LADY MARGARET HALL LIBRARY
OXFORD

Chemical Bonding and Spectroscopy in Mineral Chemistry

Chemical Bonding and Spectroscopy in Mineral Chemistry

EDITED BY

Frank J. Berry

Department of Chemistry
University of Birmingham, UK

and

David J. Vaughan

Department of Geological Sciences
University of Aston in Birmingham, UK

London New York

CHAPMAN AND HALL

First published 1985 by
Chapman and Hall Ltd
11 New Fetter Lane, London EC4P 4EE

Published in the USA by
Chapman and Hall
733 Third Avenue, New York NY 10017

Printed in Great Britain by
J. W. Arrowsmith Ltd., Bristol

ISBN 0 412 25270 8

British Library Cataloguing in Publication Data

Chemical bonding.
1. Mineralogical chemistry
2. Chemical bonds 2. Spectrum analysis
I. Berry, Frank J. II. Vaughan, David J.
549′.133 QE371

ISBN 0-412-25270-8

Library of Congress Cataloging in Publication Data

Main entry under title:

Chemical bonding and spectroscopy in mineral chemistry.

Includes bibliographies and index.
1. Mineralogical chemistry. 2. Chemical bonds.
3. Spectrum analysis. I. Berry, Frank J., 1947–
II. Vaughan, David J., 1946–
QE371.C46 1984 549′.13 84-15542
ISBN 0-412-25270-8

Contents

Contributors viii
Preface ix

1 Quantum Mechanical Models and Methods in Mineralogy **1**
J.A. Tossell
1.1 Introduction 1
1.2 Full lattice calculations 5
1.3 Cluster calculations on mineral structural properties 6
1.4 Cluster calculations on mineral spectral properties 14
1.5 Cluster calculations of valence electron density distributions 19
1.6 Applications of qualitative MO theory 22
1.7 Conclusions 26
Acknowledgements 27
References 28

2 X-ray Spectroscopy and Chemical Bonding in Minerals **31**
D.S. Urch
2.1 Introduction 31
2.2 Photoelectron and X-ray spectroscopy 32
2.3 Spectroscopic techniques 36
2.4 Application of XES and XPS to bonding studies in mineral chemistry 42
2.5 Further developments 57
2.6 Conclusions 59
References 59

3 Electronic Spectra of Minerals **63**
Roger G. Burns
3.1 Introduction 63

3.2 Background 64
3.3 Techniques 72
3.4 Crystal field spectra 74
3.5 Intervalence transitions 86
3.6 Applications 89
3.7 Summary 98
References 99

4 Mineralogical Applications of Luminescence Techniques **103**
Grahame Walker

4.1 Introduction 103
4.2 The luminescence process 104
4.3 Experimental techniques 121
4.4 Luminescence centres in some common minerals 126
4.5 Some conclusions 136
References 138

5 Mössbauer Spectroscopy in Mineral Chemistry **141**
A.G. Maddock

5.1 The basis of Mössbauer spectroscopy 141
5.2 The hyperfine interactions 143
5.3 The Mössbauer factor, f, and the intensity of the absorption lines 153
5.4 ^{57}Fe Mössbauer parameters and deductions from such data 155
5.5 Experimental details 162
5.6 Mineralogical applications 168
5.7 Antimony 191
5.8 Other physical studies 191
References 191

6 Electron Spin Resonance and Nuclear Magnetic Resonance Applied to Minerals **209**
William R. McWhinnie

6.1 Electron spin resonance spectroscopy 210
6.2 Practical aspects of ESR 216
6.3 Some applications of ESR in mineral chemistry 219
6.4 Nuclear magnetic resonance spectroscopy 227
6.5 NMR of solids 231
6.6 Applications 235
6.7 High resolution NMR studies of minerals 237
6.8 Conclusion 246
Acknowledgement 246
References 246

7 Spectroscopy and Chemical Bonding in the Opaque Minerals **251**
David J. Vaughan

7.1 Introduction 251
7.2 Compositions and crystal structures of the major opaque minerals 252
7.3 Approaches to chemical bonding models 255
7.4 Experimental methods for the study of bonding 257
7.5 Chemical bonding in some major opaque mineral groups 260
7.6 Concluding remarks 289
References 290

8 Mineral Surfaces and the Chemical Bond **293**
Frank J. Berry

8.1 Introduction 293
8.2 Spectroscopic techniques 293
8.3 Applications in mineral chemistry 301
8.4 Concluding remarks 313
References 313

Index 316

Contributors

F.J. Berry
Department of Chemistry, University of Birmingham, Birmingham, UK.

R.G. Burns
Department of Earth, Atmosphere and Planetary Sciences, Massachusetts Institute of Technology, Cambridge, MA, USA.

A.G. Maddock
University Chemical Laboratory, Cambridge, UK.

W.R. McWhinnie
Department of Chemistry, University of Aston in Birmingham, Birmingham, UK.

J.A. Tossell
Department of Chemistry, University of Maryland, College Park, MD, USA.

D.S. Urch
Department of Chemistry, Queen Mary College, University of London, London, UK.

D.J. Vaughan
Department of Geological Sciences, University of Aston in Birmingham, Birmingham, UK.

G. Walker
Department of Pure and Applied Physics, UMIST, Manchester, UK.

Preface

In recent years mineralogy has developed even stronger links with solid-state chemistry and physics and these developments have been accompanied by a trend towards further quantification in the theoretical as well as the experimental aspects of the subject.

The importance of solid-state chemistry to mineralogy was reflected in a symposium held at the 1982 Annual Congress of The Royal Society of Chemistry at which the original versions of most of the contributions to this book were presented. The meeting brought together chemists, geologists and mineralogists all of whom were interested in the application of modern spectroscopic techniques to the study of bonding in minerals. The interdisciplinary nature of the symposium enabled a beneficial exchange of information from the various fields and it was felt that a book presenting reviews of the key areas of the subject would be a useful addition to both the chemical and mineralogical literature.

The field of study which is commonly termed the 'physics and chemistry of minerals' has itself developed very rapidly over recent years. Such rapid development has resulted in many chemists, geologists, geochemists and mineralogists being less familiar than they might wish with the techniques currently available. Central to this field is an understanding of chemical bonding or 'electronic structure' in minerals which has been developed both theoretically and by the use of spectroscopic techniques. The purpose of this book is to outline the fundamental concepts associated with current models of bonding and to serve as an introduction to the techniques which may be applied in this area of mineral chemistry. It is not the intention of the text to provide a laboratory manual for the techniques discussed, neither is it intended that reviews of the literature detailing applications be comprehensive. This book is a starting point from which the interested reader can progress to many of the more detailed accounts cited in the various chapters.

The book begins with a chapter on the use of quantum mechanics in producing models of chemical bonding in minerals. The next two chapters consider the application of X-ray spectroscopy and of electronic absorption spectroscopy and are followed by a chapter on the more specialist technique of luminescence spectroscopy. All of these methods involve the interaction of some form of electromagnetic radiation with the electrons in solids and provide information about the energies and distributions of the electrons. The next two chapters deal largely with the interaction of radiation with atomic nuclei in minerals as in Mössbauer spectroscopy and nuclear magnetic resonance. These techniques generally serve as more indirect, although equally powerful, probes of the electronic structures of solids. The last two chapters deal with the special problems, as regards both theory and experiment, which are posed by opaque minerals and with the applications of spectroscopic methods for the elucidation of the surface properties of minerals.

If the readers of this book are consequently led to look more deeply into the fascinating areas of bonding and spectroscopy in mineral chemistry then our objectives will have been achieved.

F.J. Berry
D.J. Vaughan
Birmingham, 1984

1

Quantum Mechanical Models and Methods in Mineralogy

J.A. Tossell

1.1 INTRODUCTION

In this chapter I will discuss the rather brief history of the application of quantum mechanical methods to mineralogical problems. I will first consider the modelling and methodological aspects of the problem and then give a number of examples of recent quantum mechanical studies of crystal structures, spectra and valence electron densities in minerals. In addition to results of quantum mechanical computations I will discuss some recent qualitative molecular orbital (MO) interpretations of mineral properties. Finally, I will indicate future productive directions for quantum mineralogical studies and suggest the role that different scientists can play in advancing quantum mineralogy.

We should first note that mineralogy is usually defined as 'the study of naturally occurring inorganic crystalline solids', a specific and rather limited subset of matter. The employment of theories and experimental techniques which have reached maturity in mathematics, chemistry and physics is common in the history of mineralogy. The utilization of quantum mechanics in mineralogy is just another example of such a transference of technique from a more basic to a more applied area. As I will show, quantum mechanics can give information useful for interpreting or simulating the crystal structure, physical properties and (to some extent) phase equilibria of minerals. It also provides the appropriate formalism for understanding almost all mineral spectral properties.

Quantum mineralogy is quite young, its development being roughly ten years behind that of quantum chemistry. Thus, we are still exploiting techniques in quantum mineralogy developed by quantum chemists in the early 1970s. In the qualitative applications of quantum theory mineralogists are even further behind chemists. Some significant books whose publication dates illustrate these points are listed in Table 1.1. Note that I have listed an early book on qualitative bonding theory by Fyfe.[1] This was a sound book but was unsuccessful in modifying the attitudes of mineralogists since the qualitative valence bond and MO ideas used could rationalize mineral properties no better than the dominant ionic model and were unfamiliar to most mineralogists.

The first quantitative MO studies which focused specifically on minerals were by Gibbs' group[2] employing extended Hückel MO theory to correlate bond distance and angle variations with Mulliken overlap populations, and my own[3,4] using approximate self-consistent-field linear combination of atomic orbitals (SCF-LCAO) and multiple scattering Xα(MS-Xα[5]) MO methods to interpret mineral X-ray spectral data. Much more accurate calculations have since been performed which support the qualitative conclusions of the original studies. We continue to generally employ Hartree–Fock based methods for structural studies and MS-Xα methods for spectral studies. Much of this work has been previously reviewed.[6–8]

As quantum mineralogy has matured it has developed a number of distinct areas. Theoretical studies using both large-scale computation and more qualitative approaches have proved valuable. In the experimental area, early emphasis upon properties previously interpreted by the classical ionic model, such as bond distances and angles, has been replaced by study of intrinsically quantum mechanical properties such as spectra, and properties showing clearly the limitations of the ionic model, such as valence electron density distributions.

Table 1.1 Comparison of publication dates of books on quantum chemistry and quantum mineralogy

	Chemistry	*Mineralogy*
First book using crystal field theory	1960 (Griffith,[69] Orgel[70])	1970 (Burns[71])
First book using qualitative MO theory	1962–1965 (Roberts,[72] Ballhausen and Gray[73])	1964 (Fyfe[1]) 1978 (Vaughan and Craig[74]) 1980 (Burdett[61])
First book on quantitative MO computations	1972 (Schaefer[11])	Not yet!

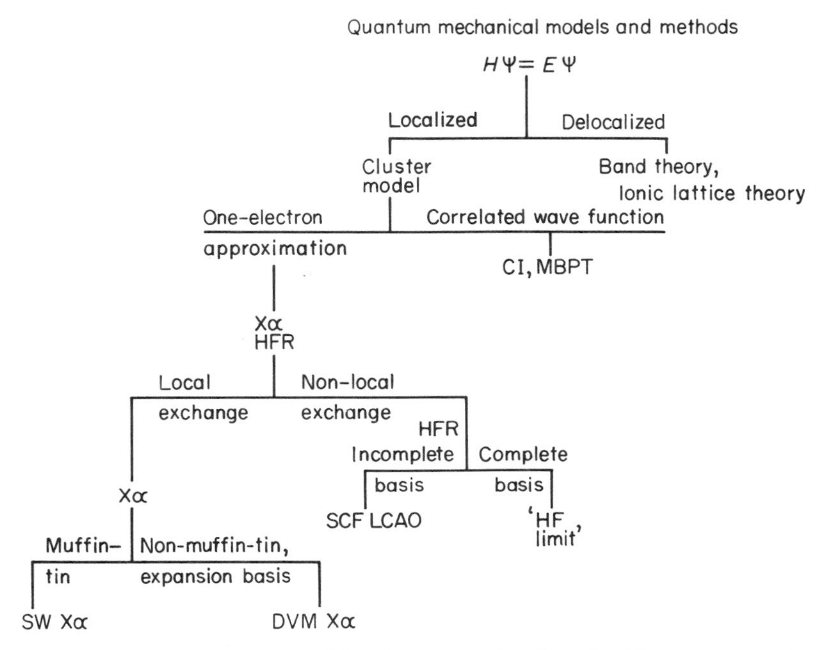

Fig. 1.1 Choices of physical model and computational method for quantum mineralogical studies.

In applying quantum mechanics to minerals, we face two distinguishable difficulties, which I will formulate as choices of 'model' and 'method'. A flow chart indicating required choices of model and method is shown in Fig. 1.1. What physical model will we use – a delocalized one, treating the whole periodic solid as in ionic lattice theory and band theory, or a localized molecular cluster approach? Although I will mention some interesting results of band theoretical and ionic lattice calculations, I will focus on the localized cluster approach which we have found to usefully describe a wide range of solid properties.

Once our physical model has been chosen we must pick a particular quantum mechanical method. All of the results which I shall discuss employ the independent electron approximation, in which each electron is assumed to move in a potential field produced by the nuclei and the averaged charge density of the other electrons. This approximation allows us to write the wavefunction for a system containing *n* electrons as an antisymmetric product of functions dependent upon the coordinates of one electron only. Antisymmetry is required so that the wavefunction satisfies the Pauli exclusion principle. However, such a wavefunction does not properly describe the *correlation* in the motion of electrons, i.e. the fact that the motion of one electron depends on the *instantaneous* positions of the other electrons, *not* simply on the charge density. This deficiency may be remedied by the techniques of configuration interaction (CI) or perturbation theory but no such calculations have yet been carried out for mineral systems. The total energy of the systems may be obtained as the quantum mechanical average value over the ground state wavefunction of the Hamiltonian operator. The most difficult terms to calculate in the Hamiltonian are those describing the electron–electron repulsions, particularly the 'exchange' terms arising from the antisymmetry requirement. To calculate the exchange contribution to the total energy we may employ the matrix Hartree–Fock or Hartree–Fock–Roothaan[9] method in which each one electron orbital is expanded as a linear combination of simple analytical functions, usually centred on the nuclei. The exchange energy can then be decomposed into a sum of integrals whose number increases as the fourth power of the number of expansion functions used. Alternatively, the exchange energy may be approximated as proportional to the one-third power of the total electron density at a point.[10] Although less accurate in theory than the Hartree–Fock method such local exchange or X_α or Hartree–Fock–Slater methods produce great computational savings and yield accurate results. Hartree–Fock–Roothaan (HFR) calculations can be further characterized by the size of the expansion basis set used (see Schafer[11] for a discussion of HFR calculations). An effectively infinite basis set is needed to obtain the true Hartree–Fock wavefunction. However, since the difficulty of the computation increases as the fourth power of the basis set size HFR calculation on most mineral model systems have employed either a minimal basis (one expansion function for each atomic orbital occupied in the free atoms from which the molecular system is formed) or a split valence basis (two expansion functions per occupied atomic valence orbital). X_α calculations may also be done with expansion basis sets[12] but most of the calculations I will describe use the muffin-tin scattered wave procedure of Johnson.[5] In this method the space of the molecular cluster is partitioned into regions and the quantum mechanical potential is simplified by various averaging procedures within these regions. Schrödinger's equation can then be solved separately in the various regions. Continuity of the wavefunction and its first derivative at the boundaries of the different regions is then required to obtain a wavefunction acceptable over all space. Although such a muffin-tin averaging procedure reduces the accuracy of the potential it makes possible a very efficient multiple scattering expansion of the wavefunction, resulting in a dramatic savings in computer time (see Appendix to Tossell and Gibbs[6]).

Thus, the calculations described herein will be quite crude by the standards of small molecule quantum chemistry. Independent electron calculations which employ large expansion basis sets (split valence plus polarization functions) and include electron correlation by configuration interaction or perturbation theory can reproduce bond distances and angles for small molecules to within experimental accuracy.[13] Such accuracy cannot be expected in quantum mineralogical studies since the large size of the mineral model systems prohibits the use of large expansion basis sets and since the approximations inherent within the molecular cluster model for a solid remain, no matter how accurate the cluster wavefunction. An inspection of the recent literature indicates that cluster models of solids are now used by a number of different groups, including solid state physics theoreticians[14,15] and theoretical chemists interested in surface properties[16] (see also Chapter 8 of this book), both of whom have addressed the questions of choice of cluster size and boundary conditions (to which we return later in this chapter).

1.2 FULL LATTICE CALCULATIONS

Let me first mention some results using the full lattice approach for a simple compound (MgO) with few electrons and high symmetry, and which is of some interest mineralogically. MgO is of interest partly because silicate minerals such as olivine break down to mixed magnesium–iron oxide and silicon dioxide (in the 6-coordinate stishovite form) at high pressures. Phase transitions in this material at high pressure were studied[17] using the full lattice modified electron gas (MEG) ionic model,[18] which evaluates the energy of an ion pair by assuming the total density to be the sum of the free ion densities, and by using energy expressions appropriate to an electron gas of uniform density. Previous studies on alkali halides by this method yielded reasonable phase transition pressures (Table 1.2). The method may be made purely non-empirical by calculating the anion wavefunction within a self-consistent stabilizing potential,[19] although early studies used free anion wavefunctions with arbitrary stabilizations. The method yielded a predicted pressure for transformation of NaCl(B1) structure MgO to the CsCl(B2) polymorph of $\sim 2560 \times 10^3$ atmospheres, which is considerably above pressures in the Earth's mantle in which MgO might occur. Even though transition pressures are very hard to predict accurately, this result suggests strongly that only NaCl structure MgO exists in the Earth.

Experimental studies by two groups on pure MgO have shown metallic conductivity at about 1 Mbar. However, two separate band calculations[20,21] indicate the MgO band gap will first increase with pressure and only become

Table 1.2 Calculated and experimental pressures (kbar) for $6 \rightarrow 8$ coordinate phase transitions in alkali halides and MgO (Cohen and Gordon[17])

Compound	*Calculated*	*Experimental*
LiCl	908	7100
NaCl	107	300
RbCl	17	5.5
MgO	2560	–

Table 1.3 Calculated and experimental phase transition pressures from MEG calculations on dihalides and dioxides (Tossell[24])

Compound and coordination numbers of transition polymorphs	*Pressure(kbar)*	
	Calculated	*Experimental*
MgF_2 6→8	200–420	330
SiO_2 4→6	125	75
SiO_2 6→8	3.9×10^3	–
TiO_2 6→8	8.3×10^2	–

zero at ~ 50 Mbar. Although the experimental data may be erroneous, it may well be that the conductive material observed is not perfectly stoichiometric single crystal MgO, but rather some reaction product, perhaps containing peroxide groups.[22] Such band theoretical studies on perfect materials may, therefore, be of limited relevance to real minerals. Nonetheless, the results of Bukowinski[20] on MgO are certainly interesting in themselves. They indicate a considerable delocalization of electron density at high pressure in MgO, reducing its 'ionic' character. Even more interesting results[23] are obtained for materials such as CaO and K at high pressure, since d-like electronic states are stabilized relative to s states leading to an electronic structure much like that of an atmospheric pressure transition metal for the K case.

My own MEG studies on more complex materials[24] indicate that structures and energies are described well for alkaline earth dihalides and somewhat less accurately for dioxides, such as SiO_2. Such studies give reasonable phase transition pressures (Table 1.3) and indicate that SiO_2 will not transform from the 6-coordinate stishovite structure to the 8-coordinate CaF_2 structure hypothesized within the Earth's mantle. Therefore, although accurate prediction of transformation pressures is very difficult, present methods can at least place some constraints upon expected high pressure behaviour. Catlow and Cormack[25] have also done interesting ionic lattice calculations using empirical ion-pair potentials. Although such semi-empirical studies can give us little direct information on the nature of bonding in solids they can accurately simulate the properties of both perfect and disordered solids which have interatomic interactions similar to those of the solids employed for the parametrization. For example, Catlow *et al.*[26] have recently predicted with reasonable accuracy how the preferred structure-type for chain silicates varies with cation identity, or cation–oxygen bond distance.

1.3 CLUSTER CALCULATIONS ON MINERAL STRUCTURAL PROPERTIES

Let me now turn to the application of cluster models in mineralogy. For some solids, definition of an appropriate cluster seems intuitively straightforward.

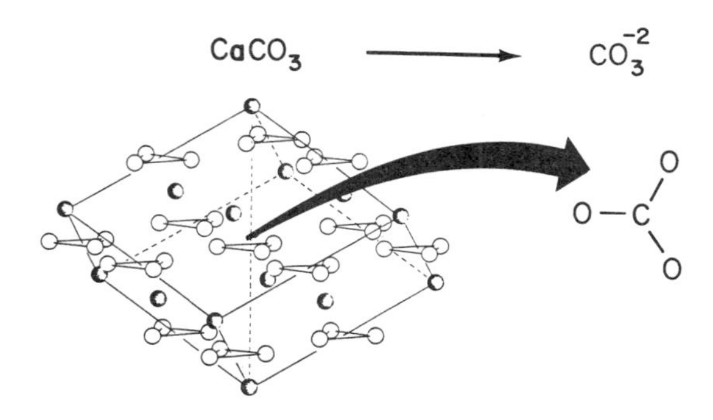

Fig. 1.2 CO_3^{2-} cluster abstracted from $CaCO_3$.

Table 1.4 Calculated and experimental C–O distances and A_1 vibrational frequencies in carbonates

	C–O distance(Å)	ν_{A1} *(cm*$^{-1}$*)*
Free CO_3^{2-} minimum basis	1.327, 1.330	1105
split valence basis	1.305	–
CO_3^{2-} in charge array, minimum basis	1.332	1064
$CaCO_3$, exp.	1.294	1082
all carbonates, experimental	1.27–1.31	–

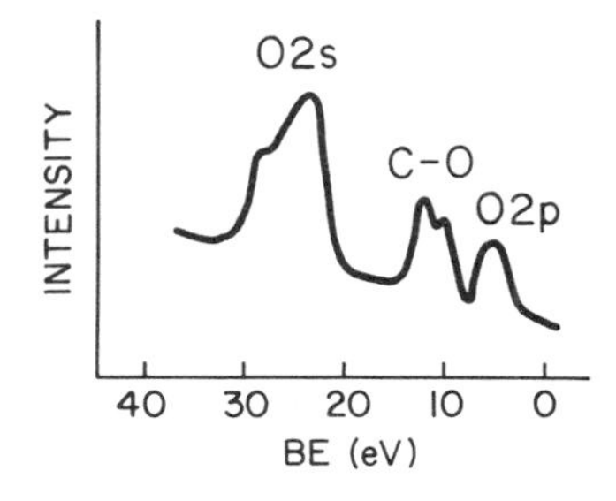

Fig. 1.3 X-ray spectra of K_2CO_3.

For example, we might describe $CaCO_3$ as a collection of discrete CO_3^{2-} and Ca^{2+} ions held together by weak ionic bonds, and detailed calculations indicate that many carbonate properties can be understood by focusing on CO_3^{2-} alone (Fig. 1.2). Rather crude HFR calculations yield minimum energy C–O distances similar to those found in carbonates (Table 1.4)[27,28] although

Table 1.5 Comparison of experimental IP's and SCF-Xα and HRF relative orbital energies (eV) for CO_3^{2-}

MO	*Experimental*	*SCF-Xα*	HFR
$1a'_2$		0	
$1e''$	0		0
$4e'$		−1.6	
$3e'$, $1a''_2$	−6.0	−5.5	−6.0
$4a'_1$	−7.7	−7.6	−8.0
$2e'$	−19.8	−18.2	−23.3
$3a'_1$	−24.1	−21.4	−27.7

incorporation of lattice effects to obtain trends from one carbonate to another have proved difficult. Ionization potentials from photoelectron spectra are found to be virtually identical for all carbonates, with the spectrum consisting of O2p non-bonding orbital, C–O bonding orbital and O2s orbital regions as shown in Fig. 1.3.[29,30] The qualitative features of the spectra have been reproduced by both HFR[31] and MS-Xα[32] cluster calculations, although quantitative discrepancies are substantial (Table 1.5). Such a comparison of calculation and experiment can be only semiquantitative since ionization potentials are not accurately equal to negatives of ground state orbital eigenvalues in either theoretical method, and since the necessary correlation, relaxation and localization corrections to the calculated ionization potentials have not been made. Many body perturbation theory studies have shown, in particular, that inner valence orbital ionizations, such as O2s, may be strongly modified by such effects.[33]

Now consider the definition of a cluster for a more difficult case, B_2O_3. The simplest cluster, BO_3^{3-} shown in Fig. 1.4, has a large negative charge and can only be formed by breaking strong covalent B–O bonds. As might be expected,

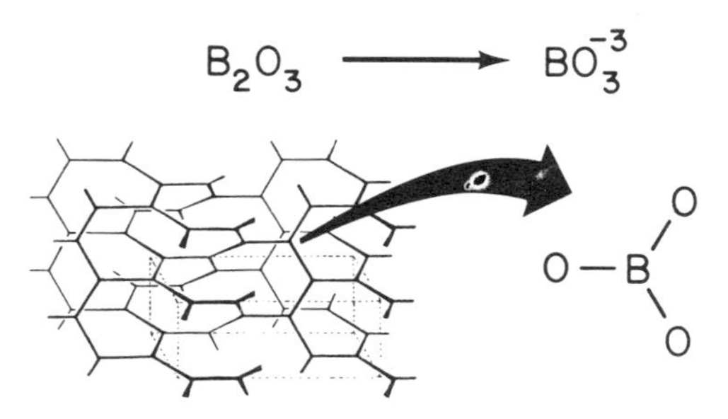

Fig. 1.4 BO_3^{3-} cluster abstracted from B_2O_3.

the calculated B–O distance for BO_3^{3-} is too large as a result of the uncompensated electron repulsions.[34] However, B $(OH)_3$ proves to model the average borate rather well and crude calculations give bond distances in good agreement with those in boric acid (Table 1.6; Gupta and Tossell, submitted). We have found in general that clusters consisting of metal atoms surrounded by nearest-neighbour oxygens yield M–O distances in good agreement with average experimental values in oxides and silicates so long as the total cluster charge is kept small by addition of terminal protons, e.g. as in the Li–N tetrahedral hydroxide series shown in Table 1.7.[35] Results of somewhat lesser accuracy have been obtained for third period metals (Table

Table 1.6 Calculated and experimental B–O bond lengths (Å)

	Minimum basis	*Split valence basis*	
	SCF calculation	*SCF calculation*	*Experimental*
BO_3^{3-}	1.419	1.435	1.34–1.40
$B(OH)_3$	1.364	1.364	1.361

Table 1.7 Experimental and calculated M–O distances (Å) for $M(OH)_2^n$, first row tetrahedral hydroxyanions

M	*n*	*Experimental*	*Calculated*
Li	−3	2.01	2.06
Be	−2	1.64	1.63
B	−1	1.48	1.48
C	0	1.40	1.42
N	+1	1.39	1.44

Table 1.8 Experimental and calculated M–O distances for second row hydroxy-anions and hydrates[8]

Central atom and coordination number	*Experimental*	*Calculated*
Na^{IV}	2.37	2.02
Na^{VI}	2.40	2.07
Mg^{IV}	1.95	1.83
Mg^{VI}	2.10	1.91
Al^{IV}	1.77	1.72
Al^{VI}	1.91	1.79
Si^{IV}	1.64	1.65
Si^{VI}	1.78	1.76

Table 1.9 Experimental and calculated M–O Distances (Å) for transition metal oxides and hydroxides

Molecular cluster	*Experimental*	*Calculated*
$Ti(OH)_6^{2-}$	1.96	1.89
CrO_4^{2-}	1.64	1.57
MnO_4^-	1.63	1.53
$Co(OH)_6^{3-}$, low spin	1.90	1.92
$Zn(OH)_4^{2-}$	1.95	1.92

1.8) by Gibbs[8] and for first series transition metals by myself (Table 1.9) and others (Jafri *et al.*[36]; Sano and Yamatera[75]). It is not yet certain why bond lengths to some central atoms (e.g. Na in $Na(OH)(OH_2)_3$ and Mn in MnO_4^-) are so different from experiment but much of the problem probably lies in the inadequacy of the minimum basis set approach (or the particular minimum

Table 1.10 Calculated and experimental bond distances (in Å) for hydroxides and fluorides with magnitude of cluster charges ⩾ 2 (Tossell[34])

	Hydroxides			*Fluorides*		
M	$R_{calc.}$	$R_{exp.}$	ΔR	$R_{calc.}$	$R_{exp.}$	ΔR
Li^{I}	1.56	1.54	0.02	1.51	1.56	−0.05
Be^{II}	1.39	–	–	1.36	1.40	−0.04
Be^{III}	1.45	1.52	−0.07	1.46	1.47	−0.01
Be^{IV}	1.66	1.62	0.04	1.59	1.58	0.01
B^{III}	1.37	1.36	0.01	1.32	1.32	0.01
B^{IV}	1.53	1.46	0.07	1.42	1.44	−0.02
C^{III}	1.25	1.26	−0.01	1.26	–	–
C^{IV}	1.42	1.40	0.02	1.33	1.32	0.01
N^{III}	1.26	1.24	0.02	1.21	1.19	0.02
N^{IV}	1.39	1.39	0.00	1.28	–	–
Na^{I}	1.94	1.82	0.12	1.92	1.93	−0.01
Mg^{II}	1.83	–	–	1.70	1.77	−0.07
Mg^{IV}	1.96	1.92	0.04	1.87	1.89	−0.02
Al^{III}	1.59	–	–	1.60	1.63	−0.03
Al^{IV}	1.75	1.74	0.01	1.66	1.72	−0.06
Si^{IV}	1.59	1.61	−0.02	1.53	1.57	−0.04
Si^{VI}	1.77	1.75	0.02	1.69	1.71	−0.02
P^{III}	1.49	–	–	1.40	–	–
P^{IV}	1.53	1.52	0.01	1.46	–	–
K^{I}	2.19	2.18	0.01	2.18	2.17	0.01
Ti^{IV}	1.79	1.76	0.03	1.72	1.73	−0.01
Ti^{VI}	1.95	1.95	0.00	1.83	1.92	−0.09

basis set used). The MEG method applied to a $M(OH)_n$ cluster also yields quite accurate M–O bond distances for non-transition metal systems (Table 1.10) at a substantially lower cost, but the calculated bond energies are too small when partial covalent character exists.[34]

Having established the capacity of the cluster model to simulate bond distances, let us now consider the simulation of bond angles in solids. Tossell and Gibbs[37] suggested that Si–O–Si bond angles in silicates be simulated by a $(SiH_3)_2O$ (disiloxane) molecule (Fig. 1.5), noting that the <Si–O–Si in disiloxane was very similar to the average <Si–O–Si in the various 4-coordinate polymorphs of SiO_2. Using the CNDO method they calculated *E* vs < curves for this molecule and a number of others of type SiH_3-A-B (Fig. 1.6). They found that the calculations reproduced observed trends in the average bridging angle and in the range of angles present in solids. For example, average Si–O–Si angles are about 145°, but the angle is quite flexible (ranging from 125° to 180°), consistent with a flat *E* vs < curve. Average Si–S–Si angles are about 110° and their range is small, consistent with a steep *E* vs < dependence. Histograms of distributions of observed angles were consistent with the calculated *E* vs < plots. Recently, *ab initio* SCF calculations at the minimum basis set level have been performed by Geisinger and Gibbs[38] for Si–O–Si and Si–S–Si linkages confirming the CNDO results. This explains why silicon oxides are more readily polymerizable and are better glass formers than are silicon sulphides – the <Si–S–Si is not deformable enough to accommodate a variety of structures.

The difference between <Si–O–Si and <Si–S–Si may be qualitatively understood by focusing upon the energy difference of highest occupied and lowest unoccupied MOs (HOMO and LUMO) for a linear B–A–B system (Fig. 1.7).[39] The highest occupied MO for the 8 valence electron case Π_u, is Ap in character and thus is less stable for the S case (due to the lower IP of the S3p orbital). The lower HOMO–LUMO energy separation for A = S gives greater stabilization of the orbital derived from the HOMO upon bending, leading to a smaller equilibrium <Si–A–Si angle.

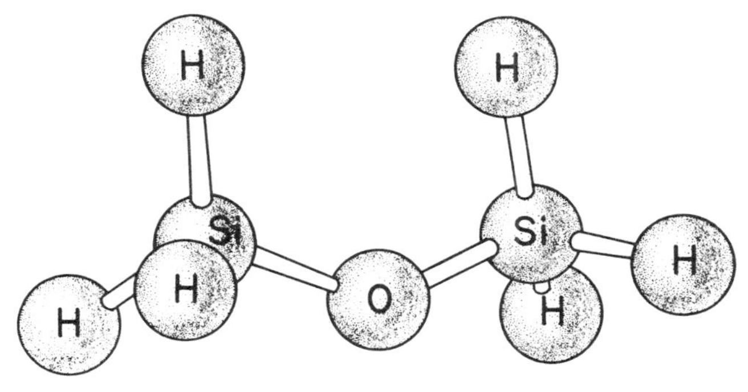

Fig. 1.5 $(SiH_3)O$ molecular model for Si–O–Si units in SiO_2.

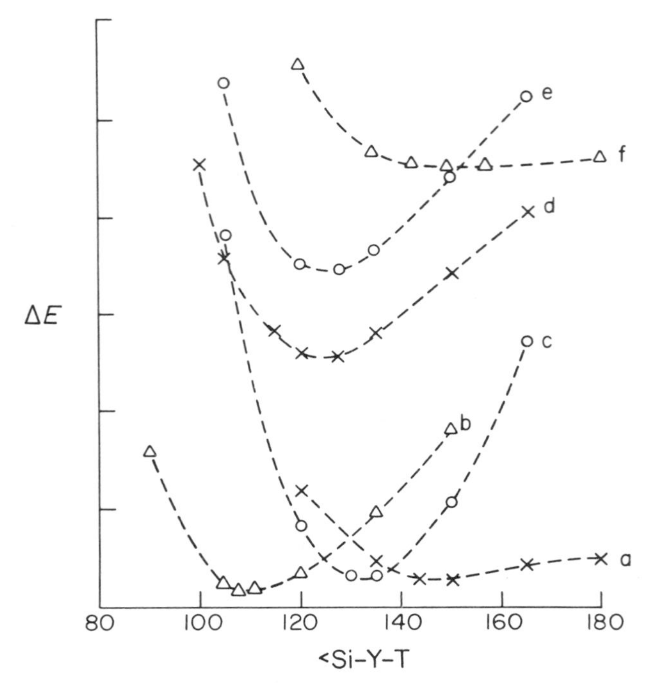

Fig. 1.6 Calculated total energy vs Si–A–B angles for: (a) SiH_3OSiH_3; (b) SiH_3SSiH_3; (c) SiH_3NHSiH_3; (d) $SiH_3OBeH_3^{-2}$; (e) $SiH_3OBH_3^{-1}$; and (f) $SiH_3OAlH_3^{-1}$. Energy divisions on vertical scale are equal to 0.01 Hartrees = 6.3 kcal/mole.

Meagher[40] has extended the CNDO calculations to the case of a two-repeat silicate chain, modelled as a $H_8Si_3O_{10}$ cluster, and has found that the calculated minimum energy chain configurations correspond to observed structures for two repeat silicates. Thus, although the particular chain silicate structure adopted might depend on the identity of the cation, the possible structures are restricted by energetic requirements of the Si–O portion. Simple molecular models seem to, therefore, give considerable information on angular trends.

Unfortunately, it is now apparent that improvement of the expansion basis set does not lead uniformly to improved angular predictions. Ernst *et al.*[41] found that for $(SiH_3)_2O$ the calculated < Si–O–Si was too small when using a minimum basis set, considerably too large using a split valence basis set and approached experiment only when O3d polarization functions were included. Similar trends (although of smaller magnitude) have been generally observed for atoms having lone pairs.[13] Thus, the close agreement of CNDO and minimum basis set SCF calculated angles with the experimental values must be considered fortuitous. Nonetheless, the trends in equilibrium angle and

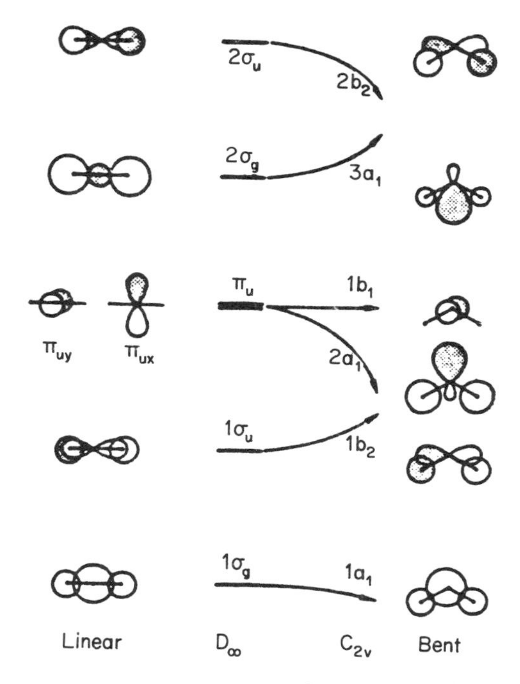

Fig. 1.7 Qualitative MO correlation diagram for linear and bent B–A–B molecule.

barrier to linearity from one system to another seem to be obtained correctly by such simple methods, even if absolute values are not.

We have so far considered the simulation of individual bond distances and angles within the cluster model. Is it possible to simulate the overall structure using this approach? The distance-least-squares (DLS) approach has been much used in the past for crystal structure simulation.[42] One chooses a set of standard bond lengths and weighting factors and varies unit cell parameters to obtain the best weighted least squares fit of simulated and standard bond distances. Quantum mechanical cluster calculations can be incorporated into this approach by using calculated bond distances and employing calculated force constants as weighting factors. Burnham *et al.*[43] have simulated SiO_2 and CO_2 structures in this way. Their results suggest that CO_2 cannot crystallize in a quartz structure even at high pressure, because the O–O distances simulated are very small. However, the coesite crystal structure of CO_2 was found to yield reasonable values for all internuclear distances. At present this method does not actually calculate a total energy but it could probably be modified to do so. Hostetler[44] has employed ion-pair potentials from HFR calculations on diatomic molecules to simulate alkali halide lattices – the corresponding melts were studied by Monte Carlo techniques. Sabelli[45] has employed SCF cluster calculations to obtain O^{2-}–O^{2-} pair potentials for a study of solid NiO. In the future, we can expect to see even more interaction between accurate cluster calculations giving improved pair potentials and full lattice crystal structure simulations for complex minerals.

1.4 CLUSTER CALCULATIONS ON MINERAL SPECTRAL PROPERTIES

Let me now turn to a discussion of the X-ray and photoemission spectral properties of some representative minerals. This topic will be further considered by Vaughan and by Urch within this volume. The prototype silicate mineral is, of course, SiO_2. An indication of its complexity is the fact that many SiO_2 band theoretical calculations have been made for the hypothetical idealized β-cristobalite structure with straight Si–O–Si angles, and that most cluster calculations have been done for SiO_4^{4-} or $Si(OH)_4$. If we consider the spectral properties of SiO_2, we find that the photoemission spectrum consists of peaks arising from O2s type orbitals, Si–O bonding orbitals, and essentially O2p non-bonding orbitals (Fig. 1.8). The nature of these peaks may be more precisely identified by correlating X-ray emission spectra with the photoemission spectra. For example, peaks in the SiK_β emission spectrum arise when an electron drops from an orbital with some Si3p character into a Si1s hole; thus the most intense SiK_β peak corresponds to a Si3p–O2p bonding orbital. SiL peaks arise from transitions with Si3d $\rightarrow$ Si2p or Si3s $\rightarrow$ Si2p character. All spectra may be readily interpreted using the molecular orbital diagram for SiO_4^{4-} obtained from a MS-Xα cluster calculation[46] (and shown in Fig. 1.9). Experimental and calculated orbital

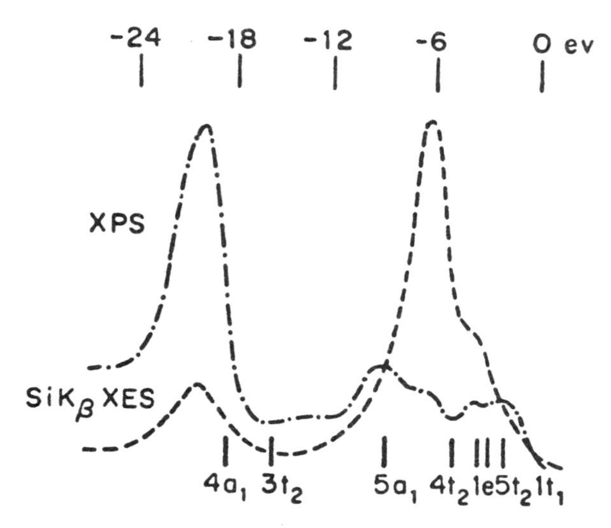

Fig. 1.8 SiO_2 X-ray photoelectron and SiK_β X-ray emission spectra with calculated orbital ionization potentials and assignments superimposed.

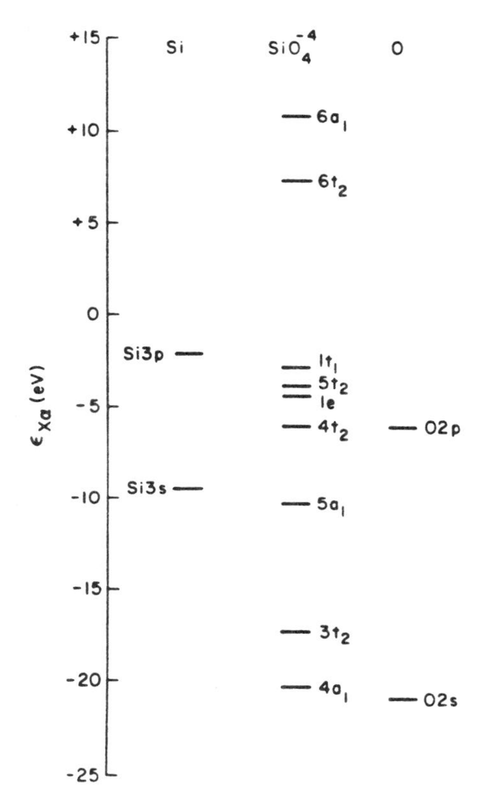

Fig. 1.9 MS-Xα SiO_4^{4-} molecular orbital diagram.

energies (in the transition state approximation;[10]) are in reasonable agreement as shown in Fig. 1.8. More sophisticated calculations have since been done using a local exchange band structure method by Chelikowsky and Schlater[47] which are in better, although still not exact, agreement with experiment. Analytic basis set or discrete variation method Xα(DVM-Xα) calculations on SiO_4^{4-} and related oxyanions[48] have also yielded ionization potentials and photoelectron and X-ray emission[49] intensities in reasonable agreement with experiment. To obtain better quantitative agreement with experiment, it will probably be necessary to use larger clusters and include the effects of correlation and hole localization (particularly for O2s ionizations). It has also been demonstrated[50] for the related PO_4^{3-} and ClO_4^- anions, that very small amounts of central atom d orbital participation give very large contributions to the LXES intensities. Thus 'd orbitals' may be important in the interpretation of some properties even though their overall participation in the wavefunction is small.

The SiO_4^{4-} calculation was originally compared to SiO_2 spectral results because few data were available for nesosilicates such as olivine,

Table 1.11 Comparison of experimental quartz and olivine relative orbital energies with SCF-Xα results for R(Si–O) = 1.609 and 1.634 Å

	Relative orbital energies (eV)					
	$4a_1$	$3t_2$	$5a_1$	$4t_2$	$1e,5t_2$	$1t_1$
SiO_2XPS	−20.8	−18.4	−7.8	−4.8	−1.9	0.
Olivine XPS and XES	−20.2		−6.4	−3.0	−1.1	0.
Difference, SiO_2 vs olivine	−0.6		−1.4	−1.8	−0.8	0.
SCF-Xα calc., R(Si − O) = 1.609 Å	−17.4	−14.5	−7.4	−3.2	−1.2	0.
SCF-Xα calc., R(Si − O) = 1.634 Å	−17.2	−14.4	−7.3	−3.1	−1.2	0.

$(Mg, Fe)_2SiO_4$, for which the SiO_4^{4-} model should be more appropriate. If the SiO_4^{4-} cluster orbital energies are compared with recently obtained data on olivines[51] agreement of calculation and experiment is even better (Table 1.11). It also appears from the calculations that the spectral differences between quartz and olivine are not a direct result of different Si–O distances, but arise mainly from the difference in degree of polymerization. An important observation of the early studies was the spectral similarity of SiO_2, $(Mg, Fe)_2SiO_4$ and the various phosphate, sulphates and perchlorates. Chemists had tended to ignore the SiO_4^{4-} group because it is found in minerals but not in compounds encountered in the chemical laboratory, while mineralogists and physicists had ignored the properties of the PO_4^{3-}–ClO_4^- series, focusing instead on crude band calculations for idealized SiO_2 structures. Recognition of the continuity of properties along the SiO_4^{4-}–ClO_4^- series would have greatly accelerated our process of understanding.

Although relaxation effects attendant upon ionization are substantial in many materials, they are particularly dramatic in certain transition

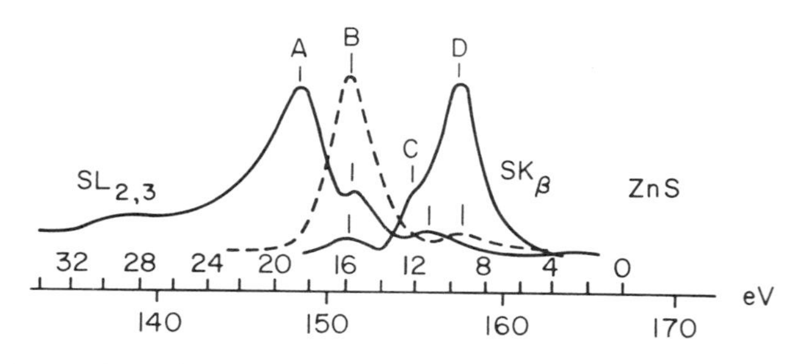

Fig. 1.10 ZnS X-ray photoelectron and X-ray emission spectra.

metal compounds. Comparison of XPS and XES for ZnS shows the Zn3d orbitals to be much more tightly bound than the S3p (Fig. 1.10). MS-Xα calculations on ZnS_4^{6-} (ref. 52; see Fig. 1.11) accurately reproduce these relative energies (Table 1.12) so long as relaxation of the Zn3d levels is incorporated using the transition state approach. It is interesting to compare the ZnS_4^{6-} cluster with isoelectronic CuS_4^{7-} (ref. 53; Fig. 1.12, see also Chapter 7). The Zn3d levels are much more tightly bound than the Cu3d, showing the dramatic and discontinuous stabilization of M3d orbitals which occurs at the end of the transition series. The low binding energies of the Cu3d orbital in Cu^{I} sulphides is also apparent from inspection of their XPS and XES, as shown for Cu_2S in Fig. 1.13.[54] The primary XPS peak, arising from Cu3d-type orbitals lies at lower IP than the main SK_β peak, arising from S3p non-bonding orbitals.

Many simple oxide and sulphide minerals have now been the subject of MS-Xα studies as discussed further by Vaughan in Chapter 7 of this book. In

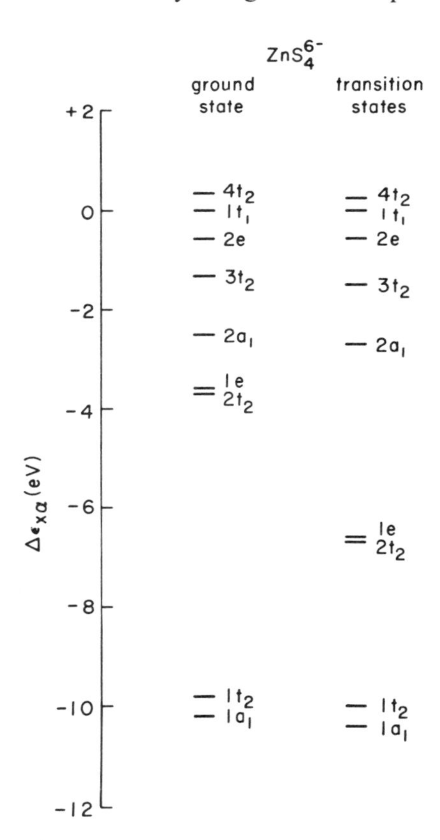

Fig. 1.11 MS-Xα ZnS_4^{6-} MO diagram for ground state and photoionization transition states (le and $2t_2$ are predominantly Zn3d orbitals).

Table 1.12 Comparison of energies of peaks in X-ray photoelectron spectra of ZnS(s) with ground state and photoionization transition state orbital energies (eV) of ZnS_4^{6-}

			Relative SCF-Xα orbital energies	
Peak	*Assignment*	*Exp. ΔE*	*Ground state*	*Transition state*
I	$4t_2, 1t_1, 2e$	0	0	0
II	$3t_2, 2a_1$	− 3.0	− 1.3, − 2.5	− 1.3, − 2.5
Zn(3d)	$1e, 2t_2$	− 7.3	− 3.6, − 3.7	− 7.6, − 7.7
III	$1a_1, 1t_2$	− 10.3	− 9.8, − 10.2	− 10

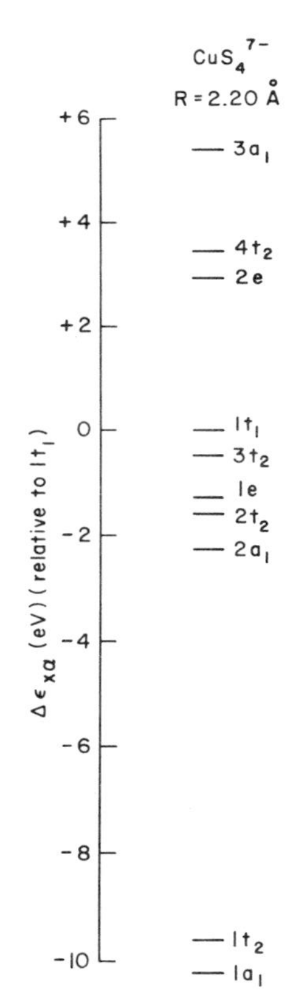

Fig. 1.12 MS-Xα CuS_4^{-7} ground state molecular orbital diagram (2e and $4t_2$ are predominantly Cu3d orbitals).

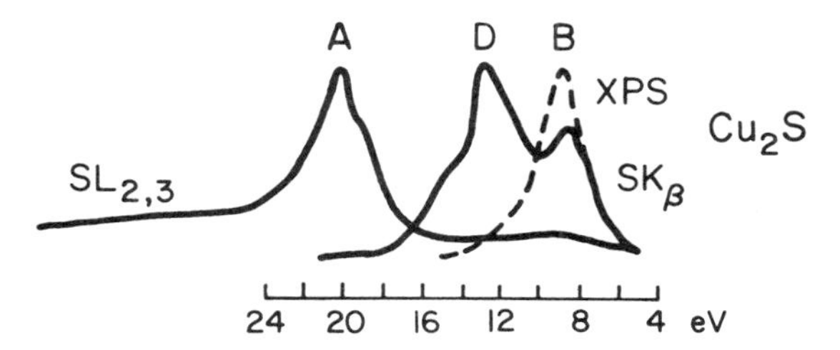

Fig. 1.13 Cu_2S X-ray photoelectron and X-ray emission spectra.

addition to their role in the interpretation of spectral properties for such minerals, the MS-Xα calculations may be used to quantify electronic structure models for more complex materials. For the case of $CuFeS_2$ (to be discussed more fully by Vaughan) X-ray spectral studies showed significant discrepancies with Fe3d orbital ionization potentials obtained from MS-Xα cluster calculations on FeS_4^{-5}. It is not clear whether the problem lies primarily in the FeS_4 cluster or in the neglect of the interaction of FeS_4 and CuS_4 polyhedra.

1.5 CLUSTER CALCULATIONS OF VALENCE ELECTRON DENSITY DISTRIBUTIONS

The electron density distribution is perhaps the most fundamental property of a material and we would expect the study of it to give substantial insights into other properties. Coppens and Stevens[55] have described experimental and theoretical methods for evluating valence electron densities and have discussed some of the early results in this field. Studies of valence electron densities in minerals have been described by Downs *et al.*,[56] Gibbs[8] and others. In most cases attention has been focused upon the difference or deformation density. $\Delta\rho$, defined as the difference between the electron density in the solid and that obtained from a super-position of free spherical neutral atom electron densities. I will briefly describe two theoretical studies of this quantity, the first

Table 1.13 Standard atomic and ionic radii and those inferred from calculated $\Delta\rho$ maps

	Atomic radii (Å)			*Crystal radii (Å)*	
Atom	*Slater*[57]	*Calculated*	*Ion*	*Shannon and Prewitt*[58]	*Calculated*
Be	1.05	1.17	Be^{2+}	0.41	0.66
B	0.85	0.79	B^{3+}	0.26	
C	0.70	0.66	C^{4+}	0.18	
N	0.65	0.54	N^{5+}	0.17	
O	0.60	0.69–0.85	O^{2-}	1.22	1.00

essentially qualitative and the second more quantitative. The qualitative study[35] considered changes in $\Delta\rho$ along the central atom – O bond for the series $Be(OH)_4^{2-}$ to $N(OH)_4^{+}$ (previously mentioned in the discussion of bond distances). For a covalent bond, one might expect the maximum in the difference density to occur at the point of contact of the covalent radii. For an ionic bond, one could define effective ionic radii in terms of the node (if present) in the difference density profile. Effective covalent and ionic radii obtained in this way from the calculated $\Delta\rho$ are compared in Table 1.13 with Slater[57] atomic radii and crystal radii from Shannon and Prewitt.[58] For B, C and N central atoms no node occurs in $\Delta\rho$, consistent with a qualitatively covalent bond. For all cases, the calculated covalent radii are fairly close to the Slater values. This study simply shows the qualitative changes which occur when one passes from one extreme bond-type to another, much as earlier quantum mechanical studies showed bond-type changes along the ten valence electron series LiF to N_2.

Consider now some quantitative studies of different density distributions.[56] For the Be–O bond in euclase, $AlBeSiO_4(OH)$, we find qualitative agreement of the observed X-ray–neutron difference density with that from a split valence RHF calculation (Fig. 1.14) although substantial discrepancies remain. The calculated map also looks fortuitously good since we have not corrected for thermal motion, which will reduce the magnitude of the bond peaks. The Si–O bond $\Delta\rho$ in euclase is described even better by extended basis set calculations,

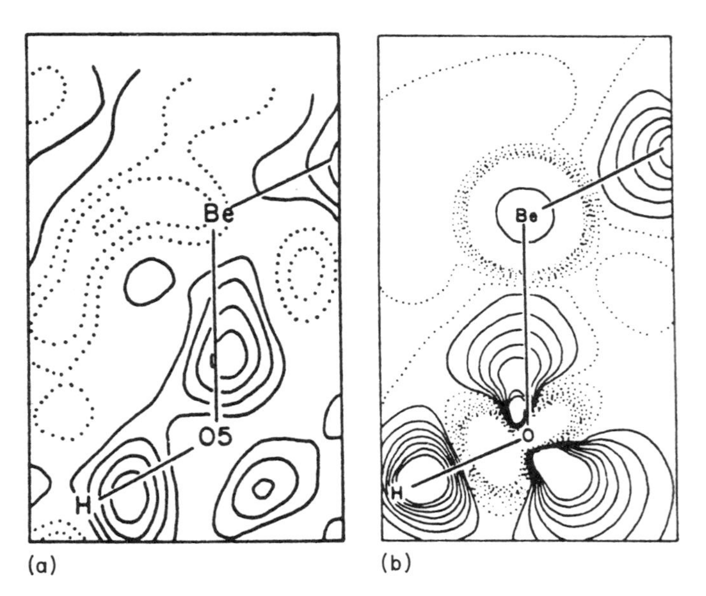

Fig. 1.14 Difference density maps through the BeOH linkage in (a) euclase and (b) $Be(OH)_4^{2-}$.

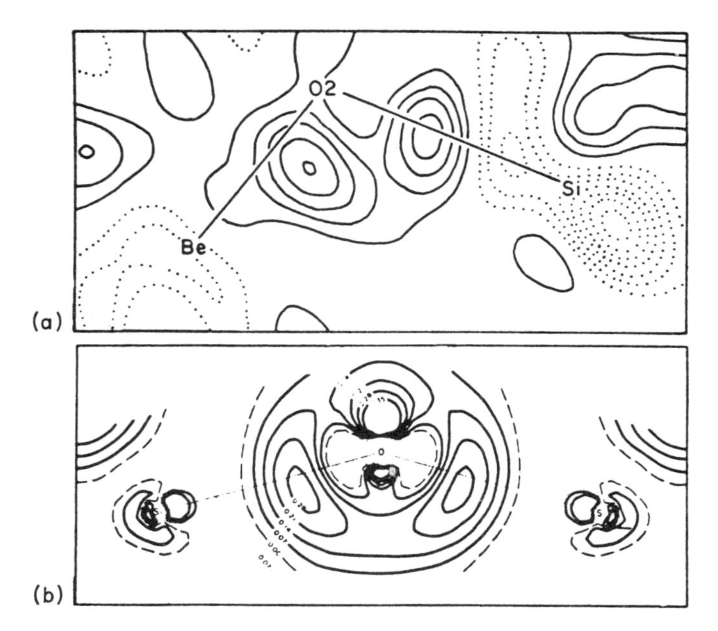

Fig. 1.15 (a) $\Delta\rho$ sections through Be–O2–Si linkage in euclase (b) $\Delta\rho$ map from $(SiH_3)_2O$ calculation using split valence plus polarization basis.

Fig. 1.16 $\Delta\rho$ map including both non-equivalent Si–O bonds in stishovite (Si at centre).

Fig. 1.17 $\Delta\rho$ map through octahedral shared edge in stishovite (O atoms at top and bottom line intersections).

using a disiloxane model (Fig. 1.15). Note that the maximum in $\Delta\rho$ is off the bond, which is therefore 'bent'. Interesting results have also been obtained for the stishovite form of SiO_2, showing difference density maxima along the Si–O bond (Fig. 1.16) higher than those in quartz (although the Si–O bond is longer, 1.78 Å compared to 1.61) and a polarization of O electron density into the edge shared between the silicons, apparently reducing their repulsions (Fig.1.17). Calculations on $Si(OH)_4(OH_2)_2$ gave essentially zero $\Delta\rho$ maxima along the Si–O bond. Density maps in agreement with experiment could be obtained only for a SiO_6H_8 model, with the H outside the centres of faces of the SiO_6 octahedron. This result illustrates the difficulty which may be encountered in the cluster modelling procedure for a solid – accurate simulation of one observable does not guarantee agreement for all.

Although the difference density studies have yielded interesting and sometimes unexpected results[56] and serve as a stringent test on the accuracy of our wavefunctions, it is not certain that $\Delta\rho$ has much physical significance. Perhaps we should be concentrating on total electron densities as suggested by Bader,[59] rather than investigating the small rearrangement of electron density which occurs during bond formation. We should also explore, both theoretically and experimentally, the potential–density relationships arising from density functional theory.[60]

1.6 APPLICATIONS OF QUALITATIVE MO THEORY

Let me now consider some qualitative MO studies of the properties of solids. Unfortunately these are much more demanding intellectually than are the 'computer experiments' previously described and the results are sometimes ambiguous or not totally convincing, so that a decision to initially focus on numerical studies was probably reasonable given the scepticism of the mineralogical audience. A leader in application of qualitative MO ideas to solids is Burdett.[61] Many of his arguments are necessarily rather involved and hard to follow without a close inspection of crystal structure pictures, but I can present here a couple of the simpler general arguments. Consider for example ZnS, which in the 4-coordinate sphalerite or wurtzite structure has just enough valence electrons per formula unit(s) to fill all the bonding and non-bonding orbitals, leaving the antibonding orbitals empty. When extra electrons are added to a system with a ZnS structure they must occupy antibonding orbitals, thus forcing structural change, breaking bonds and converting antibonding orbitals into lone pair orbitals. Such sequential bond breaking can be followed through a series of materials such as $ZnS \rightarrow GaSe \rightarrow As \rightarrow Se, Te \rightarrow I_2 \rightarrow Xe$. Similarly, an increase in the average number of valence electrons per atom breaks bonds in silicates, reducing the degree of polymerization (Table 1.14).

Bond distance and angle trends may also sometimes be understood using similar types of perturbation MO theory as Vaughan will discuss for sulphides and arsenides in his chapter. We have also been able to use qualitative theory to interpret some of our computational results. For example, in a study of Cu and Zn family compounds[62] we found that the separation of Md and Lp orbital energies was partly dependent on relativistic effects (e.g. destabilization of the Hg5d orbitals), partly upon coordination number with higher coordination number destabilizing the Md orbitals, partly upon cation charge, and partly upon atomic number (e.g. Cu^{I} vs Zn^{II} as in Fig. 1.11 and 1.12). For various metal–ligand combinations, we found that Md–Lp orbital energy differences were partly predictable from atomic orbital (AO) energies, e.g. the

Table 1.14 Breakup of silicate structures with increasing electron count[61]

System	*Number of electrons per atom*	*Examples*
SiO_2	5.33	Three-dimensional structures (with isomorphous replacement by Al, feldspars, and zeolites are found)
$(Si_2O_5)_n^{2n-}$	5.71	Double chains, e.g. gillespite ($BaFeSi_4O_{10}$), and sheets, e.g. micas
$(Si_4O_{11})_n^{6n-}$	5.87	Double chains, e.g. amphiboles and tremolite [$(OH)_2Ca_2Mg_5(Si_4O_{11})_2$]
$(SiO_3)_n^{2n-}$	6.00	Cyclic chains, e.g. benitoite ($BaTiSi_3O_9$), and linear chains, e.g. pyroxenes [$CaMg(SiO_3)_2$], and diopside
$(SiO_4)^{4-}$	6.4	Isolated tetrahedra in orthosilicates, e.g. olivines M_2SiO_4 (M mixture Mg, Fe, Mn)

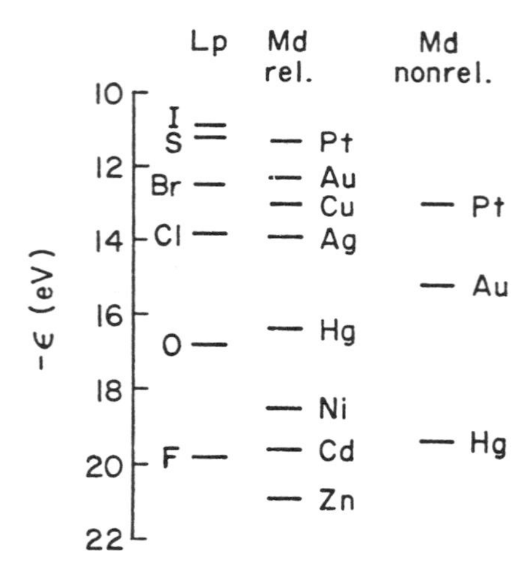

Fig. 1.18 Calculated Lp and Md atomic orbital energies.

S3p–Hg5d separation was smaller than the S3p–Zn3d separation and Cu3d was above the Zn3d and below the S3p as expected from AO energies (Fig. 1.8). Based upon the idea that interaction of completely filled orbital sets (e.g. Hg5d and S3p in Hg^{II} sulphides), was destabilizing and that the destabilization increased as the orbital energy difference decreased, we arrived at qualitative rules predicting preferred coordination number or degree of distortion for different metals and ligands. We believe this approach also has application to surface and solution species. Recently, I have extended the qualitative MO approach to skutterudite minerals,[63] the prototype being $CoAs_3$ whose structure is derived from that of ReO_3 by a distortion leading to a planar As_4^{4-} group. Qualitative MO schemes for A_4 species indicate that 20 valence electron A_4 molecules (e.g. P_4) are most stable as tetrahedra (Fig. 1.19) since the antibonding lt_1 and 3e orbitals are empty. For 24 valence electron species (e.g. P_4^{4-} or As_4^{4-}) the most stable structure has only 4 nearest-neighbour bonds (e.g. a planar geometry) and has all the σ-antibonding orbitals ($2a_{2g}$ and above) unoccupied. Such a structure may be seen as derived from the tetrahedral one (with six nearest-neighbour bonds) by the breaking of two bonds. There are also a number of other stable polymeric species based on the As^{-1} ion which is valence isoelectronic with S (which also has many polymorphs). Our previous reinterpretation[64] of the anions in $FeAs_2$ as As_2^{2-} groups suggests interactions between adjacent As_2^{2-} groups, essentially an incipient formation of As_4^{4-} species.

Using MS-Xα calculation on M centred arsenide polyhedra (Fig. 1.20) we can also qualitatively explain the instability of pure Fe and Ni skutterudites. Note that the separation of the HOMO and LUMO in As_4^{4-} is quite small (due to the large As–As distance) and that these 'frontier orbitals' generally lie

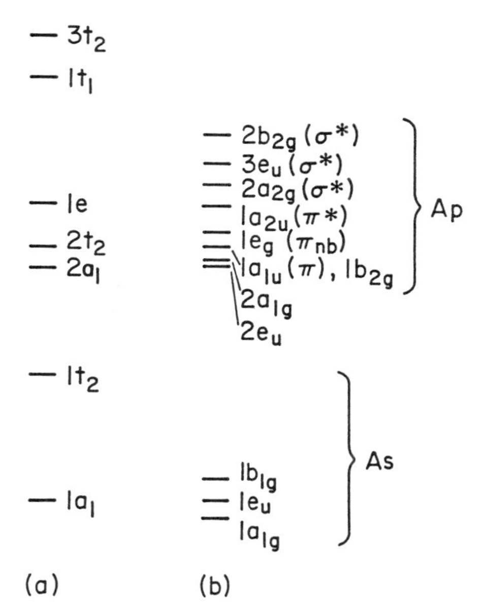

Fig. 1.19 Qualitative valence MO diagrams for A_4 species in (a) tetrahedral and (b) square planar geometry.

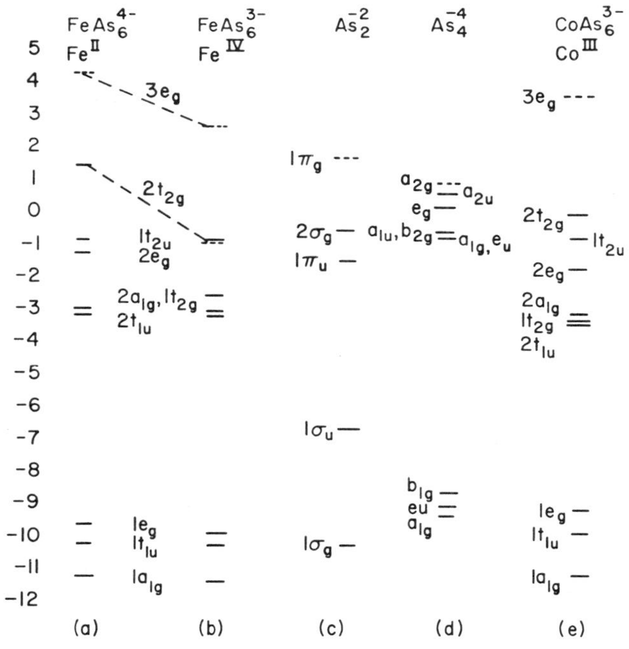

Fig. 1.20 MS-Xα orbital energies for polyhedral cluster components occurring in skutterudite minerals.

between the t_{2g} and e_gM3d orbitals. Thus, only for certain occupations of the M3d orbitals can instability with respect to M-As_4 charge transfer be avoided. A hypothetical Fe^{III} skutterudite would have a partially empty Fe3d t_{2g} orbital with a binding energy roughly midway between those shown in Fig. 1.20(a) and (b), making it more stable than the HOMO of As_4^{4-}. Charge would thus flow from As_4^{4-} to Fe^{III}, leading to a reduced Fe and a strengthened As–As bond. In a hypothetical Ni^{III} skutterudite there would be an electron in the Ni3de_g orbital with a considerably lower binding energy than that of the As_4^{4-} LUMO. In this case, electron transfer into the LUMO of As_4^{4-} (with σ^* character) would tend to break As–As bonds. When both Fe and Ni are present in the skutterudite, electron transfer can occur between them to give a Fe^{II} and Ni^{IV} configuration, without disrupting the As_4^{4-} unit.

1.7 CONCLUSIONS

We have just described a number of applications of the molecular cluster model to inorganic solids and minerals. Although the clusters chosen to model solids are to a substantial extent generated by a trial and error procedure, some general principles for choosing such clusters can now be stated. In general, bond distances and other properties may be accurately calculated using a cluster approach for those components of the crystal structure which possess the strongest bonding, e.g. the carbonate group in $CaCO_3$. Cation–anion distances and spectral properties are also found to be more strongly dependent upon cation than upon anion coordination number. Thus, the appropriate clusters are cation-centred and have a full complement of nearest-neighbour anions. For geometry prediction, it is important that the cluster be nearly neutral. Simple considerations suggest that bond distances in anions will be exaggerated by an amount proportional to the total cluster charge squared, divided by the cation formal charge.[34] However, for discussion of spectral properties it appears that stabilization of the anion by immersion inside a positively charged spherical shell yields reasonable excitation energies and relative ionization potentials and gives a simple, easily interpretable, eigenvalue spectrum. Cation–oxygen clusters with proper cation coordination numbers will usually be non-stoichiometric and negatively charged. The simplest way to neutralize such clusters is by protonation. However, the hydroxyl ions so produced are less polarizable than oxide anions and therefore yield slightly longer, less covalent bonds. The effect of the saturating hydrogens on various properties can be modified by changing the O–H distance, cation–O–H bond angle or the H1s radial wavefunction to give better agreement with experiment. Finally, cluster results can be expected to describe only the average properties of structural components in a range of solids. Thus, calculated bond distances should be compared with average values in solids, not necessarily that for any particular crystal. Even the average value will not be calculated accurately if effects arising outside the cluster are large and biased so as to change the results in a particular direction rather than in a random fashion.

Let me conclude by mentioning a few other areas where I think quantum mechanics can be usefully applied to mineralogy in the near future. Compared to our knowledge of crystalline solids, our understanding of mineral melts and solutions is meagre. New data from X-ray absorption extended fine structure[65] and near edge structure[66] can be useful with calculation[67] to help characterize solution and melt species. Similarly, we know little about defects or surface properties of minerals or about diffusion and reaction properties dependent on these phenomena. The quantum mechanical basis for the partitioning of elements between different minerals or different crystallographic sites within a mineral is little understood and many physical properties of minerals (e.g. magnetic ordering and electric conductivity) remain to be interpreted. More work, both qualitative and quantitative, can also be expected in the speculative area of high pressure mineral structures. Surface studies using cluster models for minerals or related systems presently carried out by a small number of groups[68,16] will also become more common. The use of pseudopotential approaches to eliminate core electrons will also make it possible to study metal and metal oxide clusters of varying sizes.[36] This will help us to trace the path of condensation of materials from the solar nebula.

Although chemists may contribute to mineralogy by study of specific mineral samples it is probable that more will be gained by fundamental studies of those inorganic compounds, whether in gas, liquid or solid phase, which can yield insights into mineral properties. Substantial advances in understanding minerals have already been made using information originally obtained for related gas phase molecules. Thus, further studies on mineral models such as $(SiH_3)_2O$ or $(CH_3O)_4$ Si might be of more value than studies on solid minerals themselves. Finally, it is apparent that quantum mineralogy can reach its full potential only if the practising mineralogists most familiar with the subject can be convinced to apply quantum mechanical concepts to the interpretation of mineral properties and use such concepts to guide them in the design of experiments. For these reasons, it is important for theoreticians to utilize simple qualitative quantum mechanical approaches as much as is feasible and to engage in reasoned speculation on the properties of complex minerals in addition to their rationalization of the properties of known, well-characterized minerals.

ACKNOWLEDGEMENTS

This work was supported by NSF, grant no. EAR78-01780, NATO, grant no. 1509 and the Computer Science Center of the University of Maryland.

REFERENCES

1. Fyfe, W.S. (1964) *Geochemistry of Solids*, McGraw-Hill, New York.
2. Gibbs, G.V., Hamil, M.M., Bartell, L.S. and Yow, H. (1972) *Am. Mineral.*, **57**, 1578.
3. Tossell, J.A. (1976) *J. Phys. Chem. Solids*, **37**, 1043.
4. Tossell, J.A. Vaughan, D.J. and Johnson, K.H. (1973) *Chem, Phys. Lett.*, **20**, 329.
5. Johnson, K.H. (1973) *Adv. Quant. Chem.*, **7**, 143.
6. Tossell, J.A. and Gibbs, G.V. (1977) *Phys. Chem. Mineral.*, **2**, 21.
7. Tossell, J.A. (1979) *Trans. Am. Crystall. Assoc.*, **15**, 47.
8. Gibbs, G.V. (1982) *Am. Mineral.*, **67**, 421.
9. Roothaan, C.C.J. (1951) *Rev. mod. Phys.*, **23**, 69.
10. Slater, J.C. (1972) *Adv. Quantum Chem.*, **66**, 1.
11. Schaefer, H.F. III (1972) *The Electronic Structure of Atoms and Molecules*, Addison-Wesley, Reading, MA.
12. Baerends, E.J., Ellis, D.E. and Ros, P. (1973) *Chem. Phys.*, **2**, 41.
13. Pople, J.A. (1982) *Ber-Bunsenges. Phys. Chem.*, **86**, 806.
14. Heine, V. (1980) *Solid State Physics*, **35**, 1.
15. Kenton, A.C. and Ribarsky, M.W. (1981) *Phys. Rev.*, **B23**, 2897.
16. Goddard, W.A., III, Barton, J.J., Redondo, A. and McGill, T.C. (1978) *J. Vac. Sci. Tech.*, **15**, 1274.
17. Cohen, A.J. and Gordon, R.G. (1976) *Phys. Rev.*, **B14**, 4593.
18. Gordon, R.G. and Kim, Y.S. (1972) *J. Chem. Phys.*, **56**, 3122.
19. Mulhausen, C. and Gordon, R.G. (1981) *Phys. Rev.*, **B24**, 2147.
20. Bukowinski, M.S.T. (1980) *J. Geophys. Res.*, **85**, 285.
21. Liberman, D.A. (1978) *J. Phys. Chem. Solids*, **39**, 255.
22. Freund, F. (1981) *Contr. Mineral Petrol.*, **76**, 474.
23. Bukowinski, M.S.T. (1976) *Geophys. Res. Lett.*, **3**, 491.
24. Tossell, J.A. (1980) *J. Geophys. Res.*, **85**, 6456.
25. Catlow, C.R.A. and Cormack, A.N. (1982) *Chem. in Britain*, 627.
26. Catlow, C.R.A., Thomas, J.M., Parker, S.C. and Jefferson, D.A. (1982) *Nature, Lond.*, **295**, 658.
27. Julg, A. and Letoquart, D. (1978) *Nouv. J. Cheim.*, **1**, 261.
28. Radom, L. (1976) *Austral, J. Chem.*, **29**, 1635.
29. Calabrese, A. and Hayes R.G. (1975) *J. Electron Spectrosc.*, **6**, 1.
30. Tegeler, E., Kosuch, N., Wiech, G. and Faiessler, A. (1980) *J. Electron Spectrosc.*, **18**, 23.
31. Connor, J.A., Hillier, I.H., Saunders, V.R. and Barber M. (1972) *Molec. Phys.*, **23**, 81.
32. Tossell, J.A. (1976) *J. Phys. Chem. Solids*, **37**, 1043.
33. Cederbaum, L.S., Schirmer J., Domcke, W. and Von Niesen, W. (1977) *J. Phys. Paris*, **B10**, L549.
34. Tossell, J.A. (1981) *Phys. Chem. Mineral.*, **7**, 15.

35. Gupta, A., Swanson, D.K., Tossell, J.A. and Gibbs, G.V. (1981) *Am. Mineral.*, **66**, 601.
36. Jafri, J.A., Logan, J. and Newton, M.D. (1980) *Israel J. Chem.*, **19**, 340.
37. Tossell, J.A. and Gibbs, G.V. (1978) *Acta Cryst.*, **A34**, 463.
38. Geisinger, K. and Gibbs, G.V. (1981) *Phys. Chem. Mineral.*, **7**, 204.
39. Gimarc, B.M. (1979) *Molecular Structure and Bonding: The Qualitative Molecular Orbital Approach*, Academic Press, New York.
40. Meagher, E.P. (1980) *Am. Mineral.*, **65**, 746.
41. Ernst, C.A., Allred, A.L., Ratner, M.A. *et al.* (1981) *Chem. Phys. Lett.*, **76**, 474.
42. Bauer, W.H. (1977) *Phys. Chem. Mineral.*, **2**, 3.
43. Burnham, C.W., Bish, D.L., Geisinger, K.L. and Gibbs, G.V. (1981) *Trans. Am. Geophys. Union*, **62**, 417.
44. Hostetler, C.J. (1981) *EOS Trans. Am. Geophys. Union*, **62**, 1069.
45. Sabelli, N.H. (1982) *J. Chem. Phys.*, **76**, 2477.
46. Tossell, J.A. (1975) *J. Am. chem. Soc.*, **97**, 4840.
47. Chelikowsky, J.R. and Schlater, M. (1977) *Phys. Rev.*, **B15**, 4020.
48. Sasaki, T. and Adachi, H. (1980) *J. Electron Spectrosc.*, **19**, 261.
49. Adachi, H. and Taniguchi, K. (1980) *J. phys. Soc. Japan*, **49**, 1944.
50. Tossell, J.A. (1980) *Inorg. Chem.*, **19**, 3228.
51. Tossell, J.A. (1977) *Am. Mineral.*, **62**, 136.
52. Tossell, J.A. (1977) *Inorg. Chem.*, **16**, 2944.
53. Tossell, J.A. (1978) *Phys. Chem. Mineral.*, **2**, 225.
54. Domashevskaya, E.P., Terekhov, V.A., and Marshakova, L.N. *et al.*, (1976) *J. Electron Spectrosc.*, **9**, 261.
55. Coppens, P. and Stevens, E.D. (1977) *Adv. Quantum Chem.*, **10**, 1.
56. Downs, J.W., Hill R.J., Newton, M.D. *et al.* (1982) in *Electron Distributions and the Chemical Bond* (ed. P. Coppens and M.B. Hall), Plenum Publ. Corp., New York.
57. Slater, J. (1964) *J. Chem. Phys.*, **41**, 3199.
58. Shannon, R.D. and Prewitt, C.T. (1969) *Acta Cryst.*, **B25**, 925.
59. Bader, R.F.W., Beddall, P.M. and Peslat, J. Jr. (1973) *J. Chem. Phys.*, **58**, 557.
60. Parr, R.G. (1979) in *Horizons of Quantum Chemistry* (ed. K. Fukui and B. Pullman), Reidel Publ. Co., Dordrecht, Holland, pp. 5–15.
61. Burdett, J.K. (1980) *Molecular Shapes*, Wiley-Interscience, New York.
62. Tossell, J.A. and Gibbs, G.V. (1977) *Phys. Chem. Mineral.*, **2**, 21.
63. Tossell, J.A. (1983) *Phys. Chem. Mineral.*, **9**, 115.
64. Tossell, J.A., Vaughan, D.J. and Burdett, J.K. (1981) *Phys. Chem. Mineral.*, **7**, 177.
65. Sandstrom, D.R. and Lytle, F.W. (1979) *Ann. Rev. Phys. Chem.*, **30**, 215.
66. Belli, M., Scafati, A., Bianconi, A. *et al.* (1980) *Solid State Commun.*, **35**, 355.
67. Kutzler, F.W., Natoli, C.A., Misemer, D.K. *et al.* (1980) *J. Chem. Phys.*, **73**, 3274.

68. Sauer, J., Hobza, P. and Zahradnik, R. (1980) *J. Phys. Chem.*, **84**, 3318.
69. Griffith, J.S. (1961) *The Theory of Transition Metal Ions,* Cambridge University Press, Cambridge.
70. Orgel, L.E. (1960) *An Introduction to Transition Metal Chemistry*, Butler and Tanner, Ltd., London.
71. Burns, R.G. (1970) *Mineralogical Applications of Crystal Field Theory*, Cambridge University Press, Cambridge.
72. Roberts, J.D. (1962) *Notes on Molecular Orbital Calculations*, Benjamin, New York.
73. Ballhausen, C.J. and Gray, H.B. (1965) *Molecular Orbital Theory*, Benjamin, New York.
74. Vaughan, D.J. and Craig, J.R. (1978) *Mineral Chemistry of Metal Sulfides*, Cambridge University Press, Cambridge.
75. Sano, M. and Yamatera, H. (1982) Ions and molecules in solution, in *Studies in Physical and Theoretical Chemistry*, Vol. **27**, pp. 109–116.

2

X-ray Spectroscopy and Chemical Bonding in Minerals

D.S. Urch

2.1 INTRODUCTION

The chemical bond in minerals, as in chemical compounds, can be thought of as a sharing of a pair of electrons between two or more atoms. But for more detailed and quantitative understanding of bonding it is necessary to determine the energy with which the electrons are bound and their spatial distribution. This is very much the province of theoretical chemistry which, using the techniques of quantum mechanics or of wave mechanics, shows how the Schrödinger equation can be solved in varying degrees of approximation and to varying degrees of abstraction.[1,2] The two most popular approaches are the molecular orbital (MO) theory and the valence bond theory. The former is the less sophisticated and generates wavefunctions (orbitals) for electrons that are delocalized over the network of nuclei under consideration (up to an indefinitely large number as in a metal). This theory calculates the energies of molecular orbitals (occupied and unoccupied) and their composition in terms of constituent atomic orbitals (AOs). But the delocalized nature of molecular orbitals destroys the simple concept of a pair of electrons forming a bond between two nuclei – which idea is at the heart of the valence bond method. Although these two approaches appear to give quite a different picture for electron distribution in a molecule they are in fact equivalent to each other and the set of occupied molecular orbitals can be transformed into the corresponding set of localized valence bond orbitals simply by taking suitable linear combinations.

The purpose of this chapter is to show how the predictions of such theoretical models, which have been discussed by Professor Tossell in Chapter 1, can be tested experimentally and to show how spectroscopic data can give direct insight into the electronic structure of the chemical bond. The techniques which do this most effectively are photoelectron spectroscopy[3,4] – by means of which orbital ionization energies can be measured, and X-ray emission spectroscopy[5–7] from which the bonding roles of specific atomic orbitals in individual molecular orbitals can be assessed. The first part of this chapter will, therefore, be devoted to an outline of both types of spectroscopy and to a brief description of the spectrometers used. In the remainder of the chapter, spectra, from a series of minerals of increasing structural complexity, will be discussed using the molecular orbital method so as to demonstrate the importance of these spectroscopic techniques in the determination of electronic structure.

2.2 PHOTOELECTRON AND X-RAY SPECTROSCOPY

2.2.1 Basic principles

When either radiation or electrons interact with matter, electrons can be emitted, provided that the incoming energy is great enough. If radiation of frequency ν is used, then photoemission will be initiated for electrons with ionization energies of less than E, where $E = h\nu$. If the ionization energy is E_i then the ejected photoelectron will have a kinetic energy E_k where $E_k = E - E_i$ (subject only to a very minor correction for the recoil energy of the positive ion) for a free atom in the gas phase. In a solid, further corrections must be applied to take account of the work function of the solid and the possibility of surface charging. If the solid sample is an insulator, the removal of photoelectrons will leave a positively charged surface which would then attract the emerging photoelectrons and so degrade their kinetic energy,

$$E_k = E - E_i - C \tag{2.1}$$

where C takes account of these solid state effects. The probability that an incident photon will cause the ejection of a particular electron is proportional to an integral of the type $\int \Psi_i P \Psi_f$ where $\Psi_{i,f}$ refer to wavefunctions for the initial and final states of the system and P is the transition operator. An approximation to this integral is given by replacing Ψ_i and Ψ_f by wavefunctions for the electron directly involved in the transition, and to assume that all the others are unaffected (frozen orbital approximation – see below). Thus Ψ_i would simply be the atomic orbital from which the photoelectron was ejected, ϕ_i, and Ψ_f would be replaced by ϕ_f, a plane-wave function for a free electron. If, furthermore, the electric dipole approximation is used for P, the integral becomes $\int \phi_i(er)\phi_f$. No selection rules to give forbidden or permitted transitions result from this integral (in contrast with X-ray emission) since a 'suitable' function for ϕ_f can always be found to match the parity of ϕ_i (if ϕ_i is s-like then ϕ_f should be p-like, etc.). The probability of photoemission will, therefore, depend upon the magnitude of the integral which will, in turn, depend largely upon the spatial overlap of ϕ_i and ϕ_f.[8] If ϕ_i is a valence orbital of a light atom then, within a few ångströms of the nucleus, it would roughly correspond to a function with a wavelength of about 0.1–0.3 nm. An electron represented by a plane wave of such wavelengths would have an energy of 10–100 eV. Thus, valence shell electrons will most efficiently be ejected by soft X-rays or by far ultra-violet light. If, however, radiation with an energy of more than 1000 eV is used, as in X-ray photoelectron spectroscopy (XPS), then photoelectrons from valence shell orbitals will have much shorter wavelengths and be ejected much less efficiently. More intense photoemission in XPS will come from core electrons with effective, near nucleus, 'wavelengths' of about 0.1–0.03 nm (100–1500 eV). Clearly, the number of radial nodes in an orbital function will determine its apparent 'wavelength', and the probability of photoemission from a specific orbital, as a function of irradiation energy, will vary greatly. It is found that first row elements give rise to quite weak valence band XP spectra but that even so electrons from 2s orbitals give rise to peaks an order of magnitude more intense than those from 2p orbitals. In general, it seems that for most elements core orbitals with the fewest nodes give the most intense peaks.

If the photoelectron is removed from a core orbital, then the resulting ion may relax in two possible ways; either by X-ray emission which results from electron transfer from an outer orbital (ionization energy E_j) where $\nu(\text{X-ray}) = (E_i - E_j)/h$, or by the ejection of another electron which leaves a doubly-ionized atom with vacancies in orbitals ϕ_j and ϕ_k. In the latter case, the energy of the ejected electron – the Auger electron – is approximately

$$E(\text{Auger}) = E_i - E_j - E_{k'} \tag{2.2}$$

($E_{k'}$ is the ionization energy for an electron from orbital k in the atom with an atomic number which is one greater than the atom under consideration). Because the final state after Auger emission is doubly ionized, the spectra are complex, so that whilst they can contain information of considerable importance about the nature of bonds in chemical compounds, their interpretation is a rather formidable task. Auger spectra can, however, be useful for element identification and for the determination of gross chemical properties such as valence state (see Chapter 8).

X-ray emission spectra which arise from a single electron transition lend themselves to interpretation much more easily. This is primarily because of the strong electric dipole selection rule which remains dominant for all save the shortest wavelength X-rays (i.e. $\lambda \leqslant 100$ pm). This selection rule requires that the orbital angular momentum quantum number shall change by only one unit during the transition, i.e. $\Delta l = \pm 1$.

Thus 'permitted' or 'diagram-line' X-rays will only be generated by the following transitions:

$$s \leftarrow p, \quad p \leftarrow s \text{ or } d, \quad d \leftarrow p \text{ or } f, \quad f \leftarrow d \text{ (or } g)$$

These 'atomic' selection rules can also be extended to include transitions from valence band molecular orbitals. This is because the transition is to a very localized core hole on a particular atom. In effect this very small volume acts as a probe for the electron distribution in the vicinity of that atom. X-ray spectra are thus able to give information about the atomic contributions to molecular orbitals. The intensity (I) of X-ray emission is determined by the Einstein equation which has the following form[9]

$$I = \text{constants} \times \nu^3 \times \left[\int \Psi_i P \Psi_f \right]^2 \tag{2.3}$$

where Ψ_i, Ψ_f are complete wavefunctions for the initial and final states, before and after X-ray emission and P is the transition operator. In the one electron approximation, it is assumed that only the orbitals directly involved in the transition need be considered. The effects of relaxation and electronic reorganization in other orbitals attendant upon the creation of the initial vacancy, and also in the final state after the X-ray has been emitted, are ignored. Since other orbitals are assumed unaffected this approach is also called the 'frozen-orbital' approximation. Thus, as for photoelectron spectra, Ψ_i will be replaced by the atomic orbital for the initial vacancy and Ψ_f by the orbital (atomic or molecular) with the final vacancy. When Ψ_f is to be replaced by a molecular orbital (ψ_f) the simplest form will be the Linear Combination of Atomic Orbitals (LCAO), $\psi_f = \sum a_{rf}\phi_r$, where ϕ_r is an atomic orbital at atom r. Furthermore, if P is assumed to be dominated by the electric dipole term then Equation (2.3) can be written

$$I \propto \int \phi_i(er)(\textstyle\sum a_{rf}\phi_r) \tag{2.4}$$

To be more specific, let us assume that the molecular orbital in the final state is composed of s and p orbitals from atoms A and B and that the initial vacancy is in an s orbital on atom A, written As′ to distinguish it from other s orbitals on A, the integral in (2.4) then becomes

$$\begin{aligned} &\int \phi_{As'}(er)a_{As}\phi_{As} + \int \phi_{As'}(er)a_{Ap}\phi_{Ap} \\ &+ \int \phi_{As'}(er)a_{Bs}\phi_{Bs} + \int \phi_{As'}(er)a_{Bp}\phi_{Bp} \end{aligned} \tag{2.5}$$

The first term corresponds to an electric dipole forbidden transition, $s' \leftarrow s$ on atom A, and is therefore zero. The second term is an allowed transition on A, and the third and fourth terms represent so called 'cross-over' transitions from orbitals of B to a vacancy on A. A consideration of the actual magnitudes of valence orbitals of $B(\phi_{Bs}, \phi_{Bp})$ in the region of $\phi_{As'}$ shows[10] that these integrals will be very small and can be ignored relative to the second term. Thus, it can be concluded that the intensity of an X-ray line from a valence orbital is dominated by the *atomic* transition term modified by the square of the appropriate molecular orbital coefficient, i.e.

$$I = \text{constants} \times \nu^3 \times (a_{Ap})^2 \times \left[\int \phi_{As'}(er)\phi_{Ap} \right]^2 \tag{2.6}$$

If Ap character were to be present in two or more molecular orbitals then transitions to a vacancy in an inner shell As′ orbital would give rise to two or more X-ray peaks whose relative intensities would be determined by the magnitudes of the $(a_{Ap})^2$ terms in the different orbitals. (Compared with X-ray energies the ionization energy differences for the molecular orbitals will be small so that ν can be assumed the same for all the transitions.) Relative intensities of the X-ray peaks that result from valence band–core transitions can therefore be used to measure the participation by specific atomic orbitals from individual atoms in different molecular orbitals. For example, in a bond between two second row atoms A and B the bonding roles of 3s, 3p and 3d orbitals from both atoms could be determined by an examination of the X-ray spectra given in Table 2.1.

Very often even individual spectra can reveal considerable information about the bonding in a particular compound. But much more information can be obtained about the electronic structure of a compound if all possible X-ray spectra can be assembled on a common energy scale so as to indicate the actual composition of a particular molecular orbital. To achieve this end, the ionization energies of the initial states must be determined; this then would enable the ionization energies of the final molecular orbital state to be calculated since the X-ray energies will also be known. Thus, if E_1 and E_2 are the energies of the two X-rays that result from transitions from MO's ψ_1 and ψ_2 in molecule AB to the core orbital As′ (ionization energy $E_{As'}$) then the ionization energies of ψ_1 and ψ_2 are $(E_{As'} - E_1)$ and $(E_{As'} - E_2)$ respectively. Similarly, if $E_{Bs'}$ is the ionization energy of the core s orbital on B and X-rays of energies E_3 and E_4 are observed in the K X-ray emission spectrum of B, then the corresponding final states have energies $(E_{Bs'} - E_3)$ and $(E_{Bs'} - E_4)$. If $(E_{Bs'} - E_3)$

Table 2.1

Initial vacancy	X-ray	To probe bonding role of
A1s	$AK\beta_{1,3}$	A3p orbitals
A2p	ALl, η and $AL\alpha_{1,2}$	A3s (and possibly A3d) orbitals
B1s	$BK\beta_{1,3}$	B3p orbitals
B2p	BLl, η and $BL\alpha_{1,2}$	B3s (and possibly B3d) orbitals

has, say, the same magnitude as $(E_{As'} - E_2)$, then it is reasonable to assume that the two transitions of energies E_2 and E_3 originate from the same molecular orbital. The intensities of X-rays of energies E_2 and E_3 could then be used to determine the amount of Ap character and Bp character in that MO if the corresponding atomic integrals were known. Equally important are cases where it can be established that transitions originate from different orbitals. Thus if $(E_{As'} - E_1)$ does not equal $(E_{Bs'} - E_4)$, then molecular orbital ψ_1 has Ap character but no Bp character. Arguments of this type can clearly be extended to cover other orbitals and orbital types so as to build up a complete picture of the atomic orbital structure of molecular orbitals.

The core ionization energies are, of course, measured directly using XPS and the same technique can also be used to give an indication of the ionization energies of all the various molecular orbitals. Thus, the complete set of X-ray emission data can be aligned, using core ionization energies, relative to the XP valence band spectrum. This is the spectrum which maps out on an energy scale the occupied molecular orbitals; the various X-ray spectra then indicate the composition of each one of these molecular orbitals in terms of constituent atomic orbitals. In this way, the chemical bond can be dissected to reveal its atomic components. Examples of the application of this technique to minerals will be given later in this chapter after the experimental methods for the acquisition of XPS and XES have been discussed.

2.3 SPECTROSCOPIC TECHNIQUES

2.3.1 Photoelectron spectroscopy

To be successful, photoelectron spectroscopy requires a source of monochromatic radiation, or at least a source of nearly monochromatic radiation with a clearly defined line shape. Such sources are remarkably hard to find. The noble gases have very sharp ($\sim$ 5 mV) resonance lines, e.g. He I (21.22 eV), He II (40.8 eV), Ne I (16.85, 16.67eV), etc. of which He I has been the most extensively used. The radiation is generated in a gas discharge lamp and can be used to initiate photoemission from gases and solids.[11,12] A typical spectrometer is shown in Fig. 2.1. The search for suitable radiation sources with an energy greater than 40 eV has proved quite difficult since first row elements with a Kα radiation that might be used give rise to broad Kα peaks with complex structure because the 2p orbitals are involved in chemical bonding. A core–core X-ray transition is, therefore, required and both magnesium K$\alpha_{1,2}$ (1253.6 eV) and aluminium K$\alpha_{1,2}$ (1486.6 eV) have proved suitable. The peaks are by no means monochromatic, having effective widths of 0.7 and 0.8 eV respectively and with an unresolved Kα_1–Kα_2 separation of about 0.4 eV. Even so, these line widths are not so great that the resolution of peaks of chemical interest in photoelectron spectra is compromised; also the physical properties of both elements are such that X-ray anodes can be easily fabricated.

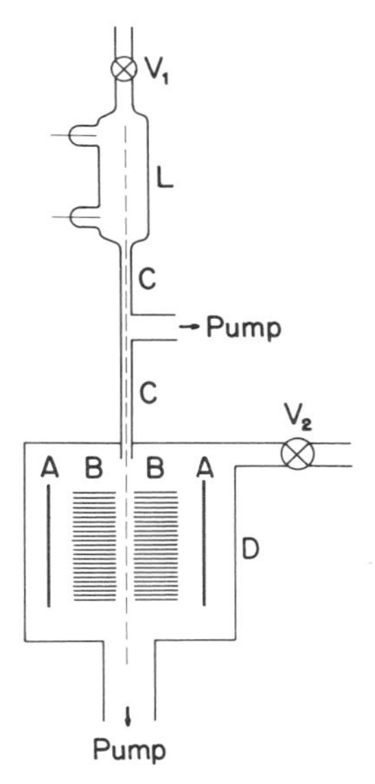

Fig. 2.1 A simple ultra-violet photoelectron spectrometer for gaseous samples. Helium (or other suitable gas) is admitted, through the upper valve (V_1) at about $10^{-4}\tau$, to the gas discharge lamp L. The light from this lamp, indicated by --- passes through the capillary C which serves both to collimate the light and to enable differential pumping between L and the sample irradiation cell D. Gaseous samples enter D through the lower valve (V_2). The kinetic energy of photoelectrons generated perpendicular to the irradiation line is measured by noting the charge received by the collecting cylinder A as a function of the retarding potential between A and the stack of discs B.

Thus two quite distinct kinds of photoelectron spectroscopy have been developed, ultra-violet photoelectron spectroscopy (UPS) – based on noble gas discharge sources, and X-ray photoelectron spectroscopy (XPS) – using (usually) magnesium or aluminium $K\alpha_{1,2}$ radiation. The latter variety of photoelectron spectroscopy is also widely known as ESCA[3] (Electron Spectroscopy for Chemical Analysis – but see below). With the increasing availability of tunable synchrotron radiation this capricious distinction may in time be lost.

A typical X-ray photoelectron spectrometer is shown in Fig. 2.2. Although gases can be handled, and although provision can be made for an ultraviolet light source, it is more usual to use such a spectrometer to study X-ray initiated photoemission from solids. And at this stage it is most important to emphasize certain fundamental features of such experiments. Electrons with relatively low kinetic energies ($\leqslant 1.5\,\text{keV}$) are being studied. Such electrons are *very* easily stopped by matter. Indeed, the maximum depth from which such electrons could be expected to emerge unscathed from a solid sample is only of the order of 10 nm. Thus, all X-ray initiated photoelectrons arise in the surface

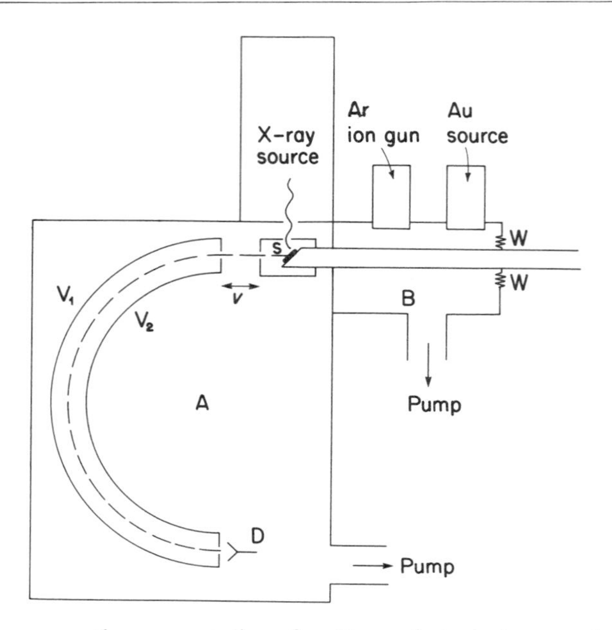

Fig. 2.2 Diagrammatic representation of an X-ray photoelectron spectrometer. The sample S is at the end of the long rod which can be moved from the analysis chamber A into the sample preparation chamber B by means of the bellows W. The sample can be treated in a variety of ways in B; two possibilities are indicated in the diagram – argon ion bombardment for sample 'cleaning' or depth profiling and gold decoration for calibration. Once in position in the analyser the sample is irradiated by X-rays (∿∿) and it emits photoelectrons (---). The energy of the photoelectrons is measured by having a retard potential (V) between the sample and the entrance to the hemi-cylindrical energy analyser. Electrons of a specific energy are brought to a focus at the detector D by the presence of a potential difference ($V_1 - V_2$) between the two plates of the analyser. Spectra are scanned by varying V with time and observing the change in the number of photoelectrons arriving at D.

layers of the sample. XPS is a surface sensitive technique. If the surface is altered or contaminated or unrepresentative of the bulk of the sample, then quite misleading conclusions can be drawn. Conversely once it is appreciated that XPS *does* monitor and analyse the top few atomic layers of a sample then it becomes the technique *par excellence* for surface studies, the method for studying the first stages of a chemical reaction, for investigating catalytic activity, for probing the nature of a surface layer of contamination, etc.

The problem that the sample may become contaminated during the measurement of the XP spectrum is a major one. At 10^{-6} torr, enough particles strike a surface in one second to generate a monolayer – were they all to stick. To obviate this difficulty, it is essential that XP spectrometers operate under high vacuum conditions; 10^{-9} torr should be regarded as a maximum operating pressure. In order that such a vacuum can be routinely and easily obtained, spectrometers are made of metal which can be baked when necessary. Facilities are also usually provided (Fig. 2.2) for samples to be pretreated before spectra are measured. Such treatment will usually consist of 'gun' for an ion bombardment (e.g. argon ions) so that surface layers can be removed. After such treatment the sample can be moved back (under high vacuum) into the spectrometer for further analysis. Whilst this is a highly effective way of removing superficial contamination (adsorbed water, carbon dioxide, dust particles, lint, etc.) or equally, a way of etching away a surface layer (surface enrichment in an alloy for example), it must be remembered that the analogy to have in mind is not that of a bacon slicer giving a smooth clean surface but more that of a sand blaster, the surface being abraded and left in a rough and fragile state. The ions will knock material from the surface, they will then enter the surface and generate dislocations and finally come to rest beneath the surface. This process will also initiate chemical reduction[13,14] to which some ions of mineralogical importance (e.g. Fe^{3+})[15,16] are particularly susceptible. 'Cleaning' samples by argon ion sputtering is, therefore, something to be undertaken prudently and delicately.

Data, from an XP spectrometer, are usually collected by measuring the rate at which photoelectrons arrive at the detector as a function of their kinetic energy. This can either be achieved in analogue manner, with kinetic energy being scanned automatically as a function of time and the corresponding fluctuations of a ratemeter displayed on a chart recorder, or data from multiple rapid scans can be collected digitally in a multichannel analyser. The latter technique facilitates subsequent data analysis (determination of peak areas for semiquantitative analysis – removal of background – peak stripping and other deconvolution procedures, etc.).

Auger electrons will also be detected in a photoelectron spectrometer. It is important that they be recognized[17] as such for two reasons, firstly, so that they are not confused with photoelectrons which could give rise to some very odd element identification and secondly, because Auger spectra are of interest in their own right. Indeed, for some elements and in some spectrometers, the Auger peaks will be more intense and more easily observed than the photo peaks (e.g. sodium and magnesium).

2.3.2 X-ray emission spectroscopy

The ions generated in the X-ray photoelectron experiment will decay either by Auger electron emission or by X-ray emission[5] and, in either case, the energy of the emitted electron or photon depends only on the atom from which it came and not upon the way in which, or energy with which, the initial vacancy was made. Thus, if interest is to be focused solely upon the energies of the emitted X-rays, there is no need for the sample to be irradiated with monochromatic radiation. X-ray tubes giving a broad 'white' X-ray spectrum as well as lines characteristic of the material of the anode are usually used in X-ray emission spectrometers. X-ray emission can, of course, also be initiated by electron bombardment.

X-ray 'fluorescence' spectrometers are widely used[18,19] for element analysis in mineralogy and the basic principles will therefore not be described in detail here. A typical spectrometer is shown in Fig. 2.3. Resolution in such a flat crystal machine is achieved by means of Soller slit collimation. In commercial spectrometers, this will be adequate to resolve most overlapping X-ray lines but for the high resolution necessary for chemical bonding investigations, special care and some modifications may be necessary:

(i) Crystals should be chosen to be as near-perfect as possible to avoid unnecessary line broadening.
(ii) Crystals should be chosen so that the peaks being studied can be observed under conditions of best resolution, i.e. at high θ angles where θ is the angle of reflection in the Bragg equation, $n\lambda = 2d \sin\theta$.
(iii) Collimation should be improved[20] so as to avoid peak tailing on the low energy (long wavelength) side.
(iv) If the overall spectrometer broadening function can be determined and if the peaks are step scanned with digital collection of the data then the

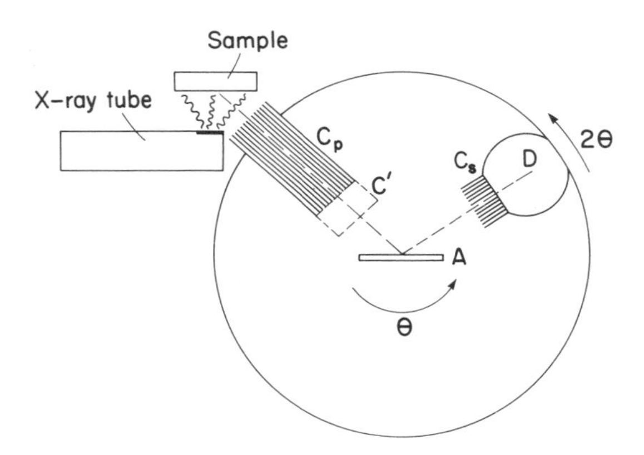

Fig. 2.3 Diagrammatic representation of an X-ray emission spectrometer. Characteristic X-rays from the sample (---) are stimulated by irradiating it with radiation from a sealed (or open-window) X-ray tube (⌇⌇⌇). A parallel beam of characteristic X-rays is generated by passage through the primary collimator Cp. Those X-rays that are diffracted by the crystal A according to Bragg's equation are detected in a gas filled proportional counter (or scintillation counter) D after having passed through the secondary collimator Cs. Spectra are scanned by rotating A through θ and D through 2θ. C' is a convenient position for extra collimation, at right angles to Cp, to improve resolution and peak profile.

broadening function can to a large extent be removed thus enhancing resolution.[21]

High resolution X-ray spectra[22] can also be obtained by using much more sophisticated equipment, e.g. a double crystal spectrometer or a curved crystal spectrometer. Another type of high resolution X-ray spectrometer uses a grating rather than a crystal as its dispersing element. Such spectrometers have distinct advantages in the 'soft' X-ray region (say $\lambda \geqslant 2$ nm) both theoretically and practically. Soft X-rays have the greatest potential for chemical bonding studies since changes of a few volts in an X-ray energy, such as might be expected for a chemical shift, can be much more easily observed as change in an X-ray energy of a few hundred rather than many thousand volts. Unfortunately, the probability that an excited ion will relax by X-ray emission diminishes as does the energy of the X-ray that would be emitted. Also working at wavelengths of greater than about 0.25 nm (which includes the soft X-ray region) requires that a spectrometer be evacuated to a pressure of $10^{-1} \sim 10^{-2}$ torr. The problems associated with soft X-ray spectroscopy and how they can be overcome can be summarized as follows.

(a) *Excitation*

An intense source of low energy photons (or electrons) is required so that suitable core vacancies can be efficiently created. This can be achieved using a specially designed X-ray tube with an aluminium or graphite anode,[23,24] or an open window gas discharge tube, or by direct bombardment with electrons (this, however, can induce chemical change, reduction, even in samples that would be thought 'stable').[25]

(b) *Dispersion*

A grating spectrometer will give the best dispersion for soft X-rays but a conventional spectrometer can be used either with crystals with large 2d spacings (e.g. OAO, 2d = 9.3 nm; OHM 2d = 6.4 nm, etc.) or with Langmuir–Blodgett films (e.g. lead stearate, 2d = 10 nm). The most exciting new development in this field is the use of multilayer devices (alternate layers of carbon and tungsten for example) which can be made with any desired effective 2d spacing.[26]

(c) *Detection*

Proportional counters with thin windows ($\leqslant 1\ \mu$m) and special gas fillings (sometimes at reduced pressure) are satisfactory.

Thus, with care reasonably well resolved X-ray spectra down to quite low energies can be obtained even with conventional equipment.

2.4 APPLICATION OF XES AND XPS TO BONDING STUDIES IN MINERAL CHEMISTRY

X-ray emission spectra of just one type from a series of different minerals or compounds can often be of interest in determining some structural or chemical feature.[27] Thus, Day[28] established that very small shifts in the Al$K\alpha_{1,2}$ for a series of aluminium compounds could be correlated with the coordination number of the aluminium (Fig. 2.4); unfortunately these shifts are at the limit of detection for most spectrometers. Similar shifts of core → core transition peaks have also been established for other elements, although for transition elements the correlation is more usually made with valence state rather than coordination number. Similar shifts of $K\beta$ peaks can be also correlated with valence state although in many cases this is more difficult to establish than for $K\alpha$, because although the $K\beta$ shifts are larger, the peak is often much broader. $K\beta$ spectra from transition metal ions show considerable intensity to the low energy side of the main peak, often as a new peak $K\beta'$. The relative intensity of $K\beta'$ to $K\beta_{1,3}$ would appear to be related in some way to the number of unpaired d shell electrons,[29,30] but this cannot be the whole story since some $K\beta'$ intensity is found for chromates and vanadates (with no electrons in the 3d orbitals) and for the ferrocyanide ($Fe(CN)_6^{4-}$) anion (six 3d electrons, all spin paired). Caution must be exercised when using this intensity ratio as shown by comparing valence state effects in manganese and iron; for manganese the $K\beta':K\beta_{1,3}$ ratio can be correlated with formal valence and used to identify the valence state of manganese in minerals,[31] but when the same approach is tried for iron it is found that the ratio has the same value to within 2% or 3% for both Fe(II) and Fe(III) (both high spin).

Valence band X-ray emission spectra (i.e. spectra that arise from valence-band → core vacancy transitions) often display considerable fine structure that

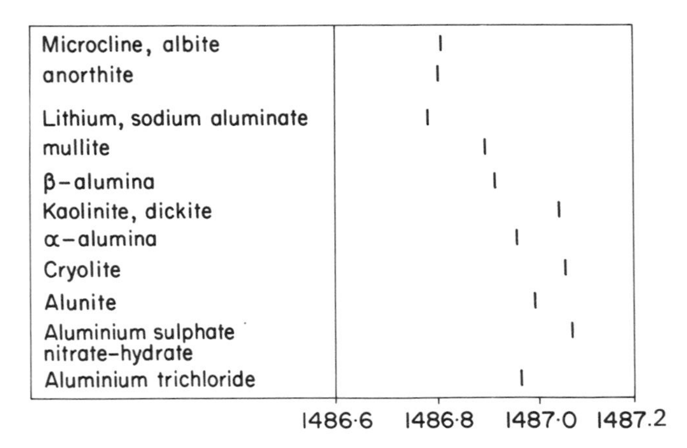

Fig. 2.4 Al $K\alpha_{1,2}$ peak positions for various minerals containing aluminium.

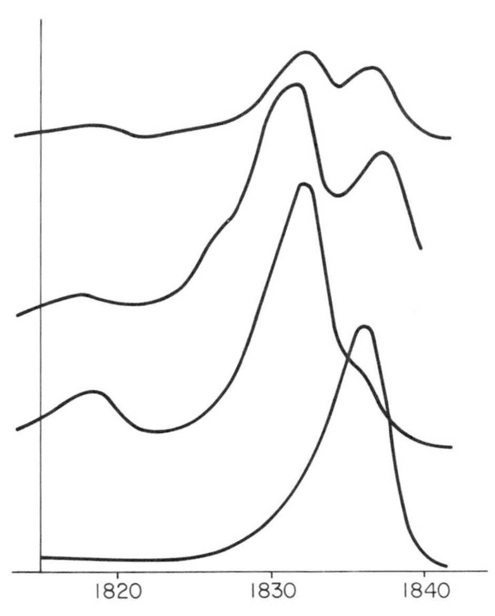

Fig. 2.5 Silicon Kβ X-ray emission spectra from various silicon compounds. From the bottom upwards, silicon, silica (SiO_2), $[Si(CH_3COCHCOCH_3)_3]^+ [HCl_2]^-$ and $[HN(C_2H_5)_3]_2^+ [Si(O_2C_6H_4)_3]^{2-}$.

can be related to particular structural features. By way of example, a series of SiK$\beta_{1,3}$ spectra is shown in Fig. 2.5.[32] The main difference between SiKβ from elemental silicon and silicon in compounds is the existence of the new peak Kβ'. The energy separation between the K$\beta_{1,3}$ and Kβ' can be related to the chemical nature of the ligand atoms bound to silicon; Si–C ~ 8 eV, Si–N ~ 11 eV, Si–O ~ 15 eV and Si–F ~ 20 eV.[33] This effect is quite general and is in no way confined to silicon. In cryolite, for example, the presence of Al–F bonds is revealed by the presence of an AlKβ' peak at 1535 eV, some 20 eV less than the energy of the main Al Kβ peak: topaz shows two Kβ' peaks at 1535 eV and 1540 eV indicative of both Al–F and Al–O bonds. (NB the origin of this Kβ' peak is quite different from that discussed above for transition metal ions.) The main SiK$\beta_{1,3}$ peak profile also shows considerable changes in different compounds. In hexa-coordinate complexes with aromatic ligands the high-energy region (1834–1839 eV) is very intense. This region also varies in intensity in glasses and in some varieties of quartz. It is possible that intensity in this region can be correlated with π bonding in the organic complexes and with Si–O–Si bond angle in the latter compounds.[34,35] Intensity on the low energy side of the main peak can, in some systems, be correlated with the presence of hydroxyl groups. Thus Myers and Andermann[36] showed that PKβ from phosphate (PO_4^{3-}) and hydrogen phosphate ($H_2PO_4^-$) differed by the presence of a new peak at the low energy side of the main K$\beta_{1,3}$ peak in the latter anion. Similarly, the MgKβ peak in brucite shows splitting of about 4 eV which can be correlated with the presence of Mg–OH bonds (see below). Gibbsite ($Al(OH)_3$) also shows considerable emission at energies just less than the main AlKβ peak. Flint and opal SiKβ spectra are broader than SiKβ for quartz; this may be due to hydration, giving rise to Si–OH groups.

Valuable as these individual spectra are in correlating particular facets of structure – physical or electronic – a more complete picture of bonding can only be obtained if more X-ray spectra, from all the elements, are gathered and aligned together using photoelectron spectra. This will now be described for a series of simple mineralogical examples.

2.4.1 Periclase (MgO)

This system was chosen[37] as being one of the simplest possible to exemplify the way in which XPS and XES can be used together to describe electronic structure. Crystallographically periclase is isomorphous with rock salt, with each magnesium octahedrally coordinated by six oxygen anions and each oxygen surrounded, at the apices of an octahedron, by six magnesium cations. All relevant spectra are assembled in Fig. 2.6. The first point to notice is that MgK$\beta_{1,3}$ exists at all, i.e. the bonding is not 100% ionic. A similar conclusion

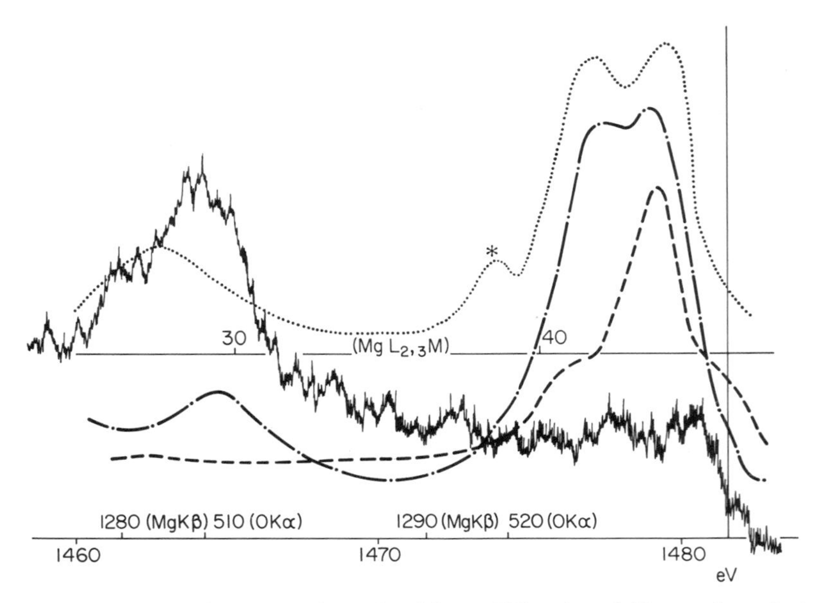

Fig. 2.6 X-ray emission (MgKβ-·-·-, $L_{2,3}$M . . . . , OKα---) and X-ray photoelectron spectra for perclase (MgO). Photoelectron kinetic energies are given on the lower scale, X-ray emission energies on the upper scales as indicated. The asterisk marks the MgL$_1$L$_{2,3}$ X-ray line.

can be drawn from the structure of the oxygen Kα peak. If it were simply an anion O^{2-} in a octahedral crystal field, all of the 2p orbitals would transform as t_{1_u}, i.e. the 2p level would not be split and would, therefore, give rise to a single sharp peak in the XE spectrum. The nature of the bonding between magnesium and oxygen can most easily be understood using just one cube of the MgO lattice. The cube has four magnesiums and four oxygens at alternate corners. If sp hybrids point from the tetrahedron of magnesium atoms to the centre of the cube, a_1 and t_2 molecular orbitals can be constructed from them which interact with a_1 and t_2 orbitals derived from p orbitals from the oxygen atoms, also pointing to the centre of the cube. Interaction can also take place between the e, t_1, t_2 orbitals from remaining magnesium p orbitals and orbitals of the same irreducible representations from the other oxygen p orbitals (oxygen 2s orbitals are omitted since they are quite tightly bound). Taken together, these interactions provide a qualitative explanation for the splitting of MgKβ (Mg3p character) into two closely-spaced peaks of nearly equal intensity and of the structure of OKα (O2p character) and $MgL_{2,3}M$ (Mg3s character). The main peak in the valence band XP spectrum is due to oxygen 2s character. This lines up with MgKβ', showing that this peak originates from the small amount of Mg3p character present in molecular orbitals that are almost wholly O2s in character.

2.4.2 Spinel ($MgAl_2O_4$)

The spinel structure is widely represented in mineralogy as a host for divalent and trivalent cations in tetrahedral and octahedral coordination. XP and XE spectra from the simplest representative of this type, Mg^{2+} (IV), Al_2^{3+} (VI), O_4 (Roman numerals to indicate coordination number) are illustrated in Fig. 2.7. A convenient model[38] with which to consider the electronic structure is based on Al_4O_4 cubes (with alternate Al and O atoms at the corners) which are linked by magnesium ions through the oxygen atoms. It is interesting to note that the structure of the AlKβ peak in spinel is similar to that of MgKβ in MgO: it is tempting to correlate such a peak profile with coordination, but more data are required to confirm this idea. The tetrahedrally coordinated Mg^{2+} gives rise to a very different peak shape – similar in fact to that found for Si(IV) and Al(IV) in quartz (Fig. 2.12) and feldspar (Fig. 2.13)(see below). The maximum of the AlKβ peak aligns with the shoulder in the OKα spectra indicating the presence in spinel of covalent Al–O bonds. The cube model for the local Al_4O_4 group suggests that there should be other Al–O orbitals based on the oxygen e, t_1 and t_2 orbitals which are derived from 2p orbitals perpendicular to the O → centre of cube direction. These orbitals should be more nearly O2p in character, hence the oxygen Kα peak at 525 eV, and the decline in AlKβ intensity in the same region. Aluminium 3s character also clearly plays a part in bonding with oxygen as shown by the peak at about 64 eV in the $AlL_{2,3}M$ spectrum. The presence of a second peak at 67.5 eV may

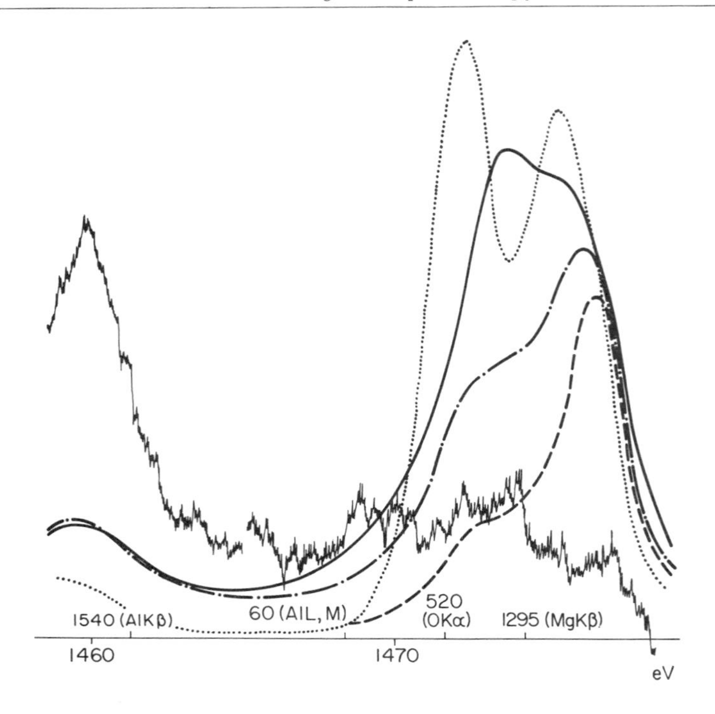

Fig. 2.7 X-ray emission (AlKβ-, $L_{2,3}$M..., MgKβ—·—, OKα---) and X-ray photoelectron spectra for spinel ($MgAl_2O_4$). The kinetic energy of photoelectrons is given on the lower scale and energy markers for the X-ray emission spectra are on the upper scale.

be due to Al3d character in molecular orbitals to oxygen. The magnesium Kβ peak profile follows that of OKα as regards peak positions. This can be attributed to the tetrahedral coordination of Mg^{2+} and the formation of bonds which are more ionic than Al–O bonds. (A bond with considerable ionic character between Mg and O, for example, would be represented by a molecular orbital with mostly oxygen and very little magnesium character. Since the interaction of the magnesium orbital is small the resulting MO will simply be a slightly perturbed oxygen orbital. A little Mg3p character thus enters the various O2p energy levels without altering them significantly. The MgKβ spectrum will, therefore, show similar structure to OKα).

2.4.3 Brucite ($Mg(OH)_2$)

Brucite is a layered material, each layer consisting of magnesium ions octahedrally coordinated by oxygens. Each oxygen carries one hydrogen, the O–H direction being more or less perpendicular to the brucite layer, and is also bound to three magnesiums; the hydrogen and the three magnesium atoms lie at the corners of a distorted tetrahedron. Again, it is interesting to note that whilst

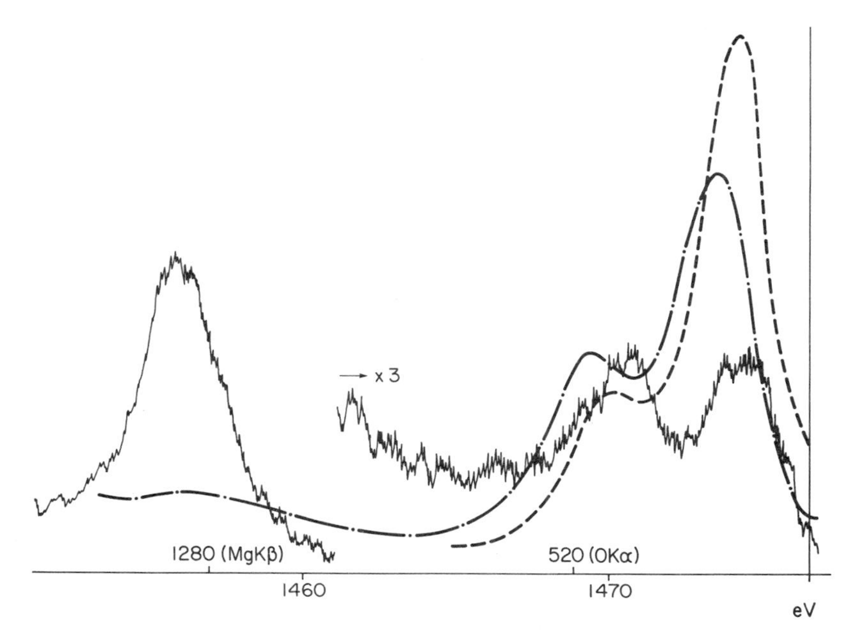

Fig. 2.8 X-ray emission (MgKβ-·-·-, OKα---) and X-ray photoelectron spectra for brucite ($Mg(OH)_2$). The kinetic energy of photoelectrons is given on the lower scale and energy markers for the X-ray emission spectra are on the upper scale. Note the change of intensity (X3) in the XPS at 1461 eV.

brucite may be thought of as ionic Mg^{2+} $(OH^-)_2$, the very existence of the MgK$\beta_{1,3}$ peaks shows that the bonding must have some covalent character. This spectrum, together with oxygen Kα and the XP valence band spectrum are presented in Fig. 2.8.[39] The spectrum would appear to be dominated by the O–H bond. The relative intensities of the component peaks of OKα correspond to calculated values for O2p character in the O–H σ bond and O2p lone pairs (perpendicular to the σ bond). The latter give rise to the main peak at ~ 525 eV, and the former, the peak at ~ 520 eV.

Magnesium 3p orbitals overlap with these two types of orbital on the oxygen atom giving the splitting observed in the MgKβ peak. What is most interesting, however, is the fact that the relative intensities of the Mg peaks are not the same as that for the oxygen peaks and also that in the XP valence band spectra the least tightly bound peaks, which must arise from the same two orbital energy levels, do not have the same relative intensity as the oxygen Kα peaks. The simplest explanation for these discrepancies would seem to lie in the presence of a small amount of oxygen 2s character in the σ bond to hydrogen in hydroxyl. Since the cross-section for photoionization by AlKα photons from O2s is about 10 times greater than that from O2p,[40] 5% O2s character in a molecular orbital would generate the same intensity peak as

50% of O2p character. Since the ls ← 2s transition is dipole forbidden (i.e. no X-ray emission from this transition), this indirect method using relative intensities of XP peaks is the only way to probe O2s character in the molecular orbitals of brucite. On the other hand, magnesium 3s could, in principle, be studied directly using the $MgL_{2,3}M$ spectrum but this has not yet been reported.

2.4.4 Gibbsite ($Al(OH)_3$), diaspore ($\alpha AlO(OH)$) and boehmite ($\gamma AlO(OH)$)

Less complete spectra are available for the hydroxides of aluminium than for magnesium, but even the $AlK\beta$ spectra show remarkable and interesting features which can be related to the structures (physical and electronic) of the minerals.[41] The spectra for aluminium hydroxide, and for both forms of the oxy-hydroxide, are shown in Fig. 2.9. The splitting of the $AlK\beta$ in gibbsite is the same as for $MgK\beta$ in brucite and presumably has the same cause. Al3p orbitals interact with the 2p σ oxygen orbital that forms the bond to hydrogen in hydroxyl, and also with the 2p π lone pair oxygen orbitals that are about 4 eV less tightly bound. Aluminium 3p character enters both these energy levels and a split $AlK\beta$ results, the relative intensities of two component peaks (1(1548 eV): ~ 2(1552 eV)) measuring the total amount of Al3p character in each level.

The spectra for the oxy-hydroxides are remarkably different, both from each other and from gibbsite. These differences can be rationalized as due to their

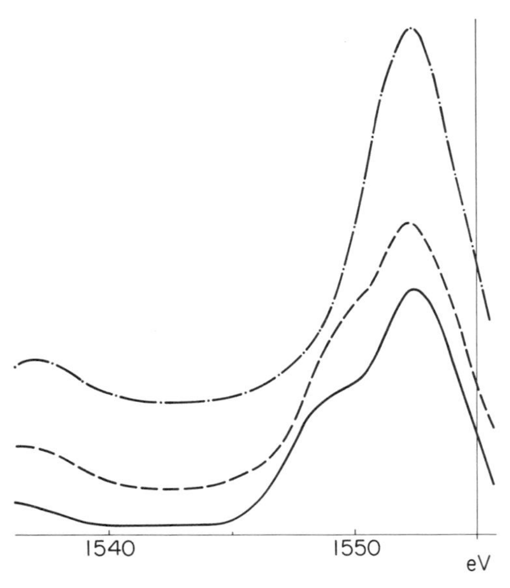

Fig. 2.9 $AlK\beta$ X-ray emission spectra for gibbsite – $Al(OH)_3$ ——, boehmite – $\alpha AlO(OH)$ --- and diaspore – $\gamma AlO(OH)$–·—·— .

very different structures. Boehmite, γ-AlO(OH), has two quite distinct types of oxygen atom, one (i) which is bridging between aluminium atoms within the layers of this mineral, and the other (ii) which is bound to two aluminium atoms and to two hydrogen atoms (one to form hydroxyl, the other by a hydrogen bond to the hydrogen of a hydroxyl in another layer). In both cases, the coordination number is four – a very distorted tetrahedron. The aluminiums are octahedrally surrounded by six oxygens, four of type (i) and two of type (ii). The aluminium Kβ spectrum will, therefore, reflect the interactions of Al3p orbitals with oxygen orbitals from both species to generate a broad peak. Structurally, diaspore is simpler in that all the oxygen atoms are similar, each one being coordinated to three aluminium atoms and to one hydrogen (for half the oxygens this will be a direct bond to hydrogen to form hydroxyl, for the other half of the oxygens it will be a hydrogen bond to the hydrogen of a nearby hydroxyl group). This is reflected in the AlKβ spectrum which shows a single peak, narrower than for boehmite, and without a shoulder.

2.4.5 Olivine (forsterite) (Mg_2SiO_4)

The simplest model for forsterite is of an ionic lattice containing magnesium cations and orthosilicate, SiO_4^{4-} anions; the bonding within the silicate anions

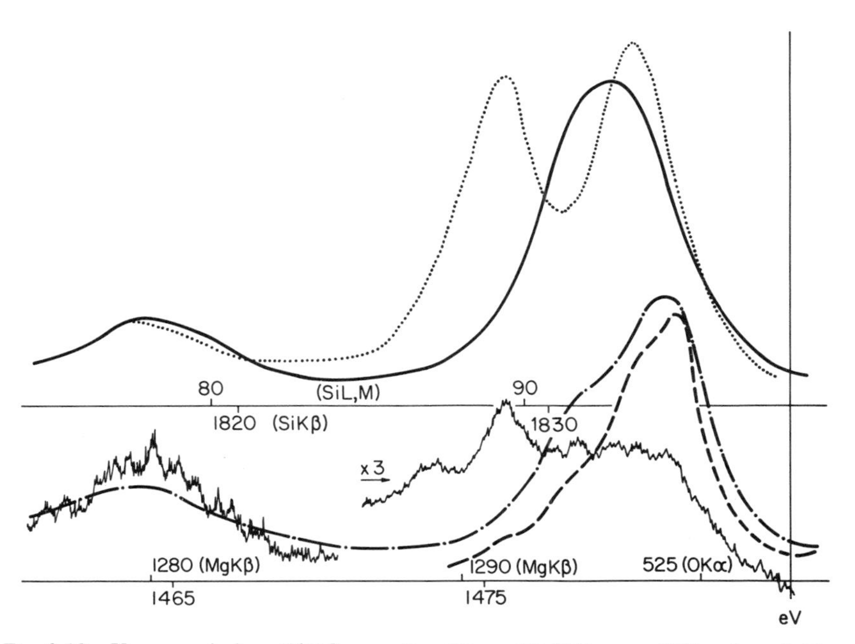

Fig. 2.10 X-ray emission (SiKβ——, $L_{2,3}$M . . ., MgKβ–·–·–, OKα---) and X-ray photoelectron spectra for forsterite (Mg_2SiO_4). X-ray photoelectron energy is shown on the lowest scale – the other scales are for the X-ray spectra, as indicated.

being covalent. Another approach would be to consider a lattice of oxygen dianions with silicon (Si^{4+}) cations in tetrahedral holes and magnesium cations (Mg^{2+}) in octahedral holes. The strongest polarization upon the oxide anions would then be exercised by the highly charged silicon leading to a model in which the Si–O bond had much more covalent character than Mg–O. In either case, the strongest covalent bonding is postulated to lie within the silicate anion. The validity of this view can be tested by comparing the XE and XP spectra for forsterite, $SiK\beta_{1,3}$, $SiL_{2,3}M$,[42,43] $MgK\beta_{1,3}$ and $OK\alpha$ together with valence band XP (Fig. 2.10), with the predictions for such spectra that can be derived from theoretical models for the bonding in SiO_4^{4-}.[44] The simplest such model is based on the one-electron Hückel molecular orbital approach, with neglect of overlap. Even so, as there are twenty five valence orbitals to be considered, the problem only becomes tractable if the simplifications, required by the tetrahedral symmetry of the anion, are utilized. The atomic orbitals may be classified according to their irreducible representations as follows:[33] Si3s, a_1; Si3p, t_2; Si3d, $e + t_2$, four oxygen 2s orbitals, $a_1 + t_2$, four oxygen 2p orbitals, orientated along the four Si–O axes, $a_1 + t_2$, the remaining eight oxygen 2p orbitals, pairs perpendicular to the Si–O axis, $e + t_1 + t_2$. Thus instead of having to solve a 25×25 determinant, the computations required break down to solutions of the following:

a_1: a 3×3 determinant
t_2: three identical 5×5 determinants, which can be approximately reduced to 3×3, 'σ' and 2×2 'π' determinants (σ and π with respect to each Si–O axis)
e: two identical 2×2 determinants
t_1: three identical 1×1 determinants (i.e. three oxygen lone pair orbitals, delocalized over the four ligand atoms).

The oxygen 2s orbitals, which enter both the a_1 and 'σ' t_2 determinants are much more tightly bound than the other orbitals. This permits further simplifications to be made. In a bond between two orbitals, ϕ_A and ϕ_B, with different Coulomb Integrals (α_A and α_B) the energies (E) of the molecular orbitals can be found by solving the secular determinant,

$$\begin{vmatrix} \alpha_A - E & \beta \\ \beta & \alpha_A - \varepsilon\beta - E \end{vmatrix} = 0 \tag{2.7}$$

where $\alpha_B = \alpha_A + \varepsilon\beta$, $\beta = \int \phi_A H \phi_B$ and ε is a constant

$$E = (\alpha_A + \alpha_B) \pm (\beta/2)(\varepsilon^2 + 4)^{\frac{1}{2}} \tag{2.8}$$

(positive sign for bonding orbital, negative for antibonding). The corresponding molecular orbitals are

$$\psi_{\text{bond}} = a_A \phi_A + (1 - a_A^2)^{\frac{1}{2}} \cdot \phi_B \tag{2.9}$$

$$\psi_{\text{antibond}} = (1 - a_A^2)^{\frac{1}{2}} \phi_A - a_A \cdot \phi_B \tag{2.10}$$

where $a_A = (\varepsilon^2 + 4 \pm \varepsilon\sqrt{(\varepsilon^2 + 4)})/2$. This result shows that when $\varepsilon = 0$ and ϕ_A and ϕ_B have the same energies the molecular orbitals are composed of equal contributions from the two atomic orbitals and that one is more and the other is less tightly bound by β than the original atomic orbitals. If two electrons enter the bonding MO a 'covalent' bond is formed. When ε is large, however, the two molecular orbitals resemble much more closely the two individual atomic orbitals, and, furthermore, the molecular orbitals are only a very little more, or less, tightly bound than the original atomic orbitals. Under these circumstances, very little stabilization would be conferred upon a pair of electrons in a tightly bound atomic orbital if that orbital were to attempt bond formation with a less tightly bound orbital. It is, therefore, not unreasonable to omit tightly bound valence shell orbitals, such as oxygen 2s, from the calculations altogether. If this is done the a_1 and 'σ' t_2 determinants are reduced further and the qualitative energy level diagram, Fig. 2.11, can be constructed.

This diagram suggests that two main transitions should be observed for $SiK\beta_{1,3}$ from $3t_2$ and $4t_2$ by virtue of Si3p character in those orbitals, and that the former should have a lower energy and be less intense than the latter. This is because $3t_2$ is more tightly bound than $4t_2$ and, being mostly oxygen 2s in character, has only a little contribution from Si3p. This corresponds to the observed $SiK\beta_{1,3}$–$SiK\beta'$ spectrum: $SiK\beta'$ aligns with the intense XP peak at 1464 eV due to oxygen 2s orbitals. The oxygen $K\alpha$ peak also shows considerable structure, with the main peak (522–525 eV) corresponding to less tightly bound orbitals than $4t_2$. From Fig. 2.11, this peak should correspond to $5t_2$, lt_1, le, oxygen lone pair orbitals, at least to a first approximation. The oxygen atoms in silicate are close enough together for these lone pair orbitals to interact so that their degeneracy will be removed. It can be shown that le will be the most tightly bound, then $5t_2$, and that lt_1 will be the least tightly bound. This will of course lead to a broadening of the main oxygen peak and possibly some structure, evidence for which can be seen in the shoulder at 523 eV. The shoulder at $\sim$ 521 eV represents O2p character in the $4t_2$ molecular orbitals since it aligns with the Si3p peak. Oxygen 2p participation in $5a_1$ should give rise to a very weak peak at lower energies, but it cannot be resolved in the data shown here ($\sim$ 518.5 eV). The weak peak at 1475 eV in the XP spectrum could be due to either a small amount of O2s character or to Si3s in $5a_1$. The presence of silicon 3s character is this orbital can, however, be detected unequivocally by the $SiL_{2,3}M$ spectrum. This spectrum is of special interest since it shows two very intense peaks and not just the one that would be expected for Si3s participation in the one orbital allowed by symmetry, $5a_1$. The other peak is at a higher X-ray energy, corresponding to a lower binding energy, where no orbitals with Si3s character are anticipated. It must, therefore, be concluded that this peak is due to Si3d character since only *s* and *d* character can initiate transitions to *p* vacancies (in the dipole approximation). Under tetrahedral symmetry 3d orbitals transform as t_2 and e. They could therefore overlap with oxygen lone pair orbitals le and $5t_2$. Such overlap

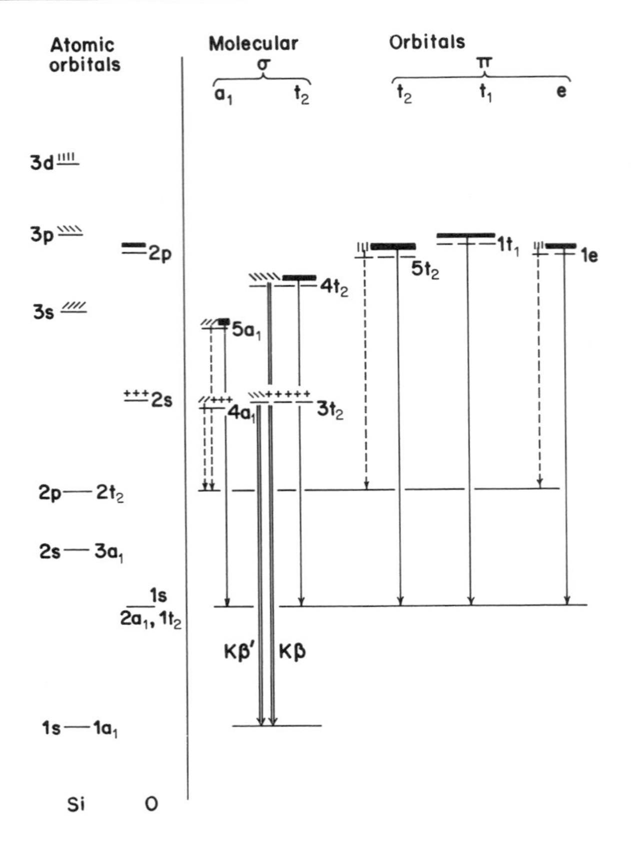

Fig. 2.11 Schematic energy level diagram for SiO_4^{4-}. The energy scale is vertical and logarithmic. Atomic orbitals are indicated on the left and molecular orbitals classified according to the tetrahedral point group (T_d) on the right. The distinction between t_2 σ and t_2 π is based on overlap considerations and not symmetry. In principle mixing between $4t_2$ and $5t_2$ is possible (see p. 52). Atomic orbitals are coded thus Si3s///, 3p \\\, 3D|||, and O2s + + +, 2p▬ so that the approximate composition of the molecular orbitals can be shown. The various possible X-ray emission transitions are shown as vertical lines, SiKβ, etc. ⇒Si $L_{2,3}$M-- →, OKα→.

would make le and $5t_2$ rather more tightly bound perhaps increasing the energy gap to the $1t_1$ level. It is significant that the peak at 93.5 in the Si$L_{2,3}$M spectrum lines up with the 523 shoulder and not with the main peak at 524.5 eV. It should also be noted that the SiKβ is broadened on its high energy side, so as to overlap with the 93.5 eV SiL peak. This may be due to interaction between $4t_2$ (Si–O σ-bonding orbitals), and $5t_2$ (Si3d–O2p π-bonding orbitals). Such σ–π mixing would allow Si3p character into the $5t_2$ orbitals, thus broadening SiK$\beta_{1,3}$ as observed. It would also stabilize $4t_2$ and destabilize $5t_2$ a little.

That the main and minor features of both the silicon and oxygen X-ray emission spectra can be rationalized in terms of a simple qualitative one electron molecular orbital model, without recourse to any consideration of interaction with magnesium, strongly supports the notion that covalent interaction is most important between silicon and oxygen. And yet, the bonding with magnesium cannot be wholly ionic because $MgK\beta_{1,3}$ can be observed. Figure 2.10 shows how very similar this spectrum is to that of oxygen $K\alpha$, in general outline and in position, although the relative intensities of the component peaks are not quite the same. It is as though the interactions between magnesium 3p and oxygen 2p orbitals ($5t_2$, $1t_1$, 1e and $4t_2$) are weak enough to lead to no major perturbation of orbital energies, but sufficient to permit some charge to be transferred to the magnesium. The presence of $MgK\beta'$ shows that a similar interaction takes place with oxygen 2s ($3t_2$, $4a_1$). In this respect the behaviour of Mg3p orbitals in forsterite is very similar to that in brucite.

Thus, an examination of the XP and XE spectra of forsterite allows a detailed, if qualitative, picture of its electronic structure to be formulated.

2.4.6 Silica (SiO_2)

The various crystalline varieties of silica (apart from stishovite) are all built up from SiO_4 tetrahedral units linked by bridging oxygen atoms which are shared between one tetrahedron and the next. The local oxygen environment of each silicon is, therefore, always the same, and indeed it is the same as in forsterite. The differences in the various types of quartz lie in the ways in which the tetrahedra are spatially disposed and, therefore, in the Si–O–Si angles. The absence of low valence cations will make the overall bonding much more covalent than in the orthosilicates but, even so, it will be of interest to see to what extent the basic SiO_4 unit can provide a model for the bonding within the whole structure.

The relevant spectra $SiK\beta$, $SiL_{2,3}M$,[47] $OK\alpha$ and XP are collected and aligned in Fig. 2.12.[48] The picture that emerges is very similar to that already described for bonding in the SiO_4^{4-} unit of forsterite. Silicon 3p character is found in both $3t_2$ ($K\beta'$ 1818.0 eV) and $4t_2$ ($K\beta_{1,3}$ 1831.5 eV) and also in $5t_2$ as evidenced by the shoulder at ~ 1835 eV. Oxygen 2p character is shown to be present in $4t_2$ by the peak at 520.5 eV, confirming this orbital as a Si–O bonding orbital. The other σ orbital, $5a_1$, can be identified with the $SiL_{2,3}M$ peak at 89 eV (Si3s), the oxygen $K\alpha$ peak at 517.5 and the XP peak at about 1468 eV kinetic energy. The most intense peak in the XP spectrum is due to orbitals that are mostly oxygen 2s in character, $4a_1$ and $3t_2$, but weak peaks in the $SiL_{2,3}M$ (77 eV) and the oxygen $K\alpha$ (508 eV) spectra point to Si3s participation in $4a_1$, and to oxygen 2p/2s mixing, respectively. The fact that the most intense feature in the XP valence band spectrum, at 1466–1478 eV, corresponds to $5a_1$, might well indicate the reciprocal presence of some oxygen 2s character in this orbital. The main oxygen peak (525 eV) is due to

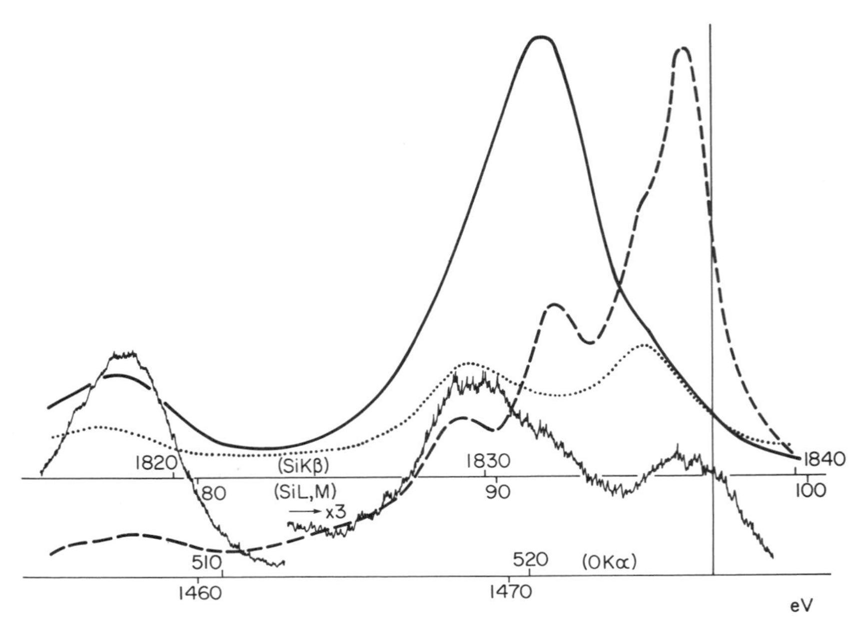

Fig. 2.12 X-ray emission (SiKβ-, $L_{2,3}$M . . . , OKα---) and X-ray photoelectron spectra for quartz (SiO_2). The lowest scale is for photoelectron kinetic energies, the others are for the X-ray spectra as indicated.

lone pair peaks $5t_2$, $1e_1$, $1t_1$, but the alignment of the shoulder at 523.5 ~ 524 eV, with both the high energy feature in SiKβ and the Si$L_{2,3}$M peak at 95 eV, shows that both Si3d orbitals and $4t_2$–$5t_2$ σ, π mixing contribute to the bonding in silica.

It is difficult, in these spectra, to identify features that can be directly related to the formation of Si–O–Si bridges from one SiO_4 unit to the next. In the SiKβ spectrum, the relative intensity of the high energy feature at ~ 1836 eV has been correlated[34] with the Si–O–Si bridging angles and the width of the SiKβ may also be related to the way in which silicate tetrahedra are linked. But more pronounced changes should be seen in the oxygen Kα spectra – after all, it is the O2p orbitals that are most actively concerned in formation of silicon–silicon bridges. It is quite possible that the structure seen in OKα (521–523–524.5 eV) reflects, not so much idiosyncratic effects within each SiO_4 unit, but rather the formation of different molecular orbitals between silicate tetrahedra. This point needs further research to establish the relationship between the OKα peak profile and the Si–O–Si bond angle.

2.4.7 Feldspar (microcline $KAlSi_3O_8$)

The close structural relationship between quartz and the feldspars is reflected in the X-ray emission spectra (Fig. 2.13).[49] The oxygen Kα has a main peak at

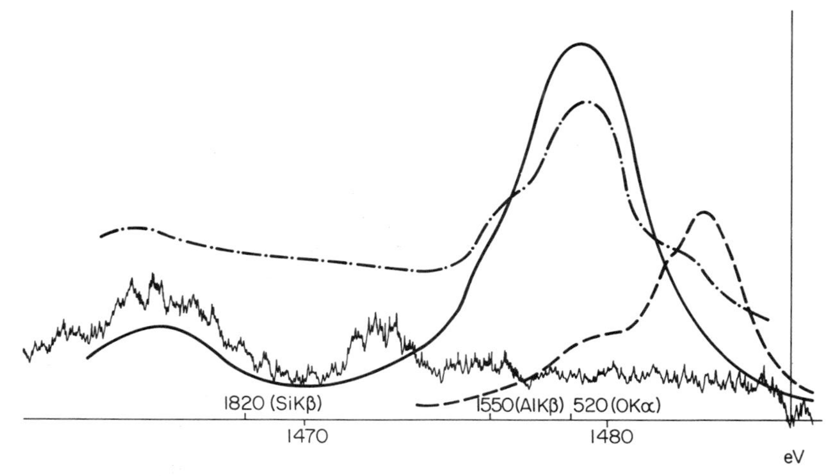

Fig. 2.13 X-ray emission (SiKβ-, AlKβ–·–·–, OKα---) and X-ray photoelectron spectra for microcline feldspar $KAlSi_3O_8$. X-ray photoelectron kinetic energy is shown on the lower scale, energy markers for the X-ray emission spectra are shown on the upper scale. The XPS peak at 1472 is due to K3p.

524.5 eV with a slight shoulder at 523 eV and a distinct plateau at 520–521 eV with a very weak emission at 518 eV. The plateau region aligns with the main Kβ peak for both silicon and aluminium, indicating that molecular orbitals with binding energies of $8 \sim 9$ eV are responsible for the Al–O and Si–O bonding in feldspars. There are interesting suggestions of further structure on the low X-ray energy side of both the aluminium and silicon Kβ peaks. If quartz is a guide, then orbitals with Si, Al3s character might be found here, but in the absence of L spectra, this cannot be confirmed. The reduction in symmetry brought about by the replacement of silicon by aluminium might well encourage 3s/3p mixing. The silicon and aluminium spectra differ on the high energy X-ray side, where aluminium shows much more prominent structure, structure which aligns with the main oxygen Kα peak.

The most important aspect of these spectra is the way in which they demonstrate the importance of the delocalized molecular orbital model for the electronic structure of feldspars. The main silicon and aluminium Kβ peaks are superimposed, indicative of the formation of orbitals with Si3p, Al3p and O2p character. There is no evidence for distinct and discrete Si–O and Al–O bonds.

2.4.8 Pyrite FeS_2

The bonding in pyrite has been extensively studied both theoretically and experimentally.[50]* The results presented here can, for example, be compared

*See also Chapter 7 of this book.

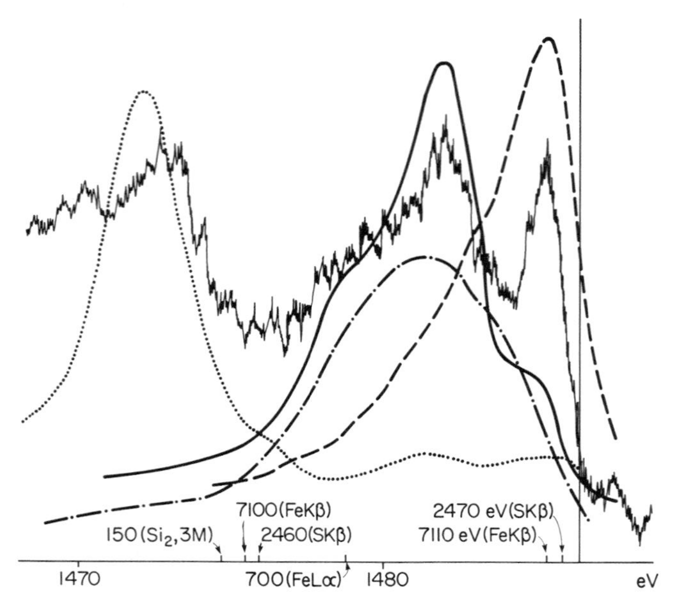

Fig. 2.14 X-ray emission (SKβ-, $L_{2,3}M\ldots$, Fe $K\beta_{2,5}$-·-·-, $L\alpha_{1,2}$---) and X-ray photoelectron spectra for iron pyrite (FeS_2). X-ray photoelectron energy is shown on the lower scale, energy markers for the X-ray emission spectra on the upper scale.

with a similar resume made by Berg *et al.*[51] All relevant XP and XE spectra are aligned and shown in Fig. 2.14. The XP peak from the least tightly bound orbitals must, since it aligns with the main $FeL\alpha_{1,2}$ peak, represent the $3d^6$ electrons on the ferrous ion. The presence of another structure in the $FeL\alpha_{1,2}$ spectrum which lines up with the $SK\beta_{1,3}$ peak, gives evidence for the formation of covalent bonds between iron and sulphur. The sulphur $K\beta$ spectrum is quite complex but can be rationalized as due to S3p character in: a sulphur–sulphur σ bond (2463 eV), sulphur–sulphur π bonds (2466 eV) and a sulphur–sulphur π^* bond(2470eV), all in the $[S–S]^{2-}$ disulphide anion. There is also evidence, from the weak peak at 2455 eV which lines up with the main $SL_{2,3}M$ peak (3s character) at 148 eV, of a very little 3s–3p mixing as part of the S–S σ bonding. Particular interest lies in the weakness of the peak at 2470 eV since the antibonding π^* orbital should be fully occupied in disulphide. The alignment with the main peak of the iron $L\alpha_{1,2}$ spectrum may, therefore, be particularly significant indicating charge delocalization from sulphur to iron. The very weak feature at 161 eV in the sulphur L spectrum also aligns with these peaks: it indicates that a little sulphur 3d character is also present in these iron–sulphur orbitals. The principal shoulder in the $FeL\alpha_{1,2}$ spectrum aligns, however, with the main sulphur $K\beta$ peak associated with π bonding in the disulphide. This indicates that ligand activity of the disulphide anion is also associated with the π orbitals of S_2^{2-} as might be expected.

An alternative view has been proposed by Tossell[52] based on $X\alpha$ calculations for the disulphide anion. These results predict that the ionization energies of the S3p σ and S3p π orbitals should be very similar, so the $SK\beta_{1,3}$ peak at 2463 eV is thought to be due to transitions from these three orbitals with the peak at 2466 eV then being associated with the two π orbitals. The weak peak at 2470 eV could then be taken as evidence for some delocalization of the iron $3d^6$ electrons.

Neither interpretation of the sulphur $K\beta$ spectrum is without problems. If the peak at 2470 eV is to be associated with π^* orbitals, then why is it so weak? But on the other hand if the peaks at 2463 eV and 2466 eV are due to $\sigma + \pi$ and π^* respectively, then why is the latter more intense than the former? More work on related disulphide compounds is required to resolve this problem.

2.5 FURTHER DEVELOPMENTS

All of the spectra that have been discussed so far have been taken from powdered samples and isotropic X-ray or photoelectron emission has been assumed. But the use of such powdered samples destroys valuable additional information which could be obtained from crystalline materials. This is because the emission of X-rays and photoelectrons is far from isotropic. This

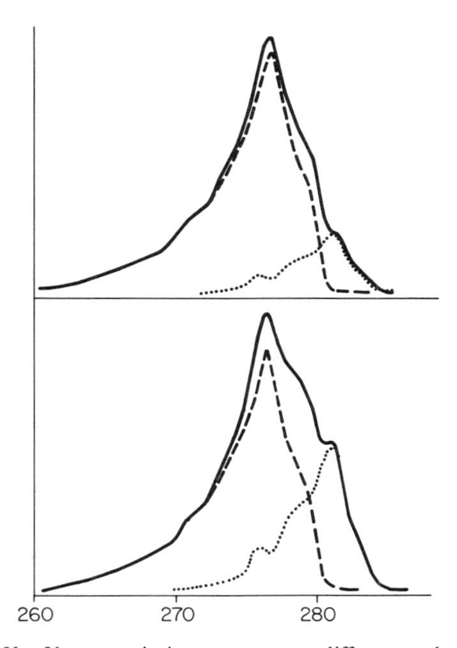

Fig. 2.15 Carbon $K\alpha$ X-ray emission spectra at different take-off angles from the graphite crystal surface; upper curves 55°, lower curves 30°. The solid line represents the experimental data, the dashed line, the σ component, the dotted line π.

point has been particularly well made by Wiech and his co-workers[53] in their studies of X-ray emission from layered materials such as graphite and from crystalline minerals such as calcite.

The electronic structure of the valence band of graphite can be built up from carbon $2p_z$ orbitals perpendicular to, and carbon $2p_x$ and $2p_y$ and 2s orbitals lying within the sheets of carbon atoms that make up the graphite structure. The π orbitals will therefore have z character and σ orbitals x, y character. Transitions from π orbitals to carbon ls vacancies will, therefore, generate X-rays that only propagate in the x–y plane (perpendicular to the electric vector in the z direction) whereas transitions $\sigma \rightarrow$ Cls will give rise to X-rays in any plane which contains the z axis. If the CKα spectrum of graphite is measured in the direction perpendicular to the plane of cleavage, only X-rays from transitions involving 2p character in σ orbitals will be observed, but as the angle of view is increased, away from the perpendicular, so increasing contributions due to 2p character in π orbitals are found.[54] This is shown in Fig. 2.15. As a result of such detailed investigations and by taking into account XP and UP spectra, it is possible to determine the electronic structure of graphite completely in terms of carbon 2s, carbon 2p σ and carbon 2p π contributions.

Using much the same technique of detailed study of anisotropic X-ray emission from crystalline materials (but not necessarily layered materials)

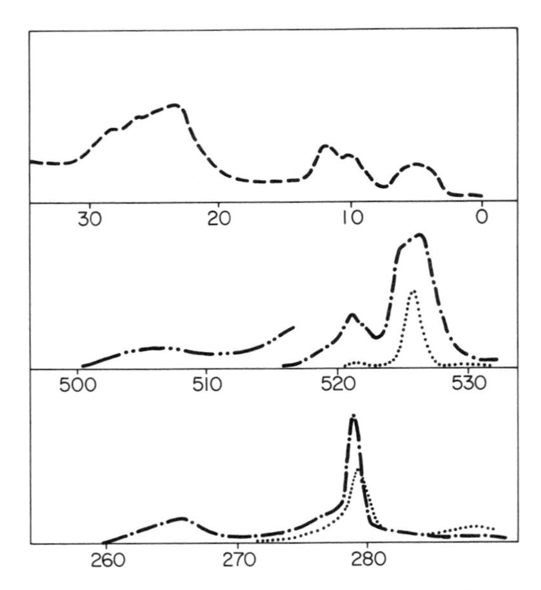

Fig. 2.16 X-ray photoelectron and X-ray emission spectra from lithium carbonate (upper XPS curve) and calcite (middle, oxygen and lower, carbon, XE spectra). An analysis of XE spectra at different take-off angles enables σ(–·–·–) and π(. . . .) contributions to the spectra to be determined (–··–··–, $\sigma \times 10$).

Tegeler *et al.*[55] have also investigated the electronic structure of the carbonate anion in calcite. The results are summarized in Fig. 2.16. As can be seen it is possible to distinguish both oxygen and carbon 2p contributions both to σ and to π bonds.

The potential of this approach to the study of the electronic structure of minerals is clearly enormous, if for no better reason than the fact that most minerals are well crystallized. And there is a further twist to the study of anisotropic X-ray emission from which even more detailed information can be obtained. The analysing crystals used to reflect X-rays at the Bragg angle within the spectrometer do so anisotropically depending upon the orientation of the electric vector of the X-rays relative to the diffracting planes of the crystal. A knowledge of the relative positions of both sample and crystal can therefore give even more detailed information about the roles played, not just by different types of orbital, but even of specific orbitals, in the electronic structures of minerals.

2.6 CONCLUSIONS

High resolution X-ray emission and photoelectron spectra can be used in combination to provide a detailed picture of the electronic structure of minerals. The chemical bond is, in effect, dissected into its component atomic parts and the relative importance of each determined. The method can be applied for all elements (save hydrogen and helium) with the sole caveat that each should be present at a unique site so as to avoid confusion due to overlapping spectra. Even if complete sets of XE and XP data are not available, peak profiles, peaks shifts, and the formation of new peaks in individual spectra, can be related to structural and chemical properties such as valence state, coordination number, chemical nature of ligand atoms, and even bond angles at the ligand. Whilst these data are best interpreted using models for electronic structure derived from sophisticated calculations (Chapter 1), the very simplest form of one electron molecular orbital theory can often give an adequate qualitative framework for the interpretation of spectra.

REFERENCES

1. Pauling, L. and Wilson, E.B. (1935) *Introduction to Quantum Mechanics*, McGraw-Hill, New York and London.
2. Murrell, J.N., Kettle, S.F.A. and Tedder, J.M. (1965) *Valence Theory*, John Wiley, London and New York.
3. Siegbahn, K., Nordling, N., Johansson, G. *et al.* (1969) *ESCA Applied to Free Molecules*, North-Holland, Amsterdam.
4. Baker, A.D., Brundle, C.R. and Thompson, M. (1972) *Chem. Soc. Rev.* **1**, 355.
5. Compton, A.H. and Allinson, S.K. (1935) *X-rays in Theory and Experiment*, Van Nostrand, New York.

6. Agarwal, B.K. (1979) *X-ray Spectroscopy*, Springer-Verlag, Heidelberg.
7. Urch, D.S. (1971) *Quart. Rev.*, **25**, 343.
8. Price, W.C., Potts, A.W. and Streets, D.G. (1972) in *Electron Spectroscopy* (Proc. Int. Conference, Asilamar, 1971) ed. D.A. Shirley, North-Holland, Amsterdam, p. 187.
9. See for example, Eyring, H., Walter, J. and Kimball, G.E. (1944) *Quantum Chemistry*, Ch. 8, John Wiley, New York.
10. Urch, D.S. (1970) *J. Phys. C, Solid State Phys.*, **3**, 1275.
11. Turner, D.W., Baker, C., Baker, A.D. and Brundle, C.R. (1970) *Molecular Photoelectron Spectroscopy*, Wiley-Interscience, London, New York.
12. Price, W.C. (1970) in *Molecular Spectroscopy*, The Institute of Petroleum, London.
13. Kim, K.S., Baitinger, W.E., Amy, J.W. and Winograd, N. (1974) *J. Elect. Spectr. Rel. Phenom.* **5**, 247.
14. Storp, S. and Holm, R. (1979) *J. Elect. Spectr. Rel. Phenom.*, **16**, 183.
15. Kasrai, M. and Urch, D.S. (1979) *J. chem. Soc. Faraday Trans. II*, **75**, 1522.
16. Brundle, C.R., Chuang, T.J. and Wandelt, K. (1977) *Surf. Sci.*, **68**, 459.
17. Urch, M.J.S. and Urch, D.S. (1980) *ESCA-Auger Tables (Al, Mg)*, Queen Mary College, Chemistry Department, London.
18. Birks, L.S. (1969) *X-ray Spectrochemical Analysis*, Wiley-Interscience, New York.
19. Jenkins, R. and de Vries, J.L. (1970) *Practical X-ray Spectroscopy*, MacMillan, London.
20. Haycock, D.E. and Urch, D.S. (1982) *J. Phys. E. Sci. Instrum.* **15**, 40.
21. Haycock, D.E. and Urch, D.S. (1978) *X-ray Spectrometry*, **7**, 206.
22. Azaroff, L.V. (ed.) (1974) *X-ray Spectroscopy*, McGraw-Hill, New York.
23. Henke, B.L. (1962) *Adv. X-ray Anal.* **6**, 361.
24. Henke, B.L. and Taniguchi, K. (1976) *J. appl. Phys.* **47**, 1027.
25. Wiech, G. (1966) *Z. Phys.* **193**, 490.
26. Barbee, T.W. (1981) in *Low-Energy X-ray Diagnostics – 1981*, (eds. D.T. Attwood and B.L. Henke) American Institute of Physics, New York (AIP Conf. Proc. No. 75), p. 133.
27. Urch, D.S. (1979) in *Electron Spectroscopy: Theory Techniques and Applications* (ed. C.R. Brundle and A.D. Baker) Vol. 3, Academic Press, London and New York, p. 1.
28. Day, D.E. (1963) *Nature Lond.*, **200**, 649.
29. Tsutsumi, K. and Nakamori, H. (1968) *J. phys. Soc. Japan*, **25**, 1418.
30. Slater, R.A. and Urch, D.S. (1972) *J. chem. Soc. Chem. Comm.*, 564.
31. Wood, P.R. and Urch, D.S. (1978) *X-ray Spectrometry*, **7**, 9.
32. Hogarth, A.J.C.L. and Urch, D.S. (1976) *J. chem. Soc. Dalton*, 794.
33. Esmail, E.I., Nicholls, C.J. and Urch, D.S. (1973) *The Analyst*, **98**, 725.
34. Wiech, G., Zopf, E., Chun, H-U and Brukner, R. (1976) *J. Non-Cryst. Solids*, **21**, 251.
35. Alter, A. and Wiech, G. (1978) *Jap. J. appl. Phys.*, **17** (supp. 2), 288.

36. Myers, K. and Andermann, G. (1972) *J. chem. Soc. Chem. Comm.*, 934.
37. Nicholls, C.J. and Urch. D.S. (1975) *J. chem. Soc. Dalton*, 2143.
38. Haycock, D.E., Nicholls, C.J., Webber, M.J. *et al.* (1978) *J. chem. Soc. Dalton*, 1785.
39. Haycock, D.E., Kasrai, M., Nicholls, C.J. and Urch, D.S. (1978) *J. chem. Soc. Dalton*, 1791.
40. Scofield, J.H. (1976) *J. Elect. Spectr. Rel. Phenom.*, **8**, 129.
41. Tolon, C. and Urch, D.S. unpublished results.
42. Wiech, G. personal communication.
43. Brytov, I.A., Dikov, Yu.P., Romashchenko, Yu.N. *et al.* (1976) *Inv. Akad. Nauk SSR, Seriya Fiz.*, **40**, 413.
44. Al-Kadier, M.A.M., Tolon, C. and Urch, D.S. (1984) *J. chem. Soc. Faraday Trans. II*, **80**, 669
45. Urch, D.S. (1979) *Orbitals and Symmetry*, 2nd printing, MacMillan, London.
46. Cotton, F.A. (1971) *Chemical Applications of Group Theory*, 2nd edn, Wiley-Interscience, New York.
47. Wiech, G. (1967) *Z. Physik*, **207**, 428.
48. Klein, G. (1973) *Proc. Int. Symp. X-ray Spectra and Electronic Structure of Matter*, Munich 1972 (ed. A. Faessler and G. Wiech) Vol. 2, p. 362.
49. Tolon, C. and Urch, D.S. unpublished results.
50. Wiech, G., Koppen, W. and Urch, D.S. (1972) *Inorg. Chim. Acta*, **6**, 376.
51. Berg, U., Drager, G., Mosebach, K., and Brummer, O. (1976) *Phys. Stat. Sol.*, **B75**, K89.
52. Tossell, J.A. (1977) *J. Chem. Phys.*, **66**, 5712.
53. Wiech, G. (1981) in *Inner-Shell and X-ray Physics of Atoms and Solids* (ed. D.J. Fabian, H. Kleinpoppen and L.M. Watson), Plenum Press, New York, p. 815.
54. Beyreuther, C., Hierl, R. and Wiech, G. (1975) *Ber. Buns. Phys. Chem.*, **79**, 1082.
55. Tegeler, E., Kosuch, N., Wiech, G. and Faessler, A. (1980) *J. Elect. Spectr. Rel. Phenom.*, **18**, 23.

3

Electronic Spectra of Minerals

Roger G. Burns

3.1 INTRODUCTION

This chapter describes origins and applications of electronic spectra of transition metal-bearing oxide and silicate minerals in the visible and nearby ultra-violet and infra-red regions. Such optical spectra are manifested by the colours and pleochroism of many minerals in transmitted light, properties used routinely by earth scientists to identify rock-forming minerals. Eye-catching, coloured, naturally-occurring inorganic compounds have appealed to mankind since antiquity in his quest for attractive pigments and beautiful gemstones; many of the earliest spectral measurements of minerals were made on large, highly-polished rare gems. However, during the past twenty years, the availability of high resolution spectrophotometers and the technology to make spectral measurements on very small crystals by using microscope accessories, have produced a vast body of experimental data for many transition metal ions in common rock-forming minerals.

Mineral structures provide a diversity of coordination sites, making it possible to measure spectra and to determine electronic energy levels of transition metal ions in coordination polyhedra having a variety of symmetries, distortions, bond-types and numbers of nearest-neighbour ligands, as well as different next nearest-neighbour metal–metal interactions. Such a diversity of coordination environments is not available in more conventional synthetic inorganic salts and organometallic compounds. Moreover, the generally high thermal stabilities of most silicate and oxide minerals make it possible to measure optical spectra of these transition metal-bearing phases at elevated temperatures and very high pressures. These features render mineral spectra interesting both to solid state physicists and chemists desiring fundamental information about transition metal cations in novel bonding situations, and to Earth and planetary scientists wanting to interpret geochemical and geophysical properties of minerals in the Earth and on surfaces of planets.[1-4] The focus of this chapter is on spectral measurements of the more common rock-forming minerals containing the most abundant transition metal ions in the region 33 000–4000 cm^{-1} (300–2500 nm), highlighting information that has been obtained in the past decade.

3.2 BACKGROUND

3.2.1 Mineralogical and geochemical constraints

The colours, and hence visible region spectra, of most minerals are directly attributable to the presence of transition metal ions in their crystal structures. Electronic transitions involving incompletely filled d or f orbitals usually predominate over other colour-producing mechanisms,[5,6] including light-scattering or diffraction phenomena (e.g. moonstone, opal) and electrons trapped in lattice defects or interstices (e.g. zircon, topaz, diamond). Cations of the first series transition elements are responsible for the colours and spectra of the common rock-forming minerals because their relatively higher abundances, appropriate valencies, and suitable ionic radii enable them to substitute readily for host Mg^{2+}, Al^{3+} or Si^{4+} ions in the crystal structures. The major minerals occurring in the Earth's crust and mantle, Moon-rocks, and meteorites include olivine, pyroxenes, garnets and dense polymorphs with spinel, corundum, ilmenite, perovskite and periclase structures. This chapter, therefore, describes optical spectra of first-series transition metal ions in these more common silicate minerals and their derivative dense oxide structure-types.

The most abundant transition metals in the Earth's crust and mantle are Fe, Ti, Cr and Mn. These metals, and transition elements in general, are renowned for their diversity of oxidation states in chemical compounds. In silicates and oxides, however, the range of valencies exhibited by first series transition elements is somewhat restricted. On Earth, $Fe^{(II)}$, $Fe^{(III)}$, $Ti^{(IV)}$, $Cr^{(III)}$, $Cr^{(VI)}$, $Mn^{(II)}$, $Mn^{(III)}$ $Mn^{(IV)}$, and perhaps $Cr^{(II)}$ and $Ti^{(III)}$ exist in natural environments, while $Fe^{(II)}$, $Ti^{(III)}$, $Ti^{(IV)}$ $Cr^{(II)}$, $Cr^{(III)}$, and $Mn^{(II)}$ occur on the Moon. Ferric iron appears to predominate on the surface of Mars. In this chapter, electronic spectra of Fe^{2+}, Fe^{3+}, Ti^{3+}, Ti^{4+} and Cr^{3+} ions only in host silicate and oxide minerals are discussed. The emphasis is on results of recent measurements since much of the earlier literature is reviewed elsewhere.[1,3,7,8]

3.2.2 Crystal field spectra

Absorption of light by transition metal-bearing minerals in the visible (400–700 nm or 25 000–14 300 cm^{-1}) and nearby infra-red (700–2500 nm or 14 300–4000 cm^{-1}) and ultra-violet (300–400 nm or 33 300–25 000 cm^{-1})

regions originate from three types of electronic transition processes: crystal field, d–d, or intra-electronic transitions; metal–metal intervalence transitions; and oxygen → metal charge transfer.

The origin of crystal field spectra is familiar to most chemists and mineralogists,[1,3] and only salient features relevant to mineral spectra described later are summarized here. Crystal field spectra originate when electrons are excited by light between incompletely filled 3d orbital energy levels within the transition metal ion. The relative energies of the 3d orbitals are controlled by different repulsive energies between anions or ligands coordinated to the cation and its electrons occupying the five 3d orbitals.

The principal parameters derived from the positions of absorption bands in crystal field spectra include: Δ, the crystal field splitting parameter which contributes to thermodynamic properties; and B and C, the Racah parameters which provide a measure of the degree of covalent character of the cation–anion bonds. Factors contributing to crystal field spectra, crystal field splittings, and Racah parameters of transition metal-bearing minerals include:

(i) The type of anion coordinated to the cation; this is oxygen in the more abundant silicate and oxide phases of the crust and mantle. However, even in these minerals the oxygen bond-type may vary from H_2O and O^{2-} in simple hydrates and oxides, to OH^- in amphiboles, micas, clays, etc., to non-bridging $Si{-}O^-$ bonds in garnets, olivines, pyroxenes, amphiboles, etc., to bridging Si–O–Si linkages in pyroxenes and amphiboles.

(ii) The valence of the cation; crystal field splittings are larger for trivalent cations (Cr^{3+}, Fe^{3+}) than for corresponding divalent ions (Cr^{2+}, Fe^{2+}).

(iii) The coordination number (N) of the cation; thus, Δ decreases in the series octahedral ($N = 6$) > cubic (8) ≫ dodecahedral (12) > tetrahedral (4). As a result, Δ and derived crystal field stabilization energies (CFCE) may be higher for octahedrally coordinated cations in common ferromagnesian silicate (olivine, pyroxenes), spinel and periclase phases, and cubic coordinated ions in garnet, than for corresponding cations in the perovskite (8–12 coordination) structure and tetrahedral sites in spinels.

(iv) The symmetry of the oxygen coordination polyhedron about the cation. Additional absorption bands appear in the crystal field spectra when the coordination polyhedra are deformed to non-cubic symmetries, such as when a cation occurs in tetragonally or trigonally distorted octahedra or rhombic distorted cubes.

(v) The cation–oxygen interatomic distance, R. For a point-charge model (i.e. oxygen anions are assumed to be point negative charges surrounding the central cation):

$$\Delta \propto 1/R^5 \qquad (3.1)$$

Since increased pressure generally compresses (shortens) interatomic distances, Δ increases and absorption bands are expected to exhibit a 'blue shift' (i.e. move to shorter wavelengths or higher energies) in high P spectra. Thus,

$$\Delta_p/\Delta_o = (R_o/R_p)^5 \tag{3.2}$$

where Δ_p, Δ_o and R_p, R_o are crystal field splittings and average cation–oxygen distances at high P and atmospheric pressure, respectively. Similarly, at elevated temperatures

$$\Delta_T/\Delta_o = (V_o/V_T)^{5/3} = [1 + \beta(T - T_o)]^{-5/3} \tag{3.3}$$

where Δ_T, Δ_o and V_T, V_o are crystal field splittings and specific volumes at temperature T and 25° C (T_o), respectively, and β is the volume coefficient of thermal expansion. Generally, $V_T > V_o$, so that a 'red shift' (movement to lower energies) is expected for absorption bands in high temperature spectra.

The intensities of absorption bands depend on several factors, including:

(i) The type of cation. Thus, spin-forbidden transitions (e.g. in Fe^{3+}) are typically two orders of magnitude less intense than spin-allowed transitions (e.g. in Fe^{2+}, Cr^{3+}, Ti^{3+}). Some cations give both spin-allowed (s.a.) and spin-forbidden (s.f.) transitions. The Fe^{2+} s.a. bands occur in the near infra-red (10 000 cm^{-1}) but Fe^{2+} s.f. peaks are found in the visible region; however, important Cr^{3+} s.f. peaks occur near intense s.a. bands in the visible region.

(ii) The symmetry of the coordination site. Cations in non-centrosymmetric environments (e.g. tetrahedral sites; olivine M2 position) may produce absorption bands 1 or 2 orders of magnitude more intense than centrosymmetric sites (e.g. octahedral sites; olivine M1 position).

(iii) At high pressures and elevated temperatures intensities of absorption bands generally increase, due to effects of increased covalency and increased vibronic coupling, respectively. Temperature intensification of crystal field s.a. bands generally dominates over pressure-include effects, however.

To illustrate some of these factors influencing energies and intensities of crystal field spectra, consider the Fe^{2+} cation octahedrally coordinated to oxygen ligands in the periclase structure. The electronic configuration of Fe^{2+}, [A]$3d^6$, is such that at normal pressures ferrous ions have four unpaired electrons corresponding to the high spin state $(t_{2g})^4(e_g)^2$ in an octahedral crystal field. This ground state configuration is represented by $^5T_{2g}$ in group theory notation, to highlight the symmetry and three-fold degeneracy (T_2) of four electrons in the three t_{2g} orbitals in the centrosymmetric (subscript g) octahedral site, and the quintet spin multiplicity (superscript 5, calculated as number of unpaired electrons plus 1). A number of excited triplet and singlet

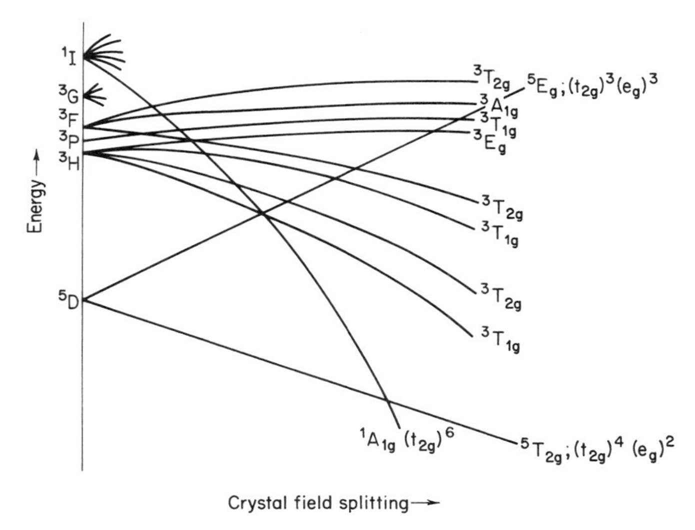

Fig. 3.1 Partial energy level diagram for Fe^{2+} in octahedral coordination.

states exist for Fe^{2+}, many of which are included in the energy level diagram illustrated in Fig. 3.1. The ^5E_g crystal field state corresponds to the quintet (superscript 5), doubly-degenerate (E), high spin configuration $(t_{2g})^3(eg^3)$, for three electrons in the two e_g orbitals, whereas the $^1A_{1g}$ state represents the unique, spherically symmetric (A) low-spin configuration $(t_{2g})^6(e_g)^0$ with zero unpaired electrons (singlet spin multiplicity) induced by large crystal field splittings (e.g. at high pressures, or by covalent ligands). The numerous triplet states depicted in Fig. 3.1 represent the large number of electronic configurations with two unpaired electrons such as $(t_{2g})^5(e_g)^1$ represented by the $^3T_{1g}$ and $^3T_{2g}$ excited states derived from the ^3H spectroscopic level.

Perhaps the most highly symmetric coordination environment available to the Fe^{2+} ion in a mineral is the periclase structure which consists of a cubic closest packed lattice of O^{2-} anions, in which all octahedral sites are filled by Mg^{2+} ions. The $[MgO_6]$ octahedra are undistorted and Mg^{2+} ions are centrosymmetric with all Mg–O distances equal to 210.6 pm. Each octahedron shares all its edges with twelve adjacent $[MgO_6]$ octahedra, and next-nearest neighbour Mg–Mg distances are 298 pm. Substitution of larger Fe^{2+} cations (octahedral radius 77 pm) for Mg^{2+} (72 pm) leads to a small expansion of the cubic cell parameter from 421.2 pm in pure MgO to 424 pm in magnesiowüstite $Mg_{0.74}Fe_{0.26}O$ (denoted as $Wü_{26}$). The absorption spectrum of this magnesiowüstite[9] illustrated in Fig. 3.2 consist of a broad asymmetric band in the near infra-red and sharper peaks in the visible region at 21 300 cm^{-1} and 26 200 cm^{-1}. The latter originate from the spin-forbidden $^5T_{2g} \rightarrow {}^3T_{1g}(H)$ and $^5T_{2g} \rightarrow {}^3T_{2g}(H)$ transitions, respectively, in Fe^{2+} (Fig. 3.1). Although one absorption band, corresponding to the spin-allowed $^5T_{2g} \rightarrow {}^5E_g$

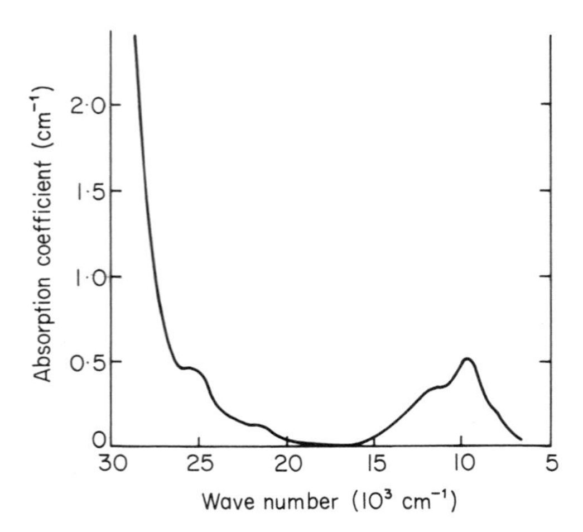

Fig. 3.2 Absorption spectrum of octahedrally coordinated Fe^{2+} ions in the periclase structure (from ref. 9). The magnesiowüstite contains 26 mol % FeO.

transition, might be expected in Fe^{2+} ions located in the regular octahedral sites in the periclase structure, the spectrum illustrated in Fig. 3.2 in fact shows two maxima at about 10 000 cm^{-1} and 11 600 cm^{-1} attributed to splitting of the degenerate ^5E_g crystal field state during the electronic transition. This dynamic Jahn–Teller effect occurs because the lifetime of the transition (approx. 10^{-13} s) is considerably smaller than the period of frequencies of vibrational modes so that the $[(Mg, Fe)O_6]$ octahedra of periclase are not actually distorted during the crystal field transition. In other mineral structures containing Fe^{2+} in distorted octahedral sites, the ^5E_g state is separated into two discrete energy levels.

The intensity scale shown in Fig. 3.2 is plotted as the absorption coefficient, α, where

$$\alpha = (\log I_o/I)/t \tag{3.4}$$

In Equation (3.4) I_o and I are the intensities of incident and emitted light incident on a crystal of thickness t. The molar extinction coefficient, ε, is related to α by

$$\varepsilon = \frac{\alpha}{C} = \frac{\alpha V}{1000X} \tag{3.5}$$

where C, V and X are concentration (moles $litre^{-1}$), molar volume (ml $mole^{-1}$), and mole fraction (e.g. 0.26 FeO in $Wü_{26}$), respectively. The value of ε calculated for the Fe^{2+} spin-allowed band located near 10 000 cm^{-1} in Fig. 3.2 is about 2 litre $mole^{-1}$ cm^{-1} which is significantly higher than intensities of crystal field transitions in octahedrally coordinated Fe^{3+} ions.

The electronic configuration of Fe^{3+}, $[A]3d^5$, coordinated to oxygen ligands is such that in the ground state the five 3d electrons occupy singly each 3d orbital, so that in this half-filled high-spin state Fe^{3+} ions have five unpaired electrons. This ground-state configuration is represented by the $^6A_{1g}$ crystal field state when Fe^{3+} ions occupy centrosymmetric octahedral sites. All crystal field transitions within Fe^{3+} ions are spin-forbidden and result in low intensity absorption bands because configurations of all possible excited states have only three (or one) unpaired electrons. A simplified energy level diagram of Fe^{3+} illustrated in Fig. 3.3 shows a number of quartet excited crystal field states originating from different orbital occupancies and symmetries of configurations with three unpaired electrons. Transitions to many of the low lying quartet states are observed in the spectra of Fe^{3+}-bearing minerals. In garnets of the grossularite–andradite solid solution series, $Ca_3(Al, Fe^{3+})_2(SiO_4)_3$, trivalent cations occupy six-coordinated sites slightly distorted from octahedral symmetry in which all six metal–oxygen distances are equal. Although these octahedra share edges with distorted cubic $[CaO_8]$ sites and share corners with tetrahedral $[SiO_4]$ sites, the $[FeO_6]$ octahedra are isolated from one another. Since garnet is a cubic mineral, absorption spectra are identical for all crystallographic orientations of polarized light through the crystals. The optical spectrum of gem quality andradite (topazolite or demantoid) illustrated in Fig. 3.4 shows absorption bands corresponding to specific Fe^{3+} crystal field transitions at 11 700 cm^{-1} ($^6A_{1g} \rightarrow {}^4T_{1g}(G)$), 17 000 cm^{-1} ($^6A_{1g} \rightarrow {}^4T_{2g}(G)$); 22 670 and 22 900 ($^6A_{1g} \rightarrow {}^4A_{1g}$, $^4E_g(G)$); 24 500 cm^{-1} ($^6A_{1g} \rightarrow {}^4T_{2g}(D)$) and 27 000 cm^{-1} ($^6A_{1g} \rightarrow {}^4E_g(D)$). The extinction coefficients of the first two absorption bands are particularly low and their large widths make them difficult to resolve.

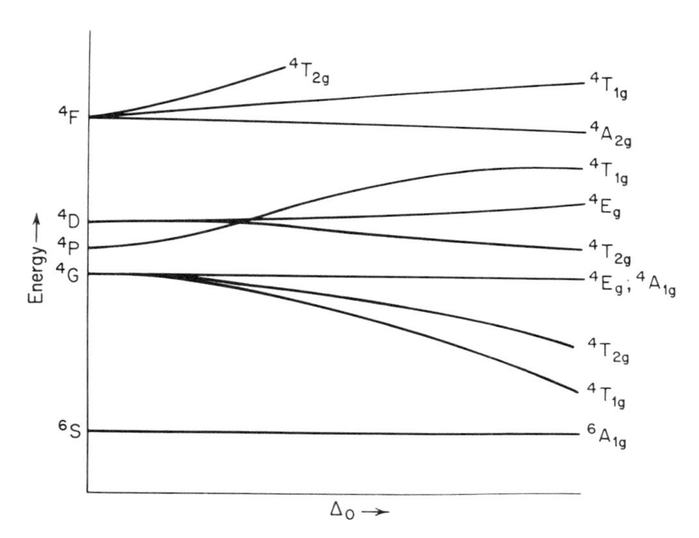

Fig. 3.3 Simplified energy level diagram for octahedrally coordinated Fe^{3+}.

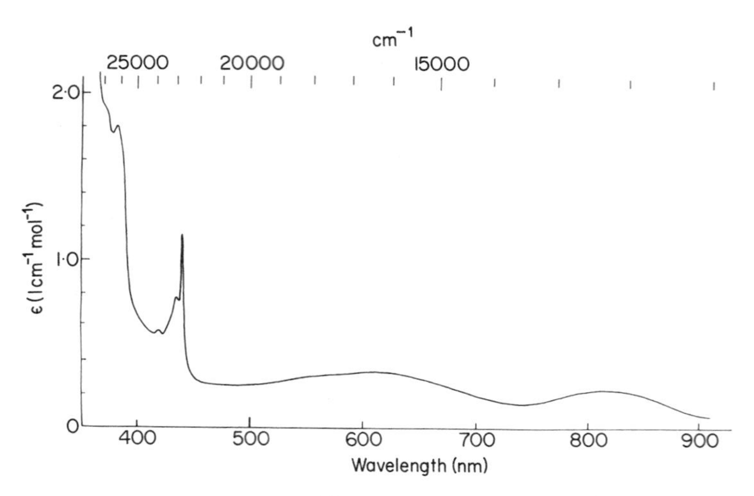

Fig. 3.4 Absorption spectrum of Fe^{3+} ions in octahedral sites of the garnet structure. The gem quality andradite contains 96 mol % $Ca_3Fe_2(SiO_4)_3$.

However, the very sharp, relatively intense peaks in the 22 000–23 000 cm^{-1} region are rather diagnostic of Fe^{3+} ions in silicates and oxides, and their appearance in the optical spectrum of a mineral usually indicates the presence of Fe^{3+} cations. Unlike other crystal field transitions discussed later, the energies of the $^6A_1 \rightarrow {}^4A_1, {}^4E(G)$ transition in Fe^{3+} are relatively unaffected by increased pressure and temperature.

3.2.3 Metal → metal intervalence transitions

Excitations of electrons between adjacent (next-nearest neighbour) transition metal ions give rise to metal → metal intervalence transitions.[10,37] Such electronic spectra originate in many mixed-valence compounds and have the following features:

(i) Depending on whether the same element or two different metals are involved, homonuclear intervalence transitions (e.g. $Fe^{2+} \rightarrow Fe^{3+}$; $Ti^{3+} \rightarrow Ti^{4+}$) are distinguished from heteronuclear intervalence transitions (e.g. $Fe^{2+} \rightarrow Ti^{4+}$).

(ii) Most intervalence transitions take place between octahedrally co-ordinated cations, although examples involving octahedral-tetrahedral (e.g. cordierite) and cubic-tetrahedral (e.g. garnet) pairs are known.

(iii) Most intervalence transitions occur between adjacent cations in edge-shared coordination polyhedra, although some examples involving face-shared octahedra (e.g. sapphire) are known.

(iv) Intervalence transitions are facilitated by short metal–metal interatomic distances.

(v) There is a strong polarization dependence of intervalence transitions. They occur only when a component of the electric vector of incident light is polarized along the metal–metal axis in the crystal structure.

(vi) Absorption bands are located principally in the visible region and generally occur at energies different from crystal field transitions.

(vii) Intensities of intervalence absorption bands depend on pressure, temperature and atomic substitution in the crystal structure. Increasing pressure intensifies intervalence transitions (analogous to crystal field spectra), while increasing temperature decreases their intensity (in contrast to crystal field spectra). Substitutional blocking of $Fe^{2+} \rightarrow Fe^{3+}$ intervalence transitions by non-transition metals such as Mg^{2+} and Al^{3+} affects the intensity (and extent of electron delocalization) in many mineral structures containing infinite chains or bands of edge-shared coordination polyhedra (e.g. glaucoplane-riebeckite series, $Na_2(Mg, Fe^{2+})_3 \cdot (Al, Fe^{3+})_2Si_8O_{22}(OH)_2$).

An excellent example of a mixed-valence mineral exhibiting a $Fe^{2+} \rightarrow Fe^{3+}$ intervalence transition in its optical spectrum is vivianite, $Fe_3 \cdot (PO_4)_2 \cdot 8H_2O$.[10,38] Freshly cleaved vivianite crystals or newly precipitated ferrous phosphate are initially pale green but turn blue when exposed

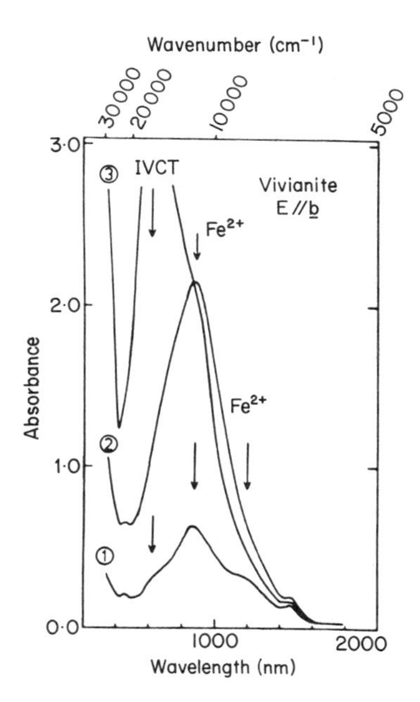

Fig. 3.5 Polarized absorption spectra of a vivianite crystal with zones of three different degrees of oxidation (from ref. 38). The arrows identify Fe^{2+} crystal field bands at 8300 cm^{-1} and 11 400 cm^{-1}, and the $Fe^{2+} \rightarrow Fe^{3+}$ intervalence charge transfer transition at 15 800 cm^{-1}.

to air. Such an intense blue coloration is atypical of pure Fe(II) or Fe(III) compounds containing Fe^{2+} or Fe^{3+} ions octahedrally coordinated to oxygen ligands. These colour variations are portrayed in the polarized absorption spectra of a zoned vivianite crystal illustrated in Fig. 3.5. The bands centred near 8300 cm^{-1} and 11 400 cm^{-1} originate from transitions to components of the ^5E_g crystal field state of Fe^{2+} ions located in distorted $[FeO_6]$ octahedra in the vivianite structure, and the weak peak at 22 200 cm^{-1} represents the Fe^{3+} spin-forbidden $^6A_{1g} \rightarrow {}^4A_{1g}, {}^4E_g$ transition. The band at about 6800 cm^{-1} represents overtones of the water stretching modes. The absorption band at 15 800 cm^{-1}, which intensifies in oxidized vivianite crystals, occurs when light is polarized along the b crystallographic axis, and originates from electron transfer between adjacent Fe^{2+} and Fe^{3+} ions separated by only 285 pm in dimers of edge-shared octahedra in the vivianite structure.

3.2.4 Oxygen → metal charge transfer

At sufficiently high energies, such as those encountered in the ultra-violet region, electronic transitions are induced between cations and nearest neighbour oxygens forming the coordination polyhedron about the metal. Features of such oxygen → metal charge transfer spectra include:[4]

(i) Absorption bands have very high intensities, perhaps 10^3–10^4 times higher than those of crystal field transitions.
(ii) Although absorption maxima are centred in the ultra-violet region, absorption edges or shoulders often extend into the visible region.
(iii) The energies of oxygen → metal charge transfer transitions depend on the cation and the symmetry of its coordination site. For octahedrally coordinated cations, oxygen → metal charge transfer energies are calculated to decrease in the order $Cr^{3+} > Ti^{3+} > Fe^{2+} > Ti^{4+} > Fe^{3+}$. Peaks for tetrahedral Fe^{2+} and Fe^{3+} cations occur at lower energies (longer wavelengths) than octahedral cations. Often, Fe^{3+}-bearing silicates have oxygen → Fe^{3+} charge transfer bands extending well into the visible region. With rising pressures and temperatures, absorption edges of oxygen → metal charge transfer bands show red shifts, absorbing increasing amounts of the visible region in high P and high T spectra.

The oxygen → Fe^{2+} absorption edge is conspicuous in the spectrum of magnesiowüstite illustrated in Fig. 3.2. Its extension into the visible region accentuates the intensity of the Fe^{2+} crystal field spin-forbidden peak near 26 000 cm^{-1}.

3.3 TECHNIQUES

Absorption spectral measurements of transparent solids are simple to perform when large single crystals are available. Such crystals are readily manipulated to produce highly polished, crystallographically orientated, mm to cm diameter surfaces of appropriate thickness. Before the advent of microscope accessories to spectrophotometers, systematic studies of electronic spectra of many common rock-forming minerals were hampered by the unavailability of suitably large crystals. Although powdered samples of small grainsize minerals can be used for diffuse reflectance spectroscopy (see Chapter 7), such spectral

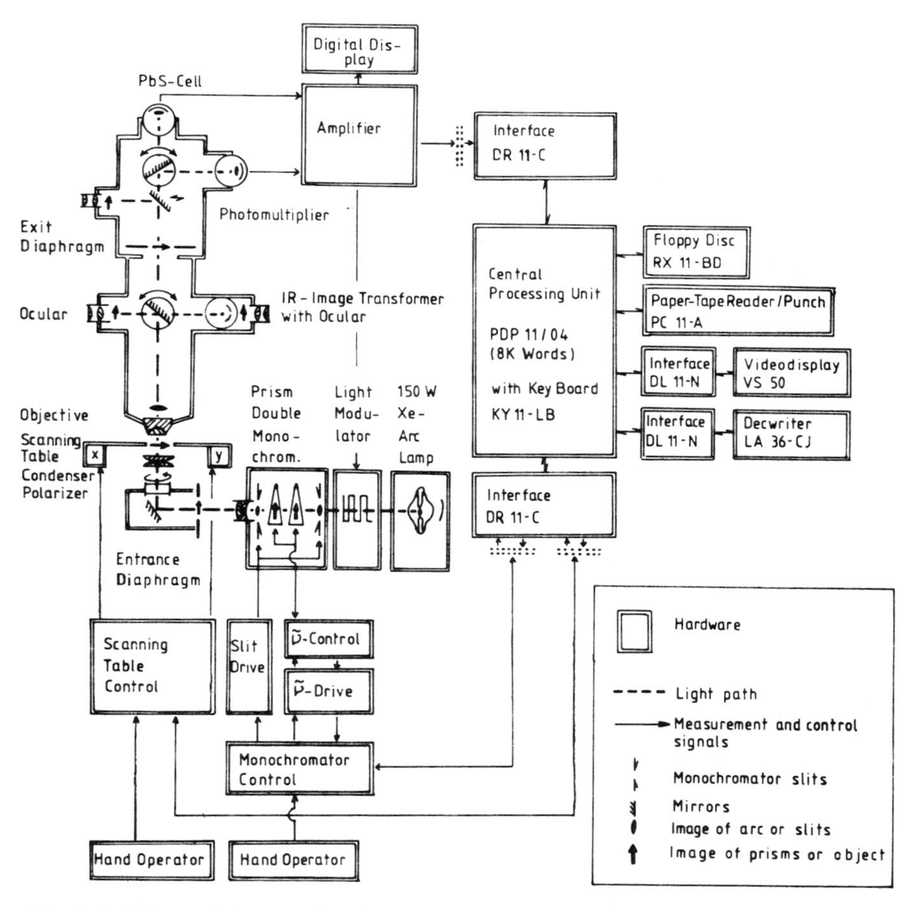

Fig. 3.6 Schematic layout of a microscope spectrophotometer system used to measure polarized absorption spectra of very small mineral crystals (from ref. 13). The computer-operated, single-beam instrument shown here comprises a polarizing microscope equipped with a stabilized light source (xenon arc lamp or tungsten lamp cover the range 250–2000 nm), a modulator which chops the light beam with a frequency of 50 Hz (the amplifier for the photodetector signals is modulated with the same phase and frequency) and a Zeiss prism double monochromator. Single crystals as small as 10 μm diameter may be measured with this system. A diamond-windowed high pressure cell can be readily mounted on the microscope scanning table for spectral measurements at very high pressures.

profiles lack the resolution and information attainable from absorption spectral measurements of single crystals in polarized light. There is also the disadvantage of diffuse reflectance spectroscopy that measurements require about a gram of powdered material, which may not be available for rare or exotic minerals occurring in moon-rocks and meteorites.

In the past decade a number of microscope absorption spectrophotometric systems have been developed[11–13] so that it is now possible to measure polarized absorption spectra of natural and synthetic crystals only tens of microns in diameter throughout the mid UV-visible-near-IR range. One such microscope spectrophotometer system is illustrated in Fig. 3.6. The evolution of microscope absorption spectroscopy has paralleled developments of high pressure diamond anvil cells, with the result that pressure variations of mineral spectra are now made routinely,[14–17] using the ruby fluorescence R_1 line for *in situ* pressure calibration. The stability of transparent diamonds to temperature has enabled diamond cells to be used in some high temperature spectral measurements.[18] One disadvantage of microscope spectrophotometry, however, is that while accurate measurements of positions (energies) of absorption bands are possible, quantitative estimates of intensities are subject to error[19] due to convergent light necessarily introduced by microscope optical systems. These errors originate from mixing of the polarization components of the various absorption bands.

3.4 CRYSTAL FIELD SPECTRA

3.4.1 Minerals containing Fe^{2+}

(a) *Periclase and spinel structures*

The optical spectrum of Fe^{2+} ions in the cubic periclase structure illustrated in Fig. 3.2 was discussed in Section 2.2. The effect of pressure on the Fe^{2+} spin-allowed transitions in the infra-red is to induce blue shifts (to higher energies) of the 10 000 cm^{-1} and 11 600 cm^{-1} bands.[20] The spin-forbidden transitions in the visible region are obscured by a red shift of the absorption edge of oxygen → iron charge transfer transitions into the visible region,[9,15] the significance of which is discussed in Sections 3.6.5(a) and 3.6.5(b).

The spinel crystal structure, like periclase, consists of a cubic closest packed lattice of O^{2-} anions, in which cations occupy half the octahedral sites and one-eight of the available tetrahedral sites. In spinel *per se*, $MgAl_2O_4$, containing small concentrations of Fe^{2+} ions, the divalent cations occur in a regular tetrahedral site in which metal–oxygen distances are all equal (193 pm). The crystal field spectrum of Fe^{2+}-bearing cubic spinel contains a broad band centred at 4830 cm^{-1} representing the $^5E \rightarrow {}^5T_2$ crystal field transition in tetrahedral Fe^{2+}, which again shifts to higher energies at high pressures.[20]

In spinel γ-Fe_2SiO_4, the high pressure polymorph of fayalite (α-Fe_2SiO_4), silicon occupies the tetrahedral sites and Fe^{2+} ions are located in trigonally-distorted octahedra. In this symmetry C_{3v} site, the octahedral $^5T_{2g}$ ground state is resolved into $^5A_1 + {}^5E'$ levels, and the ^5E_g excited state remains as $^5E''$. The optical spectrum of Fe^{2+} in cubic γ-Fe_2SiO_4 spinel[21] consists of a broad band centred at 11 430 cm^{-1}, representing the $^5A_1 \rightarrow {}^5E''$ transition. The $^5A_1 \rightarrow {}^5E'$ transition has not yet been identified, but is predicted to lie between 1000 cm^{-1} and 3000 cm^{-1} in the mid-infra-red region. Increased pressure again induces a blue shift of the 11 430 cm^{-1} band, but it becomes obscured by a strong pressure-induced red shift of the oxygen → Fe charge transfer band into the visible region.[15,22] These features are important in geophysical properties of the Earth's interior, including radiative heat transfer and electrical conductivity discussed in Section 3.6.5.

(b) *Olivines*

The electronic spectra of olivines have been studied extensively by numerous investigators. The popularity of this orthorhombic mineral to spectroscopists is attributed to (i) the availability of large euhedral, homogeneous crystals such as gem peridot from which highly polished, crystallographically orientated sections may be cut easily; (ii) the ubiquitous occurrence of olivines in igneous and metamorphic rocks, sedimentary placer deposits, meteorites and lunar samples; (iii) the importance of olivine (α-phase) and denser isochemical structure-types (e.g. spinel γ-phase) in the Earth's upper mantle; and (iv) the influence of olivines on geophysical properties of the Earth's interior, including radiative transfer of heat, electrical conductivity, and tectonic and seismic features associated with the olivine → spinel phase change.

From a crystal chemical standpoint, the olivine structure is particularly interesting because it contains Fe^{2+} ions in two six-coordinate sites which give distinguishable crystal field spectra.[1] Room-temperature Mössbauer spectra (see Chapter 5), on the other hand, resolve only one Fe^{2+} doublet. Thus, in the optical spectra illustrated in Fig. 3.7, Fe^{2+} ions in centrosymmetric, distorted M1 sites of olivine give rise to less intense broad bands near 11 000 cm^{-1} and 8000 cm^{-1} most clearly resolved in α- and β-polarized spectra, whereas Fe^{2+} ions in the non-centrosymmetric, distorted M2 sites are responsible for the sharper, more intense band at 9500 cm^{-1} in γ-polarized spectra. There are also a number of weak, sharp spin-forbidden peaks in the visible region the assignment of which has been controversial,[23] but most of which are attributable to Fe^{2+} ions. The extreme polarization dependence of the Fe^{2+} crystal field spectra in the near infra-red region is not detected by the eye; olivines are only weakly pleochroic in the visible region.

Compositional variations of the olivine spectra across the forsterite-fayalite series show that peak maxima of all Fe^{2+} crystal field bands move to lower energies with increasing Fe_2SiO_4 component.[1,23] At elevated temperatures, all bands intensify and move to lower energies.[18,24] High pressure spectral measurements of fayalite reveal blue shifts and negligible intensification of the

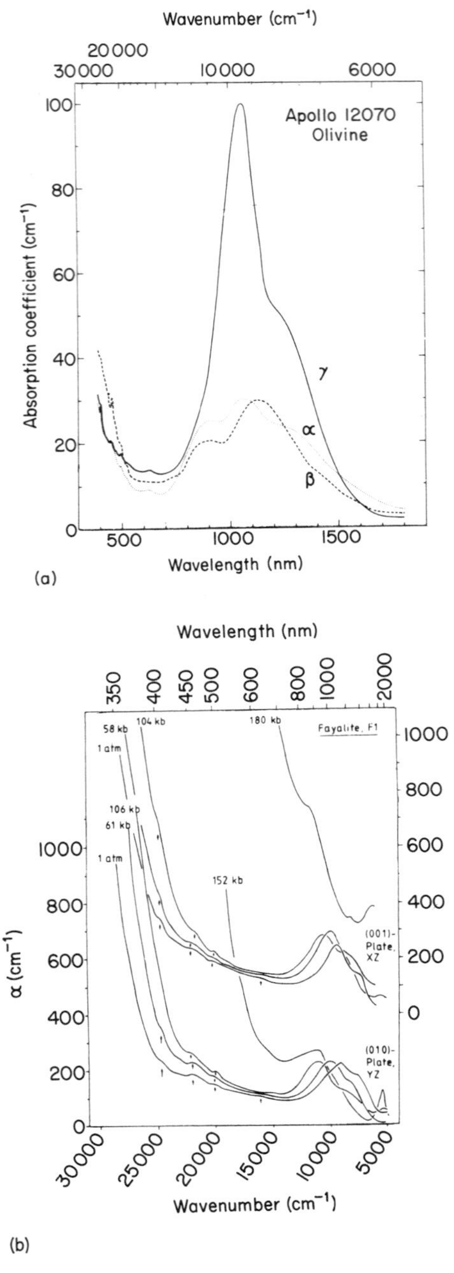

Fig. 3.7 Absorption spectra of olivines. (a) Polarized spectra of a lunar olivine containing 32 mol% Fe_2SiO_4 (from ref. 23). (b) Unpolarized spectra of orientated sections of fayalite at high pressures (from ref. 25).

Fe^{2+} crystal field bands.[14,25] However, as shown in Fig. 3.7(b) the oxygen → metal absorption edge shows a pronounced red shift into the visible region, particularly above 150 kb (15 GPa), indicating that olivines become opaque at high pressures. The significance of this in connection with heat transfer by radiation through the mantle is discussed in Section 3.6.5(a).

(c) *Pyroxenes*

Second only to olivines, the electronic spectra of pyroxenes have been studied extensively.[26] Interest in pyroxene crystal field spectra is attributable to the occurrence of this mineral as the major ferromagnesian silicate in many basic igneous rocks, granulite facies metamorphic rocks, and meteorites. Pyroxenes also predominate in Moon rocks and are responsible for the most conspicuous features measured in remote-sensed reflectance spectral profiles of the Moon's surface and lunar samples. The paragenesis, cleavages and crystal habits of pyroxenes often result in their occurrence as small crystallites. The rapid growth of pyroxene absorption spectroscopy is due in part to the development of microscope accessories to spectrophotometers and the availability of exotic chemical compositions in rocks brought back from the Moon.

The pyroxene structure is also of considerable interest to mineral spectroscopists because it again contains distinguishable coordination sites yielding distinctive Fe^{2+} crystal field spectra and two Fe^{2+} doublets in the Mössbauer spectra. Moreover, in contrast to the olivine structure, Fe^{2+} ions show strong site preferences in pyroxenes, so that there are major compositional variations in the spectra.

In orthopyroxenes, the six-coordinate M1 site approaches the configuration of a regular octahedron, but individual metal–oxygen distances are variable.[1,27] All of the oxygens bonded to cations in the M1 sites are non-bridging oxygens linked to only one silicon atom. The M2 site is very distorted, and metal–oxygen distances span a wide range. The smaller distances are to four non-bridging oxygens bound to one silicon atom. Two longer distances are to bridging oxygens bonded to two silicon atoms. The increased distortion of the M2 site, its larger average metal–oxygen distance, and the different oxygen bond-types contribute to the strong enrichment of Fe^{2+} ions in the pyroxene M2 positions. The crystal field spectra of magnesium-rich orthopyroxenes such as those illustrated in Fig. 3.8, therefore, contain absorption bands around 2350 cm^{-1}, 5400 cm^{-1} and 11 000 cm^{-1} strongly polarized in the Y, X, and Z directions, respectively originating from Fe^{2+} in M2 sites.[27] In iron-rich orthopyroxenes, the latter two bands occur at 4900 cm^{-1} and 10 700 cm^{-1}, while additional bands appearing near 8500 cm^{-1} and within the envelope spanning 10 000–13 000 cm^{-1} originate from Fe^{2+} ions in M1 sites. The positions of the Fe^{2+} M1 site crystal field bands at 8600 cm^{-1} and 13 000 cm^{-1} may be positively identified from spectral measurements of heat-treated orthopyroxenes in which disordering of Fe^{2+} ions from M2 sites into

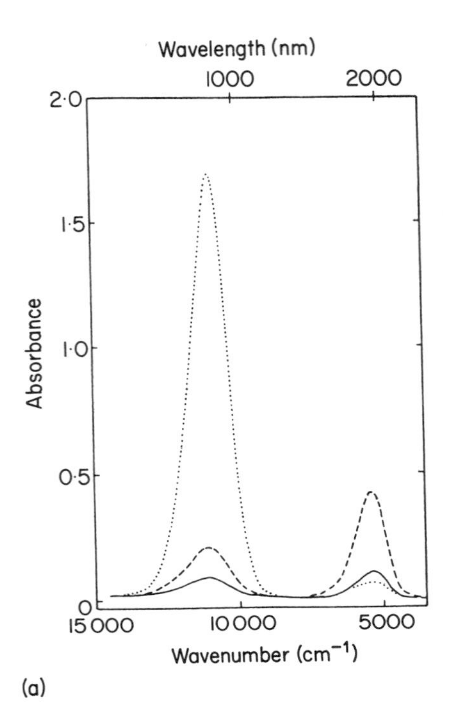

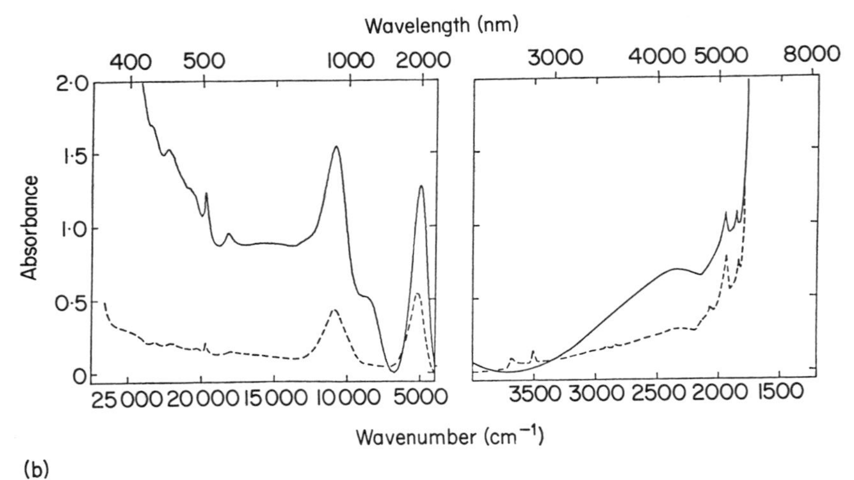

Fig. 3.8 Absorption spectra of orthopyroxenes. (a) Polarized spectra of a bronzite containing 14 mol% $FeSiO_3$ (from ref. 27). .... α-spectrum; --- β-spectrum:——γ-spectrum. (b) Spectra of two orthopyroxenes in γ-polarized light. The bands at 5400 cm^{-1} and 2350 cm^{-1} correlate in intensity and arise from Fe^{2+} in the M(2) site (from ref. 27). --- bronzite with 14 mol% $FeSiO_3$; ——hypersthene with 40 mol% $FeSiO_3$.

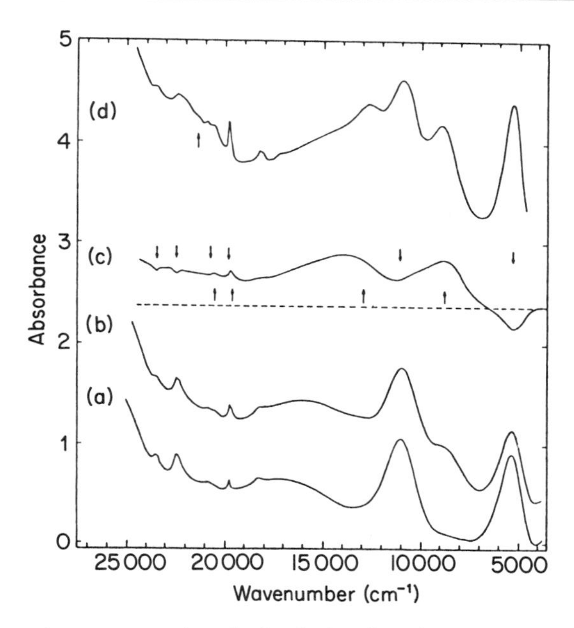

Fig. 3.9 Absorption spectra (γ-polarization) of orthopyroxenes (from ref. 19). (a) Bronzite containing 29 mol% $FeSiO_3$ before heat treatment. (b) Bronzite, Fs_{29}, after heating in a vacuum at 850° C for 24 h. (c) Difference plot obtained from subtracting unheated spectrum (a) from heated spectrum (b). Features due to Fe^{2+} in the M(1) and M(2) sites are denoted by arrows pointing up and down, respectively. (d) Hypersthene containing 40 mol% $FeSiO_3$, also heat treated, showing the Fe^{2+} M(1) absorption bands at 8600 cm^{-1} and 13 600 cm^{-1}.

M1 sites has been induced.[19] Such changes are shown in spectra illustrated in Fig. 3.9. There has been considerable debate over the group theoretical analysis of the Fe^{2+} M2 site crystal field spectra, but the energy level scheme derived by Goldman and Rossman[27] is now generally accepted. The effect of rising temperature[18] is to broaden, intensify slightly and induce a red shift of the bands at 11 000–10 700 cm^{-1} and 5400–4900 cm^{-1}. Increased pressure, however, causes these bands to move to higher energies.[14]

In monoclinic pyroxenes of the diopside-hedenbergite series, Ca^{2+} ions occupy M2 sites and Fe^{2+} ions occur in six-coordinated, distorted octahedral M1 sites, and give rise to spin-allowed bands at 10 300 cm^{-1} and 8300 cm^{-1}.[26] In augites, the deficiency of Ca^{2+} in M2 sites results in Fe^{2+} occupancy of these eight-coordinate, highly distorted sites. Such M2-site Fe^{2+} ions produce absorption bands at 9700 cm^{-1} and 4400 cm^{-1}. Examples of pyroxene absorption spectra are depicted in Fig. 3.10.[28]

The positions of the intense pyroxene M2-site Fe^{2+} crystal field bands in the 11 000–9700 cm^{-1} (1 μm) region and 5400–4400 cm^{-1} (2 μm) region serve to distinguish orthorhombic and monoclinic pyroxenes. This is illustrated in

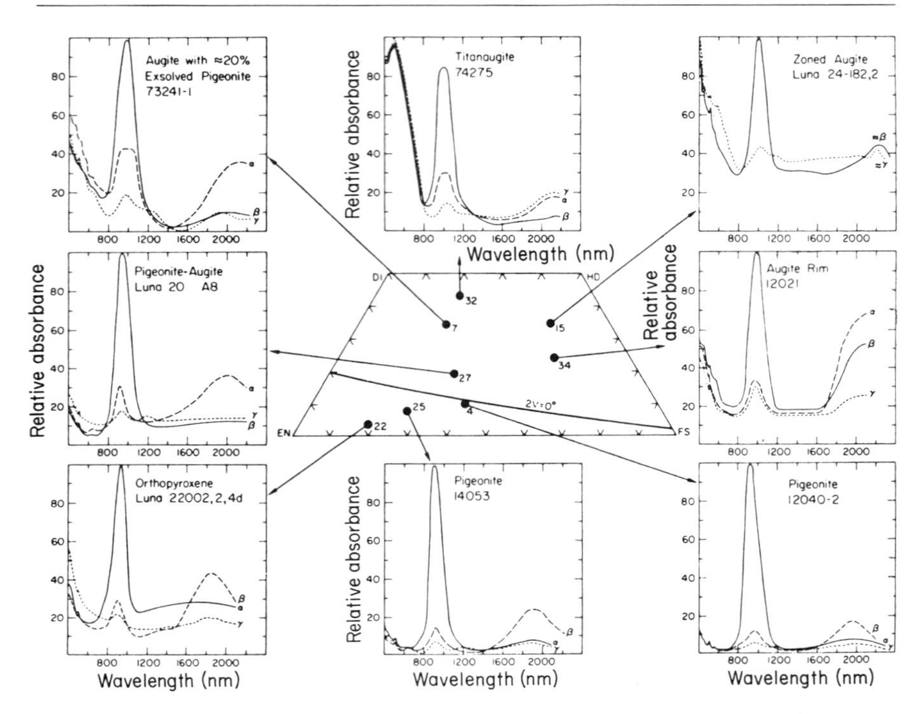

Fig. 3.10 Absorption spectra of pyroxenes from moon rocks (from ref. 28).

Fig. 3.11 in which the positions of the 1 μm and 2 μm bands are projected onto the pyroxene quadrilateral.[28] These intense pyroxene bands are utilized in planetary sciences to identify the compositions and structure-types of pyroxenes in reflectance spectra of lunar samples and remote-sensed reflectance spectral profiles of the Moon and other planetary surfaces measured through telescopes.[29] This aspect is discussed further in Section 3.6.2.

(d) *Garnets*

A distinguishing feature of cubic garnets of the pyrope-almandine series, $(Mg, Fe^{2+})_3Al_2(SiO_4)_3$, is the occurrence of Fe^{2+} ions in an eight-fold coordination site with the configuration of a distorted cube or triangular dodecahedron, which is unique to coordination chemistry. The optical spectrum of almandine[1] shows three Fe^{2+} spin-allowed bands centred in the near infra-red near 4400 cm^{-1}, 6000 cm^{-1} and 7700 cm^{-1}, and a number of spin-forbidden peaks in the visible region. The effect of increased pressure[20] is to induce blue-shifts to the spin-allowed bands, while high temperatures cause red-shifts and intensification of them.[30]

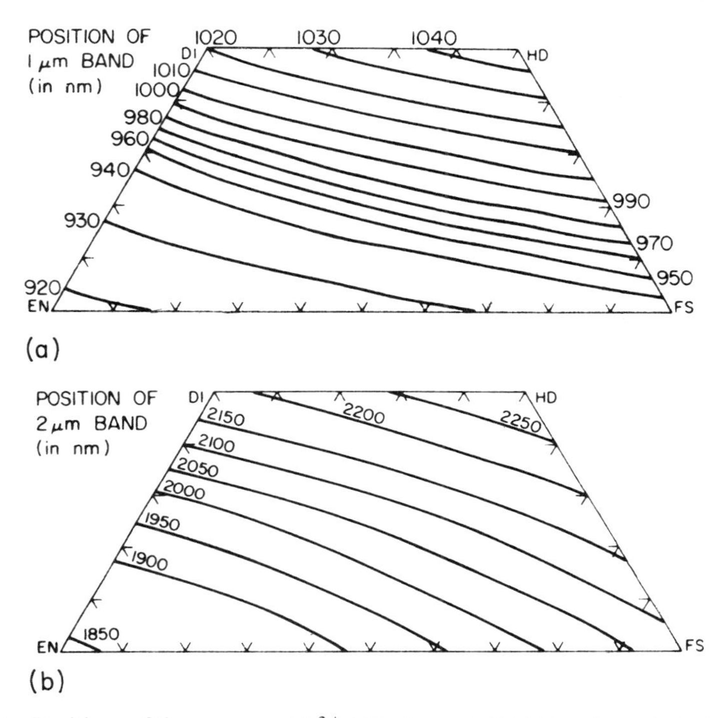

Fig. 3.11 Positions of the pyroxene Fe^{2+} M2-site crystal field bands projected on the pyroxene quadrilateral (from ref. 28). (a) The '1 μm' band (11 000–9700 cm^{-1}). (b) The '2 μm' band (5400–4400 cm^{-1}).

(e) *Other Fe^{2+} minerals*

We noted in Section 3.4.1 (d) that Fe^{2+} ions in the garnet eight-coordination sites is unique to the coordination chemistry of $Fe^{(II)}$ compounds. Another unusual coordination symmetry for Fe^{2+} ions is the square planar site of gillespite, $BaFeSi_4O_{10}$. This tetragonal mineral has been extensively studied both theoretically and under a variety of experimental conditions. Its optical spectra[1,8] show Fe^{2+} spin-allowed transitions near 20 000 cm^{-1} and 8000 cm^{-1}, the group theoretical analysis of which demonstrated that the relative energy and electronic configuration of the 3d orbitals is

$$d_{z^2}^2 < d_{xz}^1, d_{yz}^1 \ll d_{xy}^1 < d_{x^2-y^2}^1$$

The large energy separation, $\approx 20\,000$ cm^{-1}, between the d_{z^2} and $d_{x^2-y^2}$ orbitals, suggested that a high-spin to low-spin transition might be induced in Fe^{2+} at high pressures. Indeed, there is a conspicuous red to blue colour change around 2.5 Gpa. However, crystal structure refinements of the high pressure phase indicated that the square planar site becomes puckered to a distorted tetrahedron, so that Fe^{2+} ions remain in the high-spin state.

Other coordination symmetries found in common Fe^{2+}-bearing silicate minerals include a variety of distorted octahedral sites with different configurations (*cis* and *trans*) of OH^- ions and non-bridging oxygens bonded to silicon in amphiboles, micas and other phyllosilicates, and the distorted tetrahedral sites found in staurolite and melilite. Many of these Fe^{2+}-bearing minerals have been studied by Mössbauer and electronic absorption spectroscopy, and results are described elsewhere.[1,14,30]

3.4.2 Minerals containing Fe^{3+}

(a) *Corundum and haematite*

The crystal field spectra of octahedrally coordinated Fe^{3+} ions were discussed in Section 3.2.2, and a typical spectrum for Fe^{3+} ions in isolated octahedra occurring in the garnet structure is illustrated in Fig. 3.4. Much of the current interest in Fe^{3+} mineral spectra centres on the intensification of the spin-forbidden transitions by exchange interactions between next nearest-neighbour Fe^{3+} cations, such as those occurring in haematite and Fe^{3+}-bearing corundum.

The optical spectra of natural yellow sapphires and synthetic Fe^{3+}-doped Al_2O_3 crystals have been studied extensively[31] because they display incipient exchange interactions between neighbouring Fe^{3+} ions and are enhanced considerably in haematite. In the corundum structure Fe^{3+} ions substitute for Al^{3+} located in non-centrosymmetric trigonally distorted octahedral sites. Each $[AlO_6]$ octahedron shares a triangular face with another $[AlO_6]$ octahedron parallel to the *c* axis, and shares three of its edges with three adjacent $[AlO_6]$ octahedra perpendicular to the *c* axis. The metal–metal distances across the face-shared and edge-shared octahedra are 265 pm ($\parallel c$) and 279 pm (approx. $\perp c$), respectively. Although all Al^{3+} sites are crystallographically equivalent in the corundum structure, there are two magnetically inequivalent sites. All cations in a given (0001) plane of edge-shared octahedra are equivalent, but magnetically inequivalent to cations in adjacent (0001) planes separating the face-shared octahedra. In haematite such magnetic inequivalency manifests itself in antiferromagnetism. Compared with corundum, the Fe^{3+}–O distances in haematite are larger (reflecting the larger octahedral ionic radius of Fe^{3+}, 64.5 pm, relative to Al^{3+}, 53 pm), as are the Fe^{3+}–Fe^{3+} distances (289 pm $\parallel c$; 297 pm $\perp c$).

The optical spectra of Fe^{3+}-doped corundums and natural yellow sapphires[32] display pleochroism expected for Fe^{3+} ions located in trigonally distorted octahedra in a uniaxial (trigonal) mineral. Broad, weak bands at 9450 cm^{-1}, 14 350 cm^{-1}, and 18 700 cm^{-1} involve crystal field transitions to states derived from $^4T_1(G)$ and $^4T_2(G)$ (see Fig. 3.3). The $^6A_1 \rightarrow {}^4A_1, {}^4E(G)$ transition again generates a sharp peak at 22 220 cm^{-1} while peaks at 25 600 cm^{-1} and 26 700 cm^{-1} involve transitions to the $^4T_2(D)$ and $^4E(D)$ levels. The spectra of iron-bearing corundums are discussed further in Section 3.6.1 when the origin of blue coloration in sapphires is examined.

Studies of compositional variations of Fe^{3+}-doped Al_2O_3 optical spectra indicate that the peaks centred around 22 000 cm^{-1} and 26 000 cm^{-1} do not obey Beer's law, their intensities being accentuated anomalously by increasing Fe^{3+} concentration. The peaks also show unusual temperature dependencies by becoming more intense at lower temperatures.[32] These two effects are attributed to exchange interactions between pairs of Fe^{3+} ions replacing Al^{3+} in the face-shared octahedra in the corundum structure. Unusual intensifications of Fe^{3+} crystal field spectra are observed in several iron(III) minerals containing Fe^{3+} cations in adjacent coordination sites.[33] In haematite, for example, transitions are measurable in diffuse reflectance spectra of the $^4T_1$ and $^4T_2$ states only, giving rise to Fe^{3+} crystal field bands at 11 000 cm^{-1} and 19 000 cm^{-1}. The other spin-forbidden transitions at higher energies are too intense to be measured.

Exchange interactions have been invoked recently to interpret the optical spectra of a number of minerals in which crystal field transitions display unusual temperature dependencies.[34] Examples include Fe^{2+}-bearing micas and tourmalines, and blue sapphire and kyanite. More importantly, however, they set constraints on deductions about the iron oxide mineralogy of Mars' surface based on remote-sensed spectral measurements. This aspect is discussed in Section 3.6.3.

3.4.3 Minerals containing Ti

Titanium is second only to iron as the most abundant transition element. However, Ti is predominantly tetravalent on Earth and as such is stripped of 3d electrons so that Ti^{4+} ions cannot exhibit crystal field spectra. However, Ti^{4+} ions are involved in intervalence transitions, and these are discussed in Section 3.5.3.

The relatively unstable Ti^{3+} cations may occur as minor constituents in some terrestrial amphiboles and garnets. However, Ti^{3+} ions are more abundant in extraterrestrial minerals, including pyroxenes and opaque oxide phases in Moon-rocks and some meteorites, and a number of synthetic Ti(III) compounds are known. The [A]3d^1 configuration of octahedrally co-ordinated Ti^{3+} ions results in ground-state $(t_{2g})^1(e_g)^0$ and excited state $(t_{2g})^0(e_g)^1$ configurations represented by $^2T_{2g}$ and ^2E_g crystal field states, respectively. Therefore, one spin-allowed crystal field band ($^2T_{2g} \rightarrow {}^2E_g$ transition) is expected in optical spectra. However, dynamic Jahn–Teller splitting of the ^2E_g level or the occurrence of Ti^{3+} ions in distorted octahedral sites results in two absorption bands in optical spectra.

(a) *Ti^{3+}* -doped corundum

Polarized absorption spectra of synthetic Ti^{3+}-doped Al_2O_3 crystals[31] show two spin-allowed crystal field bands occurring at 18 450 cm^{-1} and 20 300 cm^{-1}. These originate from Ti^{3+} ions in the trigonally distorted, symmetry C_3, octahedral site of the corundum structure. Absorption bands at similar energies also occur in optical spectra of blue sapphire discussed in Section 3.6.1.

(b) *Pyroxene*

The green clinopyroxene in the meteorite that was discovered near Pueblo de Allende, Mexico, in 1969 is particularly interesting because it is devoid of iron and contains coexisting Ti^{3+} and Ti^{4+} ions. The polarized spectra of the Allende pyroxene, $Ca(Mg, Ti^{3+}, Ti^{4+}, Al)(Si, Al)_2O_6$, show two well-defined bands[35] around 16 000 cm^{-1} and 20 000 cm^{-1} assigned to crystal field transitions in Ti^{3+} ions located in the distorted octahedral M1 site of the clinopyroxene structure. The effect of pressure[35] is to intensify the 16 000 cm^{-1} and 20 000 cm^{-1} bands and to induce a shift to higher energies, which is consistent with their assignment as Ti^{3+} crystal field transitions. However, an additional inflexion in the optical spectra around 15 000 cm^{-1} intensifies considerably but does not change position in high pressure spectra, leading to its assignment[10,35] as $Ti^{3+} \rightarrow Ti^{4+}$ intervalence charge transfer (Section 3.5.2).

3.4.4 Minerals containing Cr^{3+}

The electronic configuration of Cr^{3+}, [A]$3d^3$, in octahedral coordination, $(t_{2g})^3(e_g)^0$, is represented by the $^4A_{2g}$ crystal field ground state. Several spin-allowed and spin-forbidden transitions may occur in Cr^{3+} in the visible region involving the $^4T_{2g}(F)$, $^4T_{1g}(F)$, $^2E_g(G)$, $^2T_{1g}(G)$ and $^2T_{2g}(G)$ excited states. The spin-allowed $^4A_{2g} \rightarrow {}^4T_{1g}(P)$ transition occurring in the ultra-violet is rarely resolved because it is obscured by the oxygen $\rightarrow$ metal charge transfer absorption edge.

Although chromium is not as abundant as iron and titanium, Cr^{3+} ions are common minor constituents of several minerals including gems[5,50] bestowing on them attractive colours. Hence, numerous spectral measurements have been made on Cr^{3+}-bearing minerals. Those extended recently to elevated temperatures and pressures are highlighted here.

(a) *Ruby*

The spectra of natural rubies and synthetic Cr^{3+}-doped Al_2O_3 crystals have been studied extensively under a variety of experimental and compositional conditions, largely as a result of their importance in ruby-laser technology. Two broad intense bands occurring around 18 000 cm^{-1} and 25 000 cm^{-1}[130]

represent respectively the spin-allowed $^4A_2 \rightarrow {}^4T_2(F)$ and $^4A_2 \rightarrow {}^4T_1(F)$ transitions in octahedrally coordinated Cr^{3+} ions, while weak peaks at 14 430 cm^{-1} and 15 110 cm^{-1} originate from spin-forbidden transitions $^4A_2 \rightarrow {}^2E(G)$ and $^4A_2 \rightarrow {}^2T_1(G)$. The polarized spectra portray the dichroism expected for Cr^{3+} in corundum's trigonally distorted (symmetry C_3) octahedral site, with small but significant differences of band maxima and intensities for light polarized parallel and perpendicular to the *c* axis. In the $E \parallel c$ spectrum band maxima are at 18 300 cm^{-1} ($^4A_2 \rightarrow {}^4A_1({}^4T_2)$) and 24 800 cm^{-1} ($^4A_2 \rightarrow {}^4A_2({}^4T_1)$) and in the $E \perp c$ spectra they occur at 18 100 cm^{-1} ($^4A_2 \rightarrow {}^4E({}^4T_2)$) and 24 500 cm^{-1} ($^4A_2 \rightarrow {}^4E({}^4T_1)$). Strictly speaking the $^4A_2 \rightarrow {}^4A_1$ transition in the $E \parallel c$ spectrum is symmetry forbidden by group theoretical selection rules, and the presence of the absorption band at 18 300 cm^{-1} is attributed to excitation of allowed vibrational modes during the electronic transition within Cr^{3+} ions. Such vibronic coupling is accentuated by increased temperature.

Measurements of ruby at elevated temperatures[30,31] have revealed that there is a general broadening, intensification (integrated areas) and shifts of band maxima to lower energies between room temperature and 900° C, the effects being most pronounced for the 18 000 cm^{-1} band (particularly in the $E \parallel c$ spectrum) which moves toward 16 000 cm^{-1} at 900° C. Above 500° C, the colour of a typical ruby becomes green. The effect of increased pressure on ruby optical spectra is to cause intensification and a blue-shift of the spin-allowed transitions at high pressures.[4] On the other hand, the ruby R_1 fluorescence line representing one peak of the spin-forbidden $^4A_2 \rightarrow {}^2E(G)$ transition doublet shows a red-shift, decreasing from 14 390 cm^{-1} at 1 atmosphere to about 13 600 cm^{-1} at 112 GPa.[36] This pressure-induced shift of the ruby fluorescence spectrum is utilized as an *in situ* pressure gauge in experiments using the diamond anvil cell.

Numerous studies have been made of the compositional variations of the optical spectra of the Al_2O_3–Cr_2O_3 solid solution series.[31] Absorption bands show a red shift with increasing Cr^{3+} concentration due to expansion of the octahedral sites as larger Cr^{3+} ions (61.5 pm) replace Al^{3+} ions (53 pm) in the corundum structure. In eskolaite or chromia, Cr_2O_3, which is green, Cr^{3+}–O distances are 197 pm and 202 pm (compared with Al^{3+}–O distances of 186 pm and 197 pm in Al_2O_3), and the Cr^{3+} spin-allowed transitions are now centred at 16 600 cm^{-1} and 21 700 cm^{-1}. The colour change from ruby red to chromia green appears between 20 and 40 mol% Cr_2O_3.

(b) *Garnets*

The optical spectra of Cr^{3+} ions in the octahedral sites of the garnet structure also show effects of site compression on the energies of the crystal field transitions. In green uvarovite, $Ca_3Cr_2(SiO_4)_3$, the spin-allowed transitions occur at 16 200 cm^{-1} and 22 700 cm^{-1}, and are produced by Cr^{3+} ions in

$[CrO_6]$ octahedra with Cr^{3+}–O distances of 203 pm. Both bands show expected blue shifts at high pressures[14] and red shifts at elevated temperatures.[30] In violet coloured chrome pyrope or knorringite, $Mg_3(Al,Cr)_2(SiO_4)_3$, in which Cr^{3+} ions substitute for Al^{3+} in $[AlO_6]$ having metal–oxygen distances of 189 pm, the two Cr^{3+} spin-allowed transitions occur at 17 600 cm^{-1} and 24 200 cm^{-1}.

3.5 INTERVALENCE TRANSITIONS

3.5.1 $Fe^{2+} \rightarrow Fe^{3+}$ charge transfer

The high abundance of iron and the common coexistence of Fe^{2+} and Fe^{3+} ions in terrestrial minerals leads to the possibility of electron transfer between iron cations located in adjacent coordination sites.[10] Such an exchange can be initiated by thermal energy or by photons. Thermally activated electron hopping usually produces opaque phases (e.g. magnetite, ilvaite, deerite) and is identified in their Mössbauer spectra. Optically-induced intervalence charge transfer is often observed in visible region spectra of mixed valence Fe^{2+}–Fe^{3+} minerals.[37,38] $Fe^{2+} \rightarrow Fe^{3+}$ charge transfer transitions are easily recognized because they usually possess different energies, intensities, widths and polarization dependencies from those of the spin-allowed or spin-forbidden crystal field transitions in the individual Fe^{2+} and Fe^{3+} ions.

These features are exemplified by the polarized absorption spectra of vivianite discussed in Section 3.2.2 and illustrated in Fig. 3.5. The $Fe^{2} \rightarrow Fe^{3+}$ intervalence band at 680 nm in vivianite is intensified considerably relative to the crystal field bands at elevated pressures,[15] but in contrast to the Fe^{2+} spin-allowed transitions in the near infra-red region, the energy of the $Fe^{2+} \rightarrow Fe^{3+}$ charge transfer band is unchanged at high pressures. The optical spectra of amphiboles of the glaucophane-riebeckite series, $Na_2(Mg,Fe^{2+})_3(Al^{3+},Fe^{3+})_2Si_8O_{22}(OH)_2$, which have also been examined in detail,[10] show how $Fe^{2+} \rightarrow Fe^{3+}$ charge transfer is affected by atomic substitution of non-transition metal ions. The crystal structure of these alkali amphiboles which contains cations in infinite bands of edge-shared octahedra, produces absorption bands in the region 16 000–18 500 cm^{-1} originating from $Fe^{2+} \rightarrow Fe^{3+}$ charge transfer. Their intensities are proportional to the product of donor Fe^{2+} and acceptor Fe^{3+} cation concentrations,[37] with the result that riebeckite is almost opaque due to electron delocalization along the chains of edge-shared octahedra. The $Fe^{2+} \rightarrow Fe^{3+}$ charge transfer spectra also show an inverse temperature dependence, because the intensities of the intervalence band decrease at elevated temperatures. Features similar to those observed in vivianite and glaucophane-riebeckite have been identified in a number of mixed valence minerals.[10,37,38] A $Fe^{2+} \rightarrow Fe^{3+}$ intervalence band centred between 16 000 cm^{-1} and 17 000 cm^{-1} is clearly distinguished in the γ-spectrum of orthopyroxene illustrated in Fig. 3.8.

3.5.2 $Ti^{3+} \rightarrow Ti^{4+}$ charge transfer

Although Ti^{3+} ions are relatively unstable on Earth, coexisting Ti^{3+} and Ti^{4+} cations are found in extraterrestrial minerals, including pyroxenes occurring in Moon-rocks and certain meteorites. The polarized spectra of the Ti pyroxene from the Allende meteorite discussed earlier, in addition to possessing Ti^{3+} crystal field bands at 16 000 cm^{-1} and 20 000 cm^{-1}, contain a $Ti^{3+} \rightarrow Ti^{4+}$ intervalence band centred near 15 000 cm^{-1} involving adjacent Ti cations separated by 315 pm in edge-shared octahedral M1 sites in the pyroxene structure. By analogy with the $Fe^{2+} \rightarrow Fe^{3+}$ charge transfer band in vivianite, the energy of this pyroxene $Ti^{3+} \rightarrow Ti^{4+}$ intervalence transition is virtually unchanged with rising pressure, but its intensity is increased considerably at high pressures.[35] $Ti^{3+} \rightarrow Ti^{4+}$ charge transfer bands have been identified in a number of synthetic Ti compounds; the weak broad band around 12 500 cm^{-1} in the polarized spectra of Ti^{3+}-doped Al_2O_3 crystals discussed later in Section 3.6.1 and Fig. 3.13 may also originate from electron transfer between Ti cations only 269 pm or 297 pm apart in face-shared or edge-shared octahedra of the corundum structure.

3.5.3 $Fe^{2+} \rightarrow Ti^{4+}$ charge transfer

Coexisting Fe^{2+} and Ti^{4+} ions frequently are present in minerals, suggesting that $Fe^{2+} \rightarrow Ti^{4+}$ charge transfer transitions may occur between the cations when they are situated in adjacent coordination sites in the crystal structure. Although $Fe^{2+} \rightarrow Ti^{4+}$ intervalence transitions have been suggested in optical spectra of several minerals,[7,10] some ambiguity exists over their assignment due to overlap with absorption bands originating from $Fe^{2+} \rightarrow Fe^{3+}$ charge transfer and crystal field transitions within Fe^{2+}, Fe^{3+} and other transition metal ions. Recent measurements[39] on the titaniferous clinopyroxene from a meteorite that fell near Angra dos Reis, Brazil, have greatly clarified spectral features of $Fe^{2+} \rightarrow Ti^{4+}$ intervalence transitions. The Mössbauer spectrum of this pyroxene indicated no detectable ferric iron, thereby eliminating complications in the visible region from $Fe^{2+} \rightarrow Fe^{3+}$ charge transfer and Fe^{3+} crystal field bands.

The polarized absorption spectra of the Angra dos Reis pyroxene illustrated in Fig. 3.12 show intense broad bands centred near 20 000 cm^{-1}, in addition to weaker broad bands around 10 000 cm^{-1} and 5000 cm^{-1} attributable to crystal field transitions in Fe^{2+} ions located in the pyroxene M1 and M2 sites. The effect of pressure[39] is to cause the normal blue shift of the Fe^{2+} crystal field bands in the near infra-red, but a red-shift and intensification of the broad band near 20 000 cm^{-1}. This pressure-induced movement of the 20 000 cm^{-1} band to lower energies, which differs from the negligible shifts observed for $Fe^{2+} \rightarrow Fe^{3+}$ and $Ti^{3+} \rightarrow Ti^{4+}$ charge transfer transitions, appears to be a distinguishing feature of $Fe^{2+} \rightarrow Ti^{4+}$ intervalence transitions, and has led to

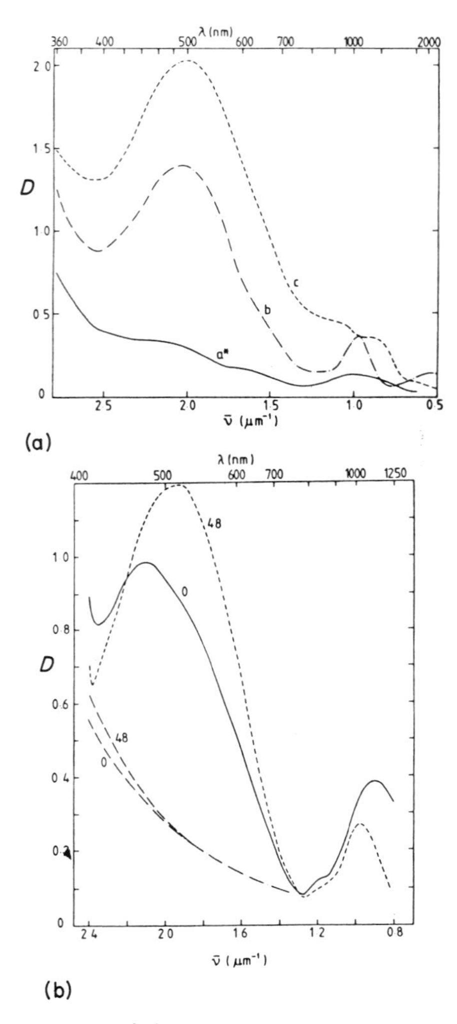

Fig. 3.12 Absorption spectra of the pyroxene from the Angra dos Reis meteorite (from ref. 39). (a) Polarized spectra measured at atmospheric pressure. (b) The E‖b spectrum measured at atmospheric pressure and at 4.8 GPa (48 kbar).

more rigorous interpretations of electronic spectra of other titanian pyroxenes. For example, the polarized spectra of an unusual blue sodic titaniferous omphacite[14] contains a broad intense absorption band at 15 400 cm^{-1}, in addition to the Fe^{3+} spin-forbidden peak at 22 700 cm^{-1} and the Fe^{2+} spin-allowed bands at 8700 cm^{-1} and 10 750 cm^{-1}. The 15 400 cm^{-1} band was originally assigned to $Fe^{2+} \rightarrow Fe^{3+}$ charge transfer. However the small pressure-induced red shift of this band in high pressure optical spectra indicates that it probably represents a $Fe^{2+} \rightarrow Ti^{4+}$ intervalence transition.[39]

3.6 APPLICATIONS

3.6.1 The sapphire blue problem

The cause of the dark blue coloration in sapphire has been a long-standing problem in mineralogy and solid-state chemistry.[5,31,40] Analyses of natural sapphires, coupled with crystal growth studies, established early that both Fe and Ti must be present but in only low concentrations to produce the intense blue colour. Natural and synthetic corundums containing only Fe^{3+} ions have yellow or pale green-blue colours, while synthetic Ti^{3+}-doped Al_2O_3 crystals are rose-pink coloured. Laboratory investigations have also revealed that heat treatment of natural and synthetic sapphires in oxidizing or reducing atmospheres induces colour changes, and that variations exist for sapphires originating from similar localities. Numerous optical spectral measurements have been brought to bear on the origin of the dark blue colour of sapphire but considerable confusion exists in the literature over assignments of absorption spectra in the visible and nearby ultra-violet and infra-red regions.

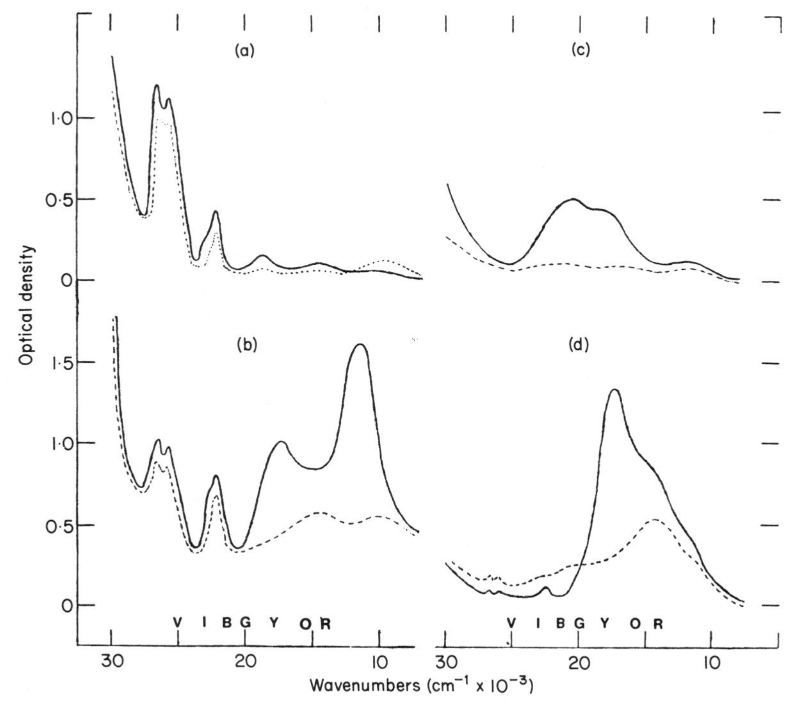

Fig. 3.13 Absorption spectra of natural and synthetic sapphires (from ref. 31). (a) Natural yellow sapphire. (b) Natural dark blue sapphire. (c) Synthetic Ti-doped Al_2O_3. (d) Synthetic Fe-Ti-doped Al_2O_3. Light polarized perpendicular (———) and parallel (---) to the *c* axis of corundum.

The optical spectra of natural blue sapphire and synthetic Fe-Ti-doped Al_2O_3 crystals are illustrated in Fig. 3.13(b) and 3.13(d), and compared with the crystal field spectra of natural Fe^{3+}-bearing yellow sapphire (Fig. 3.13(a)) and synthetic Ti^{3+}-doped Al_2O_3 crystals (Fig. 3.13(c)). It is apparent from the spectra of natural blue sapphires (Fig. 3.13(b)) that absorption minima in the violet-indigo and blue-green regions located between sharp peaks at 25 680 cm^{-1} and 22 220 cm^{-1} and broad bands spanning 17 800–14 200 cm^{-1} are responsible for the colour of blue sapphires. The latter feature is relatively less intense in the spectra of natural yellow sapphires (Fig. 3.13(a)) believed to contain negligible Ti. In synthetic Fe-Ti-doped Al_2O_3 (Fig. 3.13(d)), band maxima occur at 17 400 cm^{-1} ($E \perp c$) and 14 000 cm^{-1} ($E \parallel c$), with prominent shoulders near 12 500 cm^{-1} and 20 300 cm^{-1}. As noted in Sections 3.4.3(a) and 3.5.2, the 18 450 cm^{-1} and 20 300 cm^{-1} bands in Ti^{3+}-doped Al_2O_3 (Fig. 3.13(c)) represent crystal field transitions within Ti^{3+} ions, and the weaker band near 12 500 cm^{-1} may represent a $Ti^{3+} \rightarrow Ti^{4+}$ intervalence charge transfer transition. Also the peaks in yellow and blue sapphire spectra clustered at 22 200 cm^{-1} and around 26 000 cm^{-1} represent the spin-forbidden $^6A_1 \rightarrow {}^4A_1, {}^4E(G)$ and $^6A_1 \rightarrow {}^4T_2, {}^4E(D)$ transitions in octahedrally coordinated Fe^{3+} ions (see Fig. 3.3), intensified by exchange interactions between adjacent Fe^{3+} ion pairs in the corundum structure. Debate centres[31,40] over origins of the band maxima at 17 400 cm^{-1} and 14 200 cm^{-1}, and contributions to the broad absorption envelope spanning 9000–20 000 cm^{-1}.

The Fe^{3+} crystal field peaks near 26 000 cm^{-1} and 22 200 cm^{-1} imply that weaker contributions from the spin-forbidden $^6A_1 \rightarrow {}^4T_1(G)$ and $^6A_1 \rightarrow {}^4T_2(G)$ transitions (see Fig. 3.3) must contribute to the optical spectra at lower energies. The spectra of natural yellow sapphires (Fig. 3.13(a)) and Fe^{3+}-doped Al_2O_3 crystals suggest that they occur around 9450 cm^{-1}, 14 350 cm^{-1}, and 18 700 cm^{-1}. However, an additional intense broad band appears at 11 500 cm^{-1} in flux-grown Al_2O_3 crystals and in many blue sapphires, which appears to originate from Fe^{2+} ions, the presence of which is made possible by charge compensation by F^- or OH^- anions replacing O^{2-} ions in the corundum structure. This 11 500 cm^{-1} band has been assigned to the spin-allowed $^5T_2 \rightarrow {}^5E$ crystal field transition within octahedral Fe^{2+} ions, perhaps intensified by exchange interactions with neighbouring Fe^{3+} ions,[34] and to a $Fe^{2+} \rightarrow Fe^{3+}$ intervalence charge transfer transition between edge-shared cations perpendicular to the *c* axis.[37] The corresponding $Fe^{2+} \rightarrow Fe^{3+}$ intervalence transition across face-shared octahedra parallel to the *c* axis may occur at 9700 cm^{-1} overlapping with the $^6A_1 \rightarrow {}^4T_1(G)$ crystal field transition in Fe^{3+}. An inverse temperature dependence of the bands at 11 500 cm^{-1} and 9700 cm^{-1}, which are intensified in low temperature spectra,[37] supports their assignments to $Fe^{2+} \rightarrow Fe^{3+}$ intervalence transitions. The broad bands around 17 400 cm^{-1} and 14 200 cm^{-1} also intensify at lower temperatures,[37] suggesting that they represent $Fe^{2+} \rightarrow Ti^{4+}$ intervalence transitions across edge-shared ($\perp c$) and face-shared ($\parallel c$) octahedra, respectively. The loss of intensity of these bands when Fe-Ti-doped Al_2O_3 and some natural blue sapphires are heated in oxygen or air is attributed to oxidation of Fe^{2+} to Fe^{3+} ions and the removal of $Fe^{2+} \rightarrow Ti^{4+}$ intervalence charge transfer transitions near $17\,400\,cm^{-1}$ and $14\,300\,cm^{-1}$. The latter confine the 'windows' of minimum absorption to the blue and indigo-violet portion of the spectrum, resulting in dark blue sapphires. The absence of titanium and/or conversion of Fe^{2+} cations to Fe^{3+} ions eliminates $Fe^{2+} \rightarrow Ti^{4+}$ charge transfer, resulting in yellow or pale blue-green sapphires.

Confirmation that $Fe^{2+} \rightarrow Ti^{4+}$ charge transfer is responsible for sapphires' blue colour comes from recent Mössbauer spectral measurements of a synthetic Fe-Ti-doped Al_2O_3 crystal made from > 90% isotopically enriched $^{57}Fe_2O_3$ starting material.[31] A Fe^{2+} doublet with Mössbauer parameters

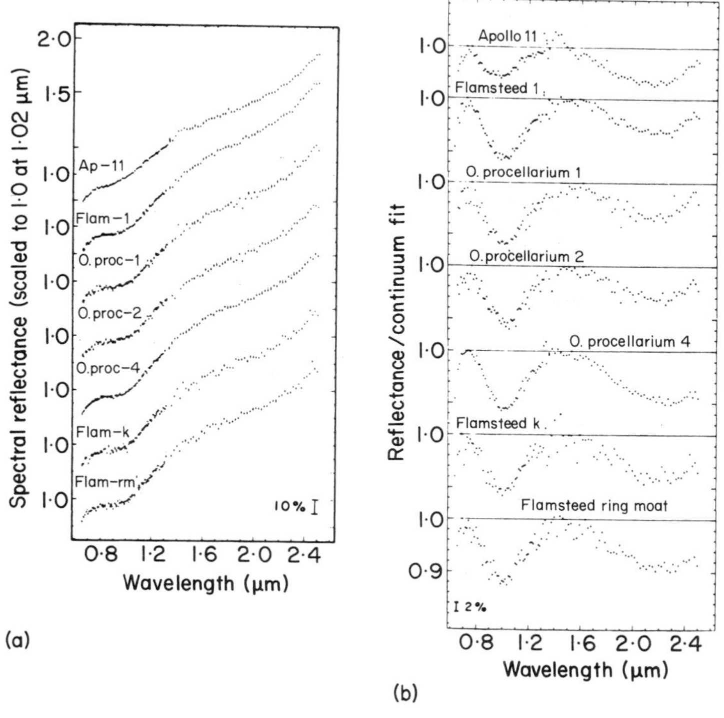

Fig. 3.14 Reflectance spectra of the Moon's surface measured through Earth-based telescopes (from ref. 41). (a) Reflectance spectral profiles of soils overlying areas of mare basalts on the Moon. All spectra are scaled to unity at 1020 nm. (b) Residual absorption features for the telescopic spectra of the soils in (a). The lunar continuum was approximated with a straight line fit to the reflectance spectrum at 730 nm and 1500 nm. Residual absorption features represent the ratio of the reflectance spectra to the estimated continuum. The spectra are dominated by Fe^{2+} M2-site pyroxene bands near 1 μm and 2 μm.

comparable with ilmenite ($FeTiO_3$) suggests that pairs of Fe^{2+} and Ti^{4+} cations in face-shared octahedra occur also in blue sapphires, facilitating the $Fe^{2+} \rightarrow Ti^{4+}$ intervalence transitions.

3.6.2 Remote sensed pyroxene spectra

The ubiquitous occurrence of pyroxenes in igneous rocks on the surfaces of terrestrial planets, coupled with the intensity of the Fe^{2+} – M2 site crystal field bands in the near infra-red spectrum (described in Section 3.4.1(c)), makes them conducive to measurement by spectral remote sensing techniques.[29] From Earth-based telescopic reflectance spectral profiles of the Moon's surface, it has been possible to identify pyroxene structure-types (orthopyroxene or pigeonite versus augite or calcic clinopyroxene), to estimate Fe/Mg ratios and Ca contents of the pyroxenes, and to map geological units from the positions of the '1 μm' band ($\approx$ 10 000 cm^{-1}) and '2 μm' band ($\approx$ 5000 cm^{-1}).

In Fig. 3.14(a), for example, remote-sensed reflectance spectra of seven areas of the Moon[41] are shown. Although the reflectance increases between 650 nm and 2500 nm, prominent inflexions are observable near the 1 μm and 2 μm regions. Removal of a background continuum (Fig. 3.14(b)) reveals the prominent pyroxene absorption bands (reflectance minima) between 950–1050 nm and 2000–2300 nm, indicating the presence of pigeonite or augite in the soils overlying the mare basalts. Subtle differences of remote-sensed reflectance spectral profiles have enabled different mare basalt-types to be mapped on the Moon's surface.[42]

3.6.3 The iron oxide mineralogy of Mars

The red colour of Mars has long been attributed to the occurrence of iron(III) oxides on its surface. On Earth, a number of polymorphs of Fe_2O_3 and FeOOH are known, and considerable debate[43–45] has centred on which, if any, of these phases are present in the Martian soil. The Viking X-ray fluorescence experiments demonstrated that the iron contents in the soil at the two Viking Lander sites are very high (*c.* 18 wt% as Fe_2O_3), while the Viking magnetic properties experiments indicated the presence of between 1 and 7% magnetic material in the soil. This was interpreted to be maghaemite (γ-Fe_2O_3), although magnetite (Fe_3O_4), feroxyhyte (δ'-FeOOH) and perhaps ferrihydrite ($Fe_2O_3 \cdot 2FeOOH \cdot nH_2O$) are also possible magnetic minerals on Mars. Confirmation that a sizable fraction of the regolith iron is in the ferric state is proved by Earth-based remote sensed spectra of the Martian surface[46] illustrated in Fig. 3.15. Features attributable to Fe^{3+} crystal field transitions (see Fig. 3.13) include the weak absorption feature centred at about 870 nm (11 500 cm^{-1}) representing the $^6A_1 \rightarrow {}^4T_1(G)$ transition; the reflectance maximum (absorption minimum) at 750 nm; a slope change near 640 nm (15 600 cm^{-1}) attributed to the $^6A_1 \rightarrow {}^4T_2(G)$ transition; and a second slope change near 530 nm (18 900 cm^{-1}) representing the onset of the $^6A_1 \rightarrow {}^4A_1$,

$^4E(G)$ transition. The latter two features are superimposed on a low energy oxygen → metal charge transfer absorption edge which is typical of Fe^{3+}-bearing minerals. Laboratory optical spectral measurements have focused on searching for oxide and clay mineral phases that most closely simulate the

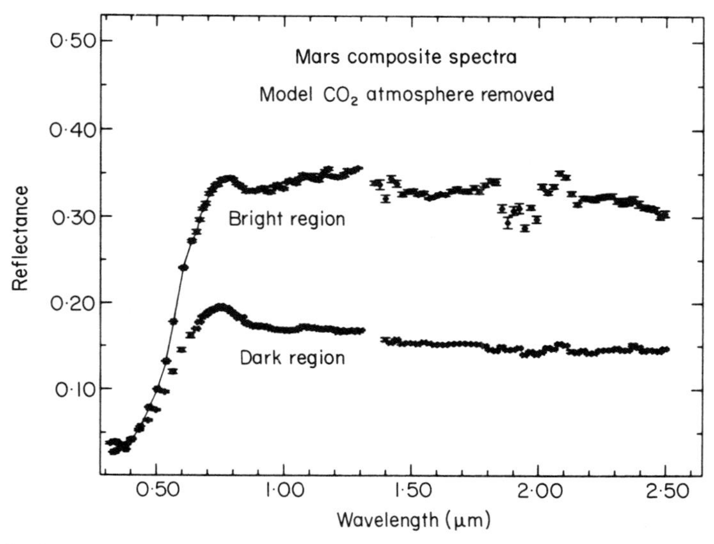

Fig. 3.15 Remote-sensed reflectance spectra of Mars (from ref. 46). A model atmospheric CO_2 spectrum has been removed to accentuate the spectra of regolith material.

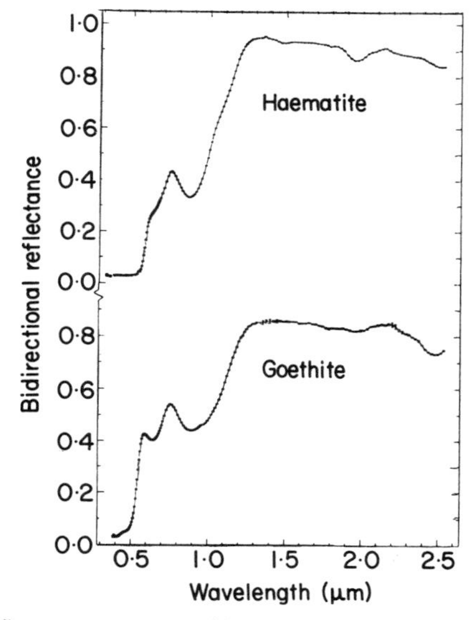

Fig. 3.16 Reflectance spectra of haematite and goethite (from ref. 43).

telescopic reflectance spectrum profiles of Mars. Most of the well-known crystalline $Fe^{(III)}$ oxides, including haematite and goethite (α-FeOOH) commonly found on Earth, are unlikely candidate phases because the intensities of the $^6A_1 \rightarrow {}^4T_2$ and $^6A_1 \rightarrow {}^4A_1$, ^4E transitions, which are accentuated by exchange coupling, are too high relative to the $^6A_1 \rightarrow {}^4T_1$ transition at 11 500 cm^{-1} compared to Mars visible-region spectra. This is illustrated by the spectra shown in Fig. 3.16. In order to account for the relatively low intensity of the absorption edge at 400–750 nm in the Martian regolith spectra (Fig. 3.15), it has been suggested[43,44] that Fe^{3+} ions occur in poorly crystalline or amorphous devitrified glass, clay or gel phases in which magnetic interactions between next-nearest neighbour Fe^{3+} ions are reduced. Since bridging hydroxyl anions also weaken cation–cation exchange interactions, the Martian soil could contain a phase with edge- or corner-shared $[Fe(O,OH,H_2O)_6]$ octahedra in which adjacent Fe^{3+} ions share OH^- anions.

3.6.4 Phase equilibria in the Earth's Mantle

The presence of transition metal ions in mineral structures modifies phase equilibria at high pressures as a result of increased crystal field stabilization energies acquired by certain cations in dense phases. Such relative CFSE may be estimated or deduced from high P and T spectral measurements. The additional electronic stabilization can influence both the depth in the Mantle at which a phase transition occurs and the distribution coefficients of transition metals in coexisting dense phases of the Lower Mantle.

(a) *The olivine → spinel transition*

Experimental phase equilibrium studies have confirmed deductions from seismic velocity data that below 400 km, olivine and pyroxene, abundant in the Upper Mantle, are transformed to more dense spinel and garnet phases. From pressure and temperature variations of the optical spectra of fayalite and γ-Fe_2SiO_4, free energy changes, ΔG_{CFS}, due to differences of crystal field splittings between Fe^{2+} in the spinel and olivine structures may be calculated[21] from Equation (3.6)

$$\Delta G_{CFS} = \Delta CFSE - T\Delta S_{CFS} \tag{3.6}$$

as functions of P and T, where ΔCFSE and ΔS_{CFS} are differences of crystal field stabilization enthalpies and electronic configurational entropies of Fe_2SiO_4 spinel and fayalite at the transition temperature and pressure. The results indicate that ΔG_{CFS} is always negative, showing that crystal field stabilization of Fe^{2+} promotes the olivine → spinel transition in Fe_2SiO_4, and expands the stability field of spinel at the expense of olivine in the system Mg_2SiO_4-Fe_2SiO_4. Because of crystal field effects, the transition pressure for the olivine → spinel transitions in Fe_2SiO_4 is lowered by about 5 GPa at 1000° C. Since olivines of the Upper Mantle contain approximately 10% Fe_2SiO_4, their transition pressures may be decreased by about 0.5 GPa due to the presence of Fe^{2+} in the crystal structures.[21] This means that the depth of the olivine → spinel transformation in a typical Upper Mantle forsteritic olivine is 15 km shallower than it would be if iron were absent from the minerals.

(b) *Partitioning of iron in the spinel → periclase + perovskite transition*

Although Upper Mantle olivine is transformed into the more dense isochemical spinel phase and delineates the onset of the Transition Zone in the Earth's interior at 400 km, the stoichiometry, $(Mg, Fe)_2SiO_4$, appears to be unstable in the Lower Mantle with respect to denser oxide structure-types. One post-spinel transformation that has been the subject of numerous investigations is the spinel to periclase plus perovskite transition:

$$\underset{\text{spinel}}{(Mg, Fe)_2SiO_4} \rightarrow \underset{\text{magnesiowüstite}}{(Fe, Mg)O} + \underset{\text{perovskite}}{(Mg, Fe)SiO_3}$$

believed to occur below 650 km. Such a disproportionation reaction raises the possibility of the breakdown products having different iron–magnesium ratios.

In investigations[47,48] of phase relations in the system: MgO-FeO-SiO_2 at high P and T compositions $(Mg_{1-x}Fe_x)_2SiO_4$ were found to yield magnesiowüstites with higher Fe/Mg ratios than coexisting perovskite. For example:

$$(Mg_{0.85}Fe_{0.15})_2SiO_4 \rightarrow (Mg_{0.74}Fe_{0.26})O + (Mg_{0.96}Fe_{0.04})SiO_3$$

The strong partitioning of iron into magnesiowüstite is the result of higher CFSE of Fe^{2+} in the periclase structure. Various estimates have been made of the CFSE of Fe^{2+} in dense oxide structures modelled as potential mantle mineral phases. All estimates indicate that octahedrally coordinated Fe^{2+} (in periclase, for example) has a considerably higher CFSE than Fe^{2+} ions in 8- to 12-coordinated sites of the perovskite structure.[4,21] The excess CFSE of Fe^{2+} in periclase-type FeO over perovskite-type $FeSiO_3$ has been calculated to be as high as 12.5 kcal mol^{-1}.[47] The large extra CFSE factor thus favours the concentration of iron in magnesiowüstite and the depletion of iron in a coexisting perovskite phase in the Lower Mantle.

3.6.5 Geophysical properties of the Earth's interior

(a) *Radiative heat transport in the Mantle*

Radiative heat transfer is an important contributor to heat flow at elevated temperatures in the Mantle. However, it has long been known that minerals absorbing radiation in the near infra-red and visible regions control this heat transport mechanism in the Upper Mantle and Lower Mantle.[4,24] As a result, numerous attempts have been made to assess the temperature- and pressure-induced variations of absorption bands in important Fe^{2+}-bearing Mantle minerals, including olivine, pyroxene, garnet, spinel and magnesiowüstite phases, on the effective radiative conductivity of the Earth's interior.

The energy transfer of photons through a grey body (i.e. one in which absorption of photons is finite, non-zero and independent of wavelength) is given by Equation (3.7)

$$K_r = 16n^2 ST^3/3\varepsilon \tag{3.7}$$

where K_r is the effective radiative conductivity, S is Stefan's constant, n is the mean refractive index, and ε is the mean absorption coefficient.

Because K_r increases proportionally to T^3, it was regarded to be one of the dominant mechanisms of heat transfer under the high temperature conditions of the Mantle. Here, temperatures range from 1500–2000° C within the Transition Zone to about 3700° C at the Core–Mantle boundary (2900 km, 136 GPa). By Stefan's law, the T^4 dependency of radiation energy flux for a black body indicates that the maximum transmission energies at temperatures 1500–2000° C and 3700° C are at about 1000 nm (10 000 cm^{-1}) and 700 nm (14 300 cm^{-1}), respectively. Therefore, it is important to assess how much radiation in this wavelength region is absorbed by electronic transitions in transition metal-bearing Mantle minerals.

The high P and T spectral data summarized in earlier sections indicate that so far as crystal field transitions are concerned, there is a general pressure-induced intensification and blue-shift of absorption bands towards the visible region. However, temperature-induced intensification appears to predominate over pressure variations. In iron-bearing olivines, silicate spinels, and magnesiowüstites, however, the red-shift of the absorption edge due to oxygen → iron charge transfer transitions, and perhaps intensification of $Fe^{2+} \rightarrow Fe^{3+}$ intervalence transitions, are likely to absorb visible region radiation at very high temperatures.

The radiative thermal conductivity K_r of forsteritic olivines increases with rising temperature and would contribute to heat flow in the Upper Mantle.[24] However, values of K_r for olivine are considered to be rather low to satisfactorily explain the dissipation of the Earth's internal heat by radiation and lattice conduction alone. Note also that Fe^{2+} crystal field transitions in almandine and pyroxenes (M2 site) absorb strongly in the wavelength range 1500–2500 nm, the region where radiation energy flux for a black body is maximum at temperatures 700–1500° C. Thus, Fe^{2+} ions in pyroxene or garnet phases of the Upper Mantle and Transition Zone would also reduce radiative heat transfer in these regions of the Earth's interior.

There is greater uncertainty about radiative heat transfer through spinel in the Transition Zone and magnesiowüstite in the Lower Mantle. The strong pressure-induced visible-region absorption in γ-Fe_2SiO_4 and magnesiowüstites at 300 kb suggests that these phases would effectively block black-body radiation at wavelengths shorter than 1500 nm. However, recent shock-compressed absorption spectral measurements[9] of two magnesiowüstites $Wü_{14}$ and $Wü_{26}$ (the iron contents of which probably span Fe/Mg ratios of the Lower Mantle periclase structure-type involved in post-spinel phase equilibria) demonstrated that absorption of visible-region radiation was considerably lower than static measurements in the diamond cell[15] inferred to contain more ferric iron. Under highly reducing conditions (i.e. absence of Fe^{3+}) in the Lower Mantle, radiative conductivity through magnesiowüstite would be comparable with phonon lattice conductivity for depths in the Earth of the order of 1000 km.

(b) *Electrical conduction in the Mantle*

Although the magnitude and mechanism of electrical conductivity of the Earth's interior are uncertain, the recent high pressure and temperature spectral measurements of minerals have provided new insights into the origin of electrical conductance by potential Mantle minerals.[4,15,49]

At a given pressure, the electrical conductivity σ is related to absolute temperature by (Equation (3.8))

$$\sigma = \sum_{i} \sigma_{oi} \exp(-E_a/kT) \qquad (3.8)$$

where E_a is activation energy, k is the Boltzmann constant, and σ_{oi} is the electrical conductivity originating from a specific conducting mechanism i. One such mechanism involves extrinsic conduction of the type oxygen → metal charge transfer.

The radial distribution of electrical conductance in the Earth has been estimated from studies of magnetic variations. There appears to be a steep increase of conductivity between depths of 400 km and 1000 km in the Mantle, apparently correlating with the olivine → spinel → post-spinel transformations throughout the Transition Zone. As a result, the Lower Mantle may have an electrical conductivity some 4 to 5 orders of magnitude higher than the Upper Mantle. These observations for the Earth may be correlated with measurements on relevant iron-bearing minerals.

The numerous studies of olivines at ambient pressures have shown that several factors affect electrical conductivity, including increasing iron contents, oxygen fugacity, and temperature.[49] Pressure also significantly influences electrical conductivities. Several studies have shown that the electrical conductivities increase significantly at high pressures.[22] Similar pressure-induced increases are observed also for the spinel γ-Fe_2SiO_4 and various magnesiowüstites.[22] These data indicate that at 300 kb, the electrical conductivity of Lower Mantle magnesiowüstite $Wü_{22}$ is six order of magnitude higher than Upper Mantle forsterite olivine, correlating with the observed variation suggested in the Transition Zone.

Temperature variations of electrical conductivities of olivines and magnesiowüstites at high pressures indicate[22] that activation energies are small, and show a systematic decrease with increasing pressure. These observations correlate with the pressure- and temperature-enhanced opacities and red shifts of the oxygen → metal charge transfer absorption edges into the visible region spectra of iron-bearing olivine, silicate spinel and periclase phases described in earlier sections. Such spectral variations, together with the rapid increases and large values of electrical conductivities as well as low values of activation energies, indicate that an extrinsic conduction mechanism of the type oxygen → iron charge transfer is important in the Lower Mantle. Thus, the very electronic transition that could inhibit radiative heat transfer in the Earth's interior may enhance its electrical conductivity.

3.7 SUMMARY

This chapter highlights features of common rock-forming minerals containing iron, titanium and chromium ions which make these naturally-occurring inorganic compounds interesting and productive phases to study in the visible and near infra-red spectral regions. The optical spectra of such minerals provide a wealth of information not otherwise available to the inorganic chemist or solid state physicist, and yield valuable data applicable to a variety of geochemical and geophysical properties of the Earth's interior and planetary surfaces. Favourable characteristics of transition metal-bearing minerals enhancing the relevancy of measuring their electronic spectra in the wavelength range 300–2500 nm include:

(i) Mineral structures provide diverse and variable atomic substitutions, so that compositional variations of absorption bands are readily measured (e.g. Fe^{2+} in the olivine series; Cr^{3+} in Al_2O_3-Cr_2O_3 solid solutions).

(ii) Mineral structures contain coordination sites with unusual coordination numbers as well as wide ranges of symmetries and distortions (e.g. square planar in gillespite; cubic in garnet; centrosymmetric and non-centrosymmetric, regular and distorted, octahedra in ferromagnesian silicates).

(iii) Such diverse coordination sites can stabilize cations with unusual coordination numbers and oxidation states (e.g. square planar and cubic Fe^{2+} in gillespite and garnet, respectively; Ti^{3+} in distorted octahedra of extraterrestrial pyroxenes; Cr^{2+} in lunar olivines).

(iv) Since silicate and oxide minerals generally have higher thermal stabilities than synthetic inorganic and organo-metallic compounds, they are readily amenable to spectral measurements well above ambient temperatures and pressures. Techniques are now available for making optical spectral measurements of minerals at elevated temperatures and very high pressures.

(v) Polarized absorption spectra, so essential for group theoretical analyses of transition metal electronic spectra, are conveniently made when petrographic microscope accessories are attached to a spectrophotometer. Such microscope spectrometer systems enable measurements to be made on typically small mineral crystals subjected to very high pressures in diamond anvil cells.

(vi) Spectral data derived from visible-near infra-red spectra of minerals are being applied to a diversity of problems in the Earth and planetary sciences. Examples include: explanations of colour and pleochroism in gems; interpretations of element partitioning and transition pressures in phase transformations in the Earth's Mantle; identifying minerals and estimating their chemical compositions from spectrum profiles of the surfaces of the Moon and Mars measured through Earth-based telescopes; and accounting for geophysical properties of the Earth's interior such as radiative heat transfer and electrical conductivity of the Mantle.

ACKNOWLEDGEMENTS

I thank Drs P.M. Bell, K. Langer, C. Pieters, G.R. Rossman and R.B. Singer for permission to use illustrations from their published papers. Research on mineral spectroscopy was supported by grants from the National Science Foundation (grant numbers EAR80-16163 and EAR83-13585) and the National Aeronautics and Space Administration (grant number NSG 7604).

REFERENCES

1. Burns, R.G. (1970) *Mineralogical Applications of Crystal Field Theory*, Cambridge University Press, Cambridge.
2. Karr, C. Jr. (ed.) (1975) *Infrared and Raman Spectroscopy of Lunar and Terrestrial Minerals*, Academic Press, New York.
3. Marfunin, A.S. (1979) *Physics of Minerals and Inorganic Materials*, Springer-Verlag, New York.
4. Burns, R.G. (1982) in *High-Pressure Researches in Geoscience* (ed. W. Schreyer), E. Schweizerbart'sche Verlagsbuchhandlung, Stuttgart, p. 223.
5. Loeffler, B.M. and Burns, R.G. (1976) *Am. Scient.*, **64**, 636.
6. Nassau, K. (1978) *Am. Mineral.*, **64**, 219.
7. Burns, R.G. and Vaughan, D.J. (1975) in *Infrared and Raman Spectroscopy of Lunar and Terrestrial Minerals*, (ed. C. Karr, Jr.,) Ch. 2, Academic Press, New York, p. 39.
8. Bell, P.M., Mao, H.K. and Rossman, G.R. (1975) in *Infrared and Raman Spectroscopy of Lunar and Terrestrial Minerals* (ed. C. Karr, Jr.), Ch. 1, Academic Press, New York, p. 1.

9. Goto, T., Ahrens, T.J., Rossman, G.R. and Syono, Y. (1980) *Phys. Earth Planet. Interiors*, **22**, 277.
10. Burns, R.G. (1981) *Ann. Rev. Earth Planet. Sci.*, **9**, 345.
11. Mao, H.K. and Bell, P.M. (1972) *Amer. Soc. Test. Mater. Spec. Tech. Publ.*, **539**, 100.
12. Langer, K. and Abu-Eid, R.M. (1977) *Phys. Chem. Minerals*, **1**, 273.
13. Langer, K. and Frentrup, K.R. (1979) *J. Microscopy*, **116**, 311.
14. Abu-Eid, R.M. (1976) in *The Physics and Chemistry of Minerals and Rocks* (ed. R.G.J. Strens), John Wiley, New York, p. 641.
15. Mao, H.K. (1976) *The Physics and Chemistry of Minerals and Rocks* (ed. R.G.J. Strens), John Wiley, New York, p. 573.
16. Frentrup, K.R. and Langer K. (1982) in *High-pressure Researches in Geoscience* (ed. W. Schreyer), E. Schweiser-bart'sche Verlagsbuchhandlung, Stuttgart, W. Germany, p. 247.
17. Smith, G.S. and Langer, K. (1982) in *High-pressure Researches in Geoscience* (ed. W. Schreyer), E. Schweiser-bart'sche Verlagsbuchhandlung, Stuttgart, W. Germany p. 259.
18. Sung, C.M., Singer R.B., Parkin, K.M. and Burns, R.G. (1977) *Proc. 8th Lunar Sci. Conf.*, Pergamon Press, Oxford, p. 1063.
19. Goldman, D.S. and Rossman, G.R. (1979) *Phys. Chem. Minerals*, **4**, 43.
20. Shankland, T.J., Duba, A.G. and Woronow, A. (1974) *J. geophys. Res.*, **79**, 3273.
21. Burns, R.G. and Sung, C.M. (1978) *Phys. Chem. Minerals*, **2**, 349.
22. Mao, H.K. and Bell, P.M. (1972) *Science*, **176**, 403.
23. Hazen, R.M., Mao, H.K. and Bell, P.M. (1977) *Proc. 8th Lunar Sci. Conf.*, Pergamon Press, Oxford, p. 1081.
24. Shankland, T.J., Nitsan, U. and Duba, A.G. (1979) *J. geophys. Res.*, **84**, 1603.
25. Smith, G.S. and Langer, K. (1982) *Am. Mineral.*, **67**, 343.
26. Rossman, G.R. (1980) in *Reviews in Mineralogy: Pyroxenes* (ed. C.T. Prewitt), Vol. 7, Ch. 3, Mineralogical Society of America, p. 93.
27. Goldman, D.S. and Rossman G.R. (1977) *Am. Mineral.*, **62**, 151.
28. Hazen, R.M., Bell, P.M. and Mao, H.K. (1978) *Proc. 9th Lunar Planet. Sci, Conf.*, Pergamon Press, Oxford, p. 2919.
29. Adams, J.B. (1974) *J. geophys. Res.*, **79**, 4829.
30. Parkin, K.M. and Burns, R.G. (1980) *Proc. 11th Lunar Planet. Sci. Conf.*, Pergamon Press, Oxford, p. 731.
31. Burns, R.G. and Burns, V.M. (1984) *Adv. Ceram.*, in press.
32. Ferguson, J. and Fielding P.E. (1971) *Chem. Phys. Lett.*, **10**, 262; (1972) *Austral. J. Chem.*, **25**, 1371.
33. Rossman, G.R. (1975) *Am. Mineral.*, **60**, 698; (1976) **61**, 398 and 933.
34. Smith G. (1978) *Phys. Chem. Minerals*, **3**, 375.
35. Mao, H.K. and Bell, P.M. (1974) *Ann. Rept. Geophys. Lab., Carnegie Inst. Washington*, Yearbk 73, p. 488.

36. Mao, H.K., Bell, P.M., Shaner, J.W. and Steinberg D.J. (1978) *J. appl. Phys.*, **49**, 3276.
37. Smith G. and Strens R.G.J. (1976) in *The Physics and Chemistry of Minerals and Rocks* (ed. R.G.J. Strens), John Wiley, New York, p. 583.
38. Amthauer, G. and Rossman, G.R. (1984) *Phys. Chem. Minerals*, **11**, 37.
39. Strens R.G.J., Mao, H.K. and Bell, P.M. (1982) in *Adv. Phys. Geochem.*, Vol. 2 (ed. S.K. Saxena), Ch. 11, Springer-Verlag, New York, p. 327.
40. Schmetzer, K. and Bank, H. (1980) *Neues Jahrb. Mineral. Abh.*, **139**, 216.
41. Pieters, C.M., Head, J.W., Adams, J.B. *et al.* (1980) *J. geophys. Res.*, **B85**, 3913.
42. Pieters, C.M. (1978) *Proc. 9th Lunar Planet. Sci. Conf.*, Pergamon Press, Oxford, p. 2825.
43. Singer, R.B. (1982) *J. geophys. Res.*, **B87**, 10159.
44. Sherman, D.M., Burns, R.G. and Burns, V.M. (1982) *J. geophys. Res.*, **B87**, 10169.
45. Morris, R. and Lauer, H.V. (1983) *Lunar Planet. Science*, **XIV**, Abstr., 524.
46. McCord, T.B., Clark, R.N. and Singer, R.B. (1982) *J. geophys. Res.*, **B87**, 3021.
47. Yagi, T. Mao, H.K. and Bell, P.M. (1978) *Phys. Chem. Minerals*, **3**, 97.
48. Bell, P.M., Yagi, T. and Mao, H.K. (1979) *Ann. Rept Geophys. Lab.*, Yearbk **78**, p. 618.
49. Shankland, T.J. (1975) *Phys. Earth Planet. Interiors*, **10**, 209.
50. Burns, R.G. (1983) *Chem. Britain*, **19**, 1004.

4

Mineralogical Applications of Luminescence Techniques

Grahame Walker

4.1 INTRODUCTION

The interest shown in the luminescence of minerals by mineralogists, petrologists and geochemists has been in the belief that it can reveal information not easily obtainable by other means. There is no doubt that cathodoluminescence can reveal detail in, for example, sedimentary rock-sections which cannot be seen under a petrological microscope whether the illumination is polarized or not. There can be a distinct difference in the luminescence properties of authigenic and detrital components of the same mineral, and such differences are themselves useful in diagenetic studies. If, however, we ask the inevitable questions as to why the luminescence varies in colour and intensity and what actually causes it, we are entering the realm of solid-state physics. Yet if we knew the answers, much valuable geochemical information could, in all probability, be revealed.

To what extent can we answer these questions and what information can the study and measurement of luminescence yield? The aim of this chapter is to indicate the present state of knowledge in this respect, to outline the background of the modern theory of luminescence as applicable to minerals, and to briefly describe the experimental techniques which have proved useful. It will be shown that these techniques can provide information not only on the actual nature of luminescence centres and their distribution in minerals but also on the site occupancy and cation ordering of known luminescent ions.

It should be pointed out at this stage that luminescence techniques are usually applicable only to relatively iron-free minerals since iron normally quenches all luminescence when present in concentrations greater than a few per cent.

Luminescence, or the emission of light other than by incandescence, can be 'excited' or brought about in a number of ways and this is reflected in the terminology used. For example, cathodoluminescence (CL) occurs when the source of excitation is electrons or 'cathode rays' as in a TV tube. If, however, ultra-violet or visible light can be used to excite luminescence in a particular material the resulting emission is termed photoluminescence (PL).

Thermoluminescence (TL) is thermally stimulated luminescence which can occur when energy stored in a solid from a previous excitation is released by heating, usually in the dark.

Unfortunately, some confusion surrounds the use of the older terms fluorescence and phosphorescence to describe luminescence. These terms are often crudely used to differentiate between 'prompt' emission (fluorescence) which decays very rapidly after the removal of the source of excitation, and a much slower decay of several milliseconds or more, which is traditionally called phosphorescence. In fact, as will be described in Section 4.2, phosphorescence in inorganic solids such as minerals involves an additional temperature-dependent process to that of simple fluorescence. The confusion arises when, in some minerals, even the fluorescence has a very long decay time.

In general, luminescent minerals exhibit both fluorescence and phosphorescence. However, we shall now proceed to discuss the theory of luminescence and its terminology in some detail.

4.2 THE LUMINESCENCE PROCESS

Luminescence emission occurs as a result of a radiative electronic transition in which an electron 'jumps' from a higher energy state to a lower one, the difference in energy being released as a photon. In contrast to the light emission from incandescent solids, luminescence occurs at temperatures down to absolute zero. Any detectable emission is obviously the result of a large number of more-or-less identical transitions. Clearly the electrons must first be excited into higher energy states by some means, for example, using an electron beam or UV light as already mentioned in the introduction.

Quite often the transition which gives rise to the actual emission is between the lowest excited state and the ground state of a localized centre in the structure. Excitation into any excited state is usually enough to produce luminescence since the electron cascades down the energy levels of the centre non-radiatively to the emitting state; the energy released in such non-radiative transitions produce lattice vibrations or 'phonons'. The efficiency of luminescence excitation in solids varies enormously but in naturally occurring minerals it is usually quite low, seldom exceeding 1% and often one or even two orders of magnitude lower. It is primarily determined by the competing non-radiative processes which occur where most of the energy of excitation is again dissipated in phonon creation.

The probability of a radiative transition from one electronic state to another, k_f, depends on the square of the magnitude of the transition dipole moment for the two states; by far the most common type of transition is an electric dipole one.[1,2] If the initial and final electronic states are characterized by wavefunctions Ψ_i and Ψ_f respectively then the electric dipole transition moment $\mathbf{M}_{if}$ is given by

$$\mathbf{M}_{if} = \int \Psi_i^* (\Sigma e\mathbf{r}_n) \Psi_f \, dV \tag{4.1}$$

where $\Sigma \mathbf{r}_n$ is the sum of the position vectors of all charges e in the system, and the integration is performed over all space. $\mathbf{r}_n$ can be replaced by cartesian coordinates to find the components of $\mathbf{M}_{if}$ in say x, y and z directions.

Note that if Ψ_f is replaced by Ψ_i in this expression we merely have the permanent dipole moment of the initial state of the system (which may, of course, be zero). However, since the transition dipole moment involves two states of different energies, it may oscillate at a frequency ν_{if} which would be that of the emitted photon corresponding to the energy difference between the states. If the amplitude of the transition moment is zero then no photon can be emitted and the transition is said to be strictly 'forbidden'. It can, of course, be forbidden in say x and y directions but 'allowed' in say the z direction. Such a situation would produce photon emission polarized in the z direction. The probability of a spontaneous radiative transition between the two states is given by

$$k_f = \frac{64\pi^4}{3hc^3} \nu_{if}^3 |\mathbf{M}_{if}|^2 \tag{4.2}$$

The actual calculation of the transition moment is seldom attempted since it is much simpler to use symmetry methods to determine whether a transition is forbidden or not.[3,4]

Just as the permanent dipole moment of, say, a water molecule is a fundamental property of that system so is the transition dipole moment of any two states of such a system; it is, therefore, invariant to symmetry operations performed on the system. Let us take as a simple example the symmetry operation of inversion in a centre of symmetry for two d-orbitals of a transition-metal ion. Since d-orbitals have even parity with respect to such an inversion Ψ_i and Ψ_f would not change sign although the position vectors or coordinates of the charges would; consequently $\mathbf{M}_{if}$ would change sign. This it cannot do and therefore it must be zero. A selection rule is, therefore, established where electric dipole transitions are only allowed between states of opposite parity (Laporte rule). Considerations of symmetry can, therefore, lead to a set of such selection rules which describe under what conditions a transition is 'forbidden' or 'allowed'. Remember that $\mathbf{M}_{if}$ must remain totally symmetric.[3,4] Perhaps the most important of these rules is that the electron-spin quantum number must not change during a transition. The derivation of this rule, however, assumes there is no electron spin-orbit coupling (caused by interaction between the electron spin and orbital magnetic moments). The presence of even a small amount of spin-orbit coupling, which mixes spin and spatial parts of the wavefunction, leads to a breakdown of 'forbiddenness'. However, although luminescence transitions between states of different spin do often occur, spin-forbiddenness has a very significant effect on the lifetime or decay time of the process. In the reverse transition the absorption would be very weak, since the radiative probability would be small. As already indicated other selection rules depend on the spatial symmetry of the electronic states involved but there are various ways in which the symmetry of a state can be modified, e.g. by interaction with a low symmetry crystal field (the surrounding ions in the lattice may not be symmetrically placed about the luminescence centre) or by an asymmetrical vibrational mode of the surrounding ions or ligands. Again the primary effect of forbiddenness on luminescence is to lengthen the decay time. The transition which gives rise to emission from the Mn^{2+} ion in various minerals is both spin-and-symmetry-forbidden but the decay time can be as long as several tens of milliseconds; a fully-allowed transition would be expected to have a decay time of around 10 ns or less.

There are other factors which affect the decay time, however. One is due to metastable electron trapping states which effectively lengthen the decay time; the other is due to competing non-radiative processes, which shorten the measured decay time. We will consider the latter first. The experimentally observed decay time of the luminescence is given by $\tau = 1/(k_f + k_i)$ where k_f is the probability of radiative decay and k_i the probability of non-radiative decay processes from the same state. If the decay of the excited state is the only rate-determining step in the luminescence process (which is not always true) then the luminescence intensity I will decay exponentially with time t after the removal of the source of excitation according to $I = I_o \exp(-t/\tau)$. Clearly if k_i is much larger than k_f not only will the decay time be shortened appreciably but the luminescence will be very weak. Non-radiative deactivation within a centre occurs by interaction with the vibrating lattice which depends on temperature; k_i is therefore temperature dependent but also depends to some degree on the energy of the transition which determines the number of phonons which must be created simultaneously – the larger the number the smaller the probability. However k_i may be increased significantly by the resonance transfer[5] of energy to another centre with a short-lived excited state where the probability of non-radiative decay dominates (i.e. a quenching centre). For long-lived emission at room temperature from Mn^{2+} centres, the non-radiative decay probability within the centre is often negligible but resonance transfer to quenching centres such as Fe^{2+} can reduce the intensity of luminescence appreciably. A large variation of luminescence intensity with temperature is indicative of a large non-radiative decay probability at higher temperatures.

An additional delay between excitation and emission is often introduced by metastable electron states known as electron traps which are filled during excitation. Once an electron has become trapped in such a state it requires a certain energy (the binding energy E) to release it but this can be provided thermally. Emission which follows the emptying of such traps at a fixed temperature is known as phosphorescence although this term is often loosely used to describe any luminescence with a decay time longer than a few milliseconds whether or not trapping states are involved. (Fluorescence is the corresponding term for relatively short-lived luminescence where no trapping states are involved between excitation and emission). At a particular temperature there is a certain probability p that a trap will be emptied where $p \propto \exp(-E/kT)$. Thermoluminescence occurs when such traps in a solid are progressively emptied by heating in the dark following previous excitation.

Luminescence emission in general occurs at longer wavelengths (lower energy or wavenumber) than the corresponding transition in absorption and this is referred to as the Stokes shift. Moreover, with the exception of rare-earth centres and the Cr^{3+} ion, the emission band is usually quite broad even at low temperatures. In fact, both the size of the Stokes shift and the width of the broad band are quantified by an important parameter S known as the Huang–Rhys factor. This factor is a measure of the interaction between the electronic states of the centre and the vibrating lattice (often referred to as electron–phonon coupling).

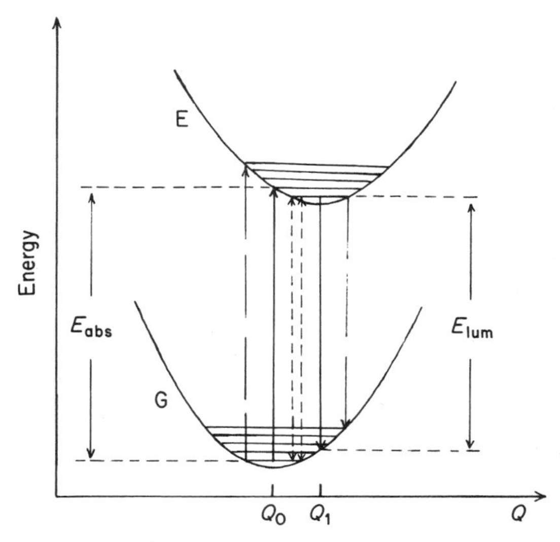

Fig. 4.1 Configurational coordinate diagram (see text). Continuous vertical lines indicate the most probable and therefore the most intense transitions in absorption and emission. Short-dashed vertical lines indicate the no-phonon transition between the lowest vibrational levels of the two electronics states. Long-dashed lines indicate the energy limit of the spectral band in absorption and emission; the difference in length of these lines and those representing the no-phonon transition is a measure of the width of the spectral band.

The situation is best explained in terms of a configurational coordinate diagram (Fig. 4.1). There the potential energy of the electronic states of the centre is plotted as a function of a spatial coordinate Q which represents the average distance to the nearest-neighbour ions in the lattice. G and E are the curves for the ground state and first excited state of the centre; the quantized vibrational or phonon levels are shown for each state. One can imagine the surrounding ions in a 'breathing mode' moving outwards and inwards about some equilibrium position Q_0. Note that the equilibrium coordinates Q_0 and Q_1 are in general different for the two states.

We assume that an electronic transition between the two states occurs so quickly that the surrounding ions do not change their position significantly during transition; such a transition is therefore vertical on the diagram or very nearly so (the Frank–Condon principle). At low temperature the initial state prior to the transition is the lowest vibrational energy state for that particular electron level. If Q_0 and Q_1 are appreciably different then the final state after the transition is likely to be a higher vibrational state of the final electron level, i.e. most of the transitions involve an increase in vibrational energy (or the creation of phonons). However, provided the difference between Q_0 and Q_1 is not too large it is possible to have a 'no-phonon' transition which is purely electronic and does not involve any change in vibrational energy; in this case the transition occurs at the same wavelength in absorption and emission. Following a more common absorption transition which creates phonons, there is a rapid thermal relaxation of the centre to a lower vibrational state of the excited electronic level prior to emission which, in general, will produce a similar number of phonons to that created during absorption. Hence the Stockes shift is

$$E_{\mathrm{abs}} - E_{\mathrm{lum}} = 2\hbar\omega S \tag{4.3}$$

where $\hbar\omega$ is the energy of a phonon. Hence S is the average number of phonons created in absorption or emission. Figure 4.1 also shows how the difference $(Q_1 - Q_0)$ influences the width of the spectrum in absorption and emission. Note that if the curves are similar in shape then the absorption and emission spectra for transitions between these two states are mirror images about the no-phonon or zero-phonon line, which is sharp since no phonons are involved. The width W of the broad band due to transitions in which phonons are created is given by

$$W^2 = 5.6(\hbar\omega)^2 S \tag{4.4}$$

What is more, the Huang–Rhys factor S also determines the fraction of the total intensity in the no-phonon transition since this is equal to $\exp(-S)$. As might be expected, S is temperature-dependent but values usually quoted are for temperatures near absolute zero. This however means that at normal temperatures S is effectively larger and the no-phonon transition is then often too small to be detected. If S is greater than about 6 it is usually difficult or impossible to detect the no-phonon line. In addition, other vibrational levels besides the lowest are also populated at higher temperatures.

Figure 4.2 shows the observed emission and absorption (or excitation) transitions between the lowest split level of the $^4T_1$ (G) and the ground state of the Mn^{2+} ion in enstatite (from the Khor Temiki meteorite) at 5 K.[6] The Huang–Rhys factor is about 5 which is typical for many luminescence centres in minerals. Note the repetitive interval of peaks separated by 165 cm^{-1}; these are the so-called phonon-replicas of the no-phonon line where one, two, or more phonons of this particular vibrational frequency are created during the transition. The peak at 16 660 cm^{-1} which appears only in absorption is probably the no-phonon line of the next split component of the $^4T_1$ level. Its absence from the emission spectrum suggests that the luminescence occurs predominantly from the lowest level. At room temperature the spectrum in absorption and emission is broad and featureless on account of the interaction with a more strongly vibrating lattice.

There are, of course, centres for which the emission and absorption bands are broad and featureless even at very low temperatures. Such centres show a correspondingly large Stokes shift and may have a Huang–Rhys factor as high as 20 or 30. The well-known F-centres in rocksalt and other alkali halides are in this category. (An F-centre is an electron trapped at an anion vacancy.) On the other hand there are centres where the luminescence emission spectrum, even at room temperature, is dominated by the sharp no-phonon transition (e.g. Cr^{3+} in periclase, ruby, spinels, or trivalent rare earths such as Eu^{3+} in some apatites). These centres clearly have a Huang–Rhys factor which is rather small.

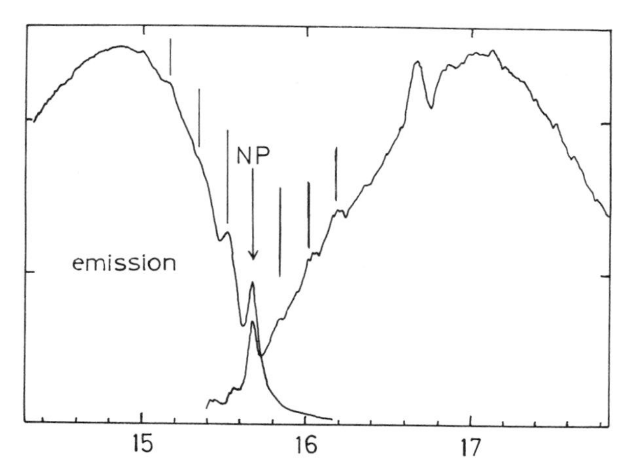

Fig. 4.2 The overlap of emission and excitation spectra of Mn^{2+} in orthoenstatite (from the Khor Temiki meteorite) measured at 5 K. The no-phonon line is indicated and other vertical lines show the position of the phonon replicas. The wavenumber scale is labelled at 1000 cm^{-1} intervals. Spectra are corrected for instrumental response and excitation intensity. (Green and Walker, unpublished work.)

A weak phonon sideband is sometimes visible but only one phonon replica of each vibrational mode frequency is present.

To sum up we can, therefore, roughly classify the luminescence emission spectra of localized centres into three types: (i) those where the no-phonon transition is dominant ($S \ll 1$); (ii) those where the no-phonon transition is visible only at low temperatures along with a multi-phonon sideband ($1 \lesssim S \lesssim 6$) and (iii) those which remain broad and featureless even at very low temperatures ($S \gtrsim 6$). All these spectra are a consequence of the centre being situated in a vibrating lattice; the differences are due to the magnitude of the influence on the electron(s) in the centre exerted by the surrounding ions.

We have so far discussed only luminescence transitions within a localized centre of atomic or small molecular dimensions, i.e. intra-centre transitions. However, in insulators, which most minerals are, most luminescence transitions are likely to be of this nature particularly in predominantly ionic crystals. The discussion has necessarily been brief and simplistic but the reader is referred to the more detailed analyses by Imbusch[7], Rebane[8] and Stoneham.[9]

Many models of the luminescence process have used the band theory of solids to explain luminescence transitions following the theory as applied to semiconductors. In semiconductors where the energy bandgaps between the delocalized electronic levels of the valence and conduction bands are separated by less than about 3 eV, visible luminescence due to transitions involving these delocalized levels is observed (see Fig. 4.3). Inter-centre transitions between levels of different localized centres (such as donor–acceptor pairs) are also commonly observed and, at low temperatures, emission can occur due to the recombination of 'excitons' (bound electron-hole pairs), particularly those trapped at impurities (bound-excitons).

Are these types of luminescence transitions therefore likely to be observed in minerals? Only the sulphides among common minerals could be classified as semiconductors although intra-centre transitions such as occur in Mn^{2+} are still a likely source of luminescence, e.g. in zinc blende. Although a very broad emission band in diamond has been ascribed to donor–acceptor transitions with varying spatial separation between donor and acceptor atoms, most luminescence bands appear to be due to transitions either within a localized centre or between a localized centre and the valence or conduction band.[10] The latter type of transition, often referred to as a 'free-to-bound' or 'bound-to-free' transition, may well occur in some minerals (such as quartz?). Simultaneous measurements of photoluminescence excitation spectra and photo-conductivity may help to identify such transitions.

Luminescence from mineral crystals will often be polarized, except of course from randomly orientated microcrystalline samples. This polarization results from the fact that the transition may not be allowed in all directions in the centre. The direction of polarization reflects the symmetry of the luminescence centre in non-cubic crystals. Luminescence from cubic crystals

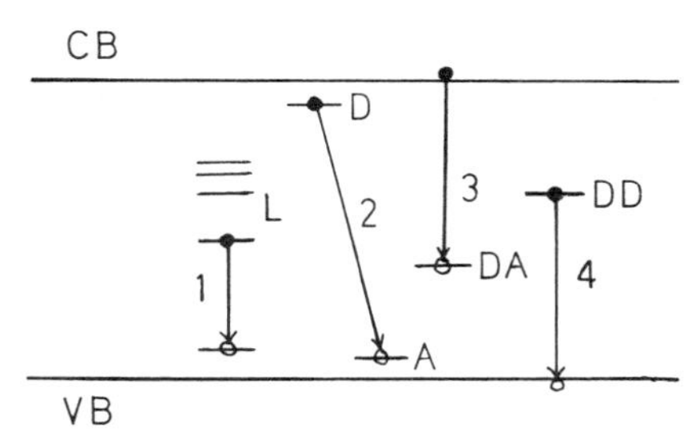

Fig. 4.3 Possible luminescence transitions in minerals. 1. Internal transition in a localized centre L with energy levels between the valence and conduction bands (the most likely). 2. Donor–acceptor transition in which an electron leaves a neutral donor D for a neutral acceptor A when donor and acceptor atoms are reasonably close together in the crystal. The transition leaves both donor and acceptor atoms ionized. 3. Conduction band (CB) to deep acceptor (DA) transition. A 'free' electron is trapped by an acceptor atom and ionizes it. 4. Deep donor (DD) to valence band transition. An electron on a neutral donor atom recombines with a 'free' hole.

can also be polarized if centres of non-cubic symmetry are preferentially orientated in certain directions in the crystal (e.g. along dislocations) or if the crystal is anisotropically stressed.

4.2.1 Historical background

The luminescence properties of certain minerals have been known for a very long time. Indeed minerals such as willemite and fluorspar (from which the term 'fluorescence' originates) were among the earliest known luminescent materials. Nevertheless, knowledge and understanding of the luminescence centres in common rock-forming minerals have only begun to be developed in a coherent manner during the last twenty years. In 1965 three important papers were published[11–13] which suggested that cathodoluminescence might be a very useful analytical tool, particularly in sedimentary petrology. In fact about this time much interest was stimulated in the use of both cathodoluminescence and thermoluminescence as possible diagnostic techniques in mineralogy and petrology. In 1966 a NATO Advanced Research Institute on the application of Thermoluminescence to Geological Problems was held at Spoleto in Italy and the proceedings subsequently published.[14] Thermoluminescence bibliography was already extensive by this time although the nature of the luminescence centres involved were often not known and sometimes not even discussed. Medlin[15,16] had shown that Mn^{2+} was responsible for the orange-red emission in calcites and dolomites, and Geake and Walker[17,18] had demonstrated that the red cathodoluminescence from meteoritic enstatite was also correlated with the presence of manganese. However, the nature of the emission centres in two of the most common luminescent mineral components, namely quartz and feldspars, was not known at this time (1968). Even today the nature of the various luminescence centres in quartz is still not known with any certainty.

Sippel[19] investigated the luminescence properties of quartz in sandstones and although recently Zinkernagel[20] has studied quartz cathodoluminescence in greater detail the precise nature of the dominant blue and red emitting centres is a problem which remains unsolved. However, better progress has been made in the elucidation of luminescence centres in feldspars. Additional impetus was provided by the fact that plagioclase was found to be the main luminescent component of lunar soils, rocks and breccias.[21,22] Plagioclase was found to exhibit three different luminescence bands and the nature of the centres responsible for two of these was suggested by the work of Geake *et al.*;[22,23] unequivocal confirmation was provided later by Telfer and Walker[24,25] using the techniques of luminescence excitation spectroscopy (see Section 4.3.2). Following Medlin's early work, Sommer[26] made a detailed study of the cathodoluminescence of carbonates and about the same time (1972) Graves and Roberts[27] analysed the spectral emission of thermoluminescent calcite at different temperatures. The article by Nickel[28] reviewed the state of cathodoluminescence as a tool in sedimentology up to 1977.

About the same time as the upsurge of interest in cathodoluminescence in minerals came the early applications of crystal field theory to explain the electronic absorption spectra of transition-metal ions, most notably Fe^{2+}, in minerals. The monograph by Burns[29] reviewed the early work and gave a succinct account of how such measurements of crystal or ligand-field spectra could give information about cation ordering in mineral structures. Ligand-field theory was first applied to the interpretation of the spectra of transition-metal luminescence centres such as Mn^{2+} by Orgel[30] as early as 1955 but Medlin was probably the first to apply it to luminescence centres in minerals. By the early 1970s it had become clear that the Mn^{2+} ion was responsible for luminescence in a large number of minerals producing emission colours ranging from green to deep red. However, Mariano and Ring[31] demonstrated that rare earths such as europium may also be responsible for luminescence emission in some natural minerals including apatite, although strong yellow emission in this mineral is usually due to Mn^{2+}.

In 1979, a translated and up-dated version of a book by Marfunin,[32] first published in Russian in 1975, revealed that a considerable amount of work on the luminescence of minerals has been done in the USSR. Much of the work reported corroborates that done in Europe and America although some is unique.

Cathodoluminescence photography has produced many beautiful pictures of minerals in rock sections luminescing brightly in different colours and shades. The nature of the luminescence centres responsible for these different colours however may not be known, nor easily determined. The article by Remond[33] for example shows that although some conclusions can be drawn by comparison with pure synthetic analogues carefully doped with specific impurities (a technique used previously by Geake *et al.*[23]), many interesting cathodoluminescence microphotographs cannot easily be explained without prior knowledge of the nature of the centre likely to be present.

4.2.2 Types of luminescence centres in minerals

Luminescence centres are generally of two types, impurity centres and defect centres. In impurity centres, the luminescence transition involves the lattice-perturbed electronic levels of the impurity ion and clearly transition-metal ions and rare earths fall into this category. However, many defect centres occur as a result of the presence of an impurity, and sometimes the differentiation between these two types of centre seems a little artificial although in other instances it is quite marked, e.g. an F-centre is quite definitely a defect centre.

We shall, however, for the present purposes consider transition-metal and rare-earth centres separately and consider all other centres, impurity-related and otherwise, under the heading of defect centres.

(a) *Transition-metal ion centres*

These luminescence centres are the most well defined and best understood. Divalent or trivalent transition-metal ions present as impurities substitute for normal metal cations in the structure, e.g. Mn^{2+} for Ca^{2+} in calcite, anorthite, or wollastonite. Mn^{2+} is, in fact, one of the most common luminescence centres in minerals where it readily substitutes for Ca^{2+}, Mg^{2+} or Zn^{2+} in carbonates, silicates and a host of other structures. The isoelectronic d^5 ion Fe^{3+} also gives rise to luminescence particularly in feldspars[24,25] where it is thought to substitute for Al^{3+} although the emission is mainly in the infra-red with the tail of the band sometimes visible in the red. Cr^{3+} is, of course, well known in ruby where its sharp red line emission was used to produce the first solid-state laser. However, Cr^{3+} often substitutes for octahedral Al^{3+} in, for example, grossular garnets and spinels where it gives rise to a deep red emission; it is not usually found in tetrahedral coordination.

The application of ligand-field theory to the spectra of transition metal ions in minerals has been dealt with at length elsewhere[29,34] (see also Chapter 3). Only a few brief comments about the more common d^5 luminescent ions will therefore be given here.

The ligand-field splitting parameter Δ or $10\,Dq$ is a measure of the strength of the field due to the surrounding anions or ligands and represents the energy difference between the $e_g(d_{x^2-y^2}, d_{z^2})$ and $t_{2g}(d_{xy}, d_{yz}, d_{zx})$ orbitals in any cubic field. For d^5 ions in minerals the ground state has one electron in each d-orbital with all spins parallel (Hund's rule); it is therefore a spatially symmetric sextet ($^6A_{1g}$) state and is not split by the ligand field. The lower excited states derive from the ^4G state of the free ion; some of these ($^4A_1$, ^4E) have the same

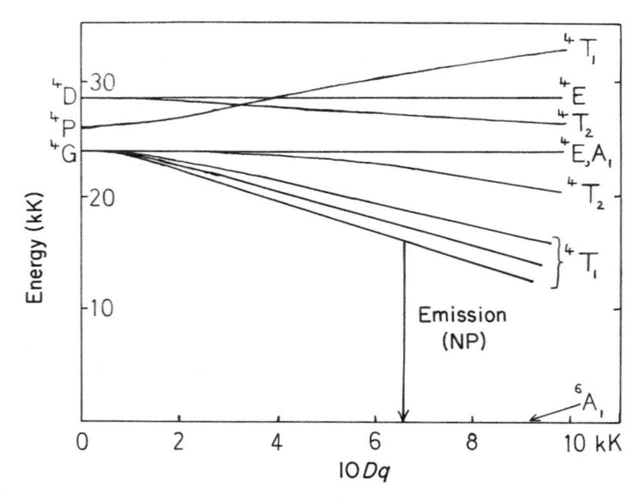

Fig. 4.4 Orgel energy level diagram for a d^5 ion. This is specially drawn for Mn^{2+} in orthoenstatite showing a large splitting of the $^4T_1(G)$ level. Splitting of other levels is smaller and, in the interests of clarity, is not shown. The luminescence no-phonon transition is shown. No-phonon transitions in absorption to all levels will of course occur at the same value of 10 Dq. The free-ion terms are shown on the left. (The unit 1 kK $= 1000\,cm^{-1}$.)

electron configuration as the ground state ($t_{2g}^3\,e_g^2$), although two of the electron spins are now paired, and are not affected to any appreciable extent by the field. The positions of the remaining $^4T_1$ and $^4T_2$ triply degenerate states are however more dependent on the size of the ligand field (see Fig. 4.4). (The nomenclature is that used for irreducible representations in group theory and describes the symmetry, degeneracy and spin-multiplicity of the state.) Nevertheless, the energy difference between these levels and the ground state still owes rather more to inter-electron repulsion terms than to the ligand field. The inter-electron repulsion parameters (the Racah parameters) B and C can be determined directly from the spectral positions of the $^4E(G)$ and $^4E(D)$ states and a best fit value of Δ or 10 Dq can then be evaluated from the positions of the ligand-field dependent states. Energies of states of the same symmetry are given by an energy matrix containing the parameters B, C and $10\,Dq$. These matrices are obtained by starting either from the weak field[35] or strong field[36] point of view, i.e. assuming either $B \gg Dq$ or vice versa and then introducing either the ligand field or the electron repulsion terms as a perturbation.

Transition-metal d^5 centres can be easily and unambiguously identified as responsible for particular luminescence bands by the technique of luminescence excitation spectroscopy (see Section 4.3.2(b)) which reveals the characteristic ligand-field bands of these ions (Figs. 4.5 and 4.6).

Figure 4.4 shows the Orgel energy-level diagram applicable to d^5 ions such as Mn^{2+} and Fe^{3+} showing the luminescence transition from the lowest component

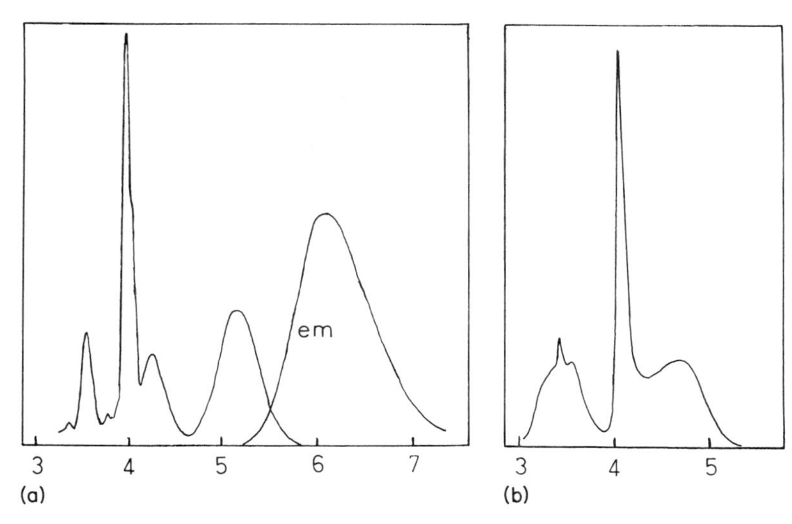

Fig. 4.5 (a) Emission and excitation spectra of Mn^{2+} in a calcite (Clitheroe, Lancashire). (b) Excitation spectrum of Mn^{2+} in a typical plagioclase (laboradorite.) The wavelength scale is marked at 100 nm intervals.

of the ${}^4T_1(G)$ level; the three-fold degeneracy of this level is often removed by distortions from cubic point symmetry which occur in many cation sites in minerals. Note that the wavelength of the emission not only depends on the magnitude of the crystal field but can also be influenced by the splitting of the ${}^4T_1(G)$ level if this is appreciable (as in enstatite, for example). Figures 4.5 and 4.6 show how the different crystal environments affect the ligand-field spectra of Mn^{2+} luminescence centres. From these spectra, obtained by luminescence excitation spectroscopy, can be extracted a great deal of information besides the obvious deduction that the centres producing the emission are quite definitely Mn^{2+} ions. As already indicated, it is possible to evaluate the Racah parameters B and C, the ligand field parameter $10\,Dq$ and, in some cases, to get some idea of deviations from cubic symmetry of the site in which the Mn^{2+} ion resides. Table 4.1 gives ligand-field parameters for Mn^{2+} in various mineral structures determined from such spectra. Bearing in mind that the size of Racah parameters compared with their known free-ion values give a measure of the degree of covalency or orbital overlap with the ligands and taking into account the strength of the ligand field it is not too difficult to decide what site the Mn^{2+} ion is occupying. For example, the spectra of Mn^{2+} in forsterite and enstatite show that the ion occupies only one (M_2) of the two possible cation sites whereas in diopside it occupies both sites. Moreover, the differences in the degree of distortion between M_2 sites in forsterite and enstatite is clearly evident in the spectra. As we shall see later it is possible to carry out a complete analysis of the effective pseudosymmetry of the site from polarized luminescence excitation spectra on single crystals.

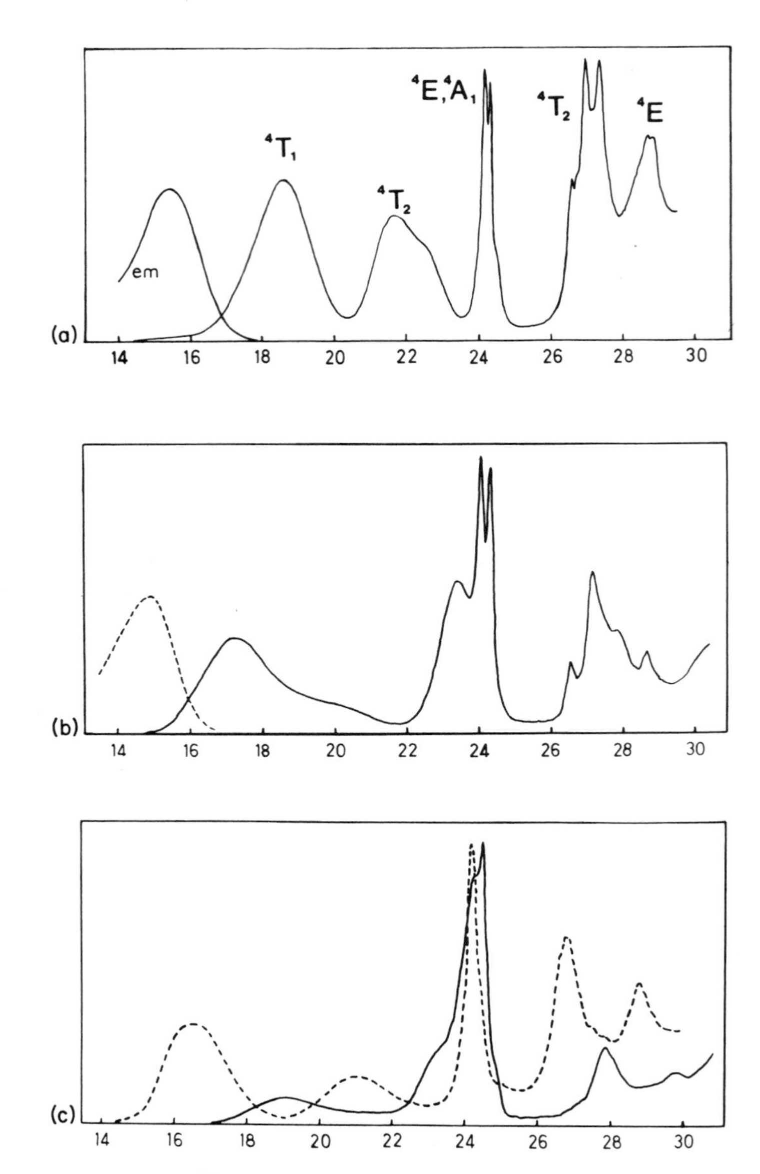

Fig. 4.6 Spectra of Mn^{2+} measured at room temperature. (a) Emission and excitation spectra of Mn^{2+} in a synthetic forsterite. Transitions to ligand-field energy levels are indicated. (b) Emission and excitation spectra of Mn^{2+} in orthoenstatite (from the Khor-Temiki meteorite). Note the large splitting of the lowest $^4T_1$ level. (c) Excitation spectra of Mn^{2+} in the M_1 site (broken line) and the M_2 site (continuous line) of diopside (BM40397) (see Fig. 4.12 for emission spectrum of diopside). The wavenumber scale is marked at 2000 cm^{-1} intervals, and all spectra are corrected for instrumental response and excitation intensity. (Green and Walker, unpublished work.)

Table 4.1 Ligand-field parameters† for Mn^{2+} in various minerals

Mineral	*Average cation–anion distance (nm)*	*10 Dq* (cm^{-1})	*Racah parameter B* (cm^{-1})	B/B_O
Forsterite (M_2 site)	0.2133	7500 ± 200	794 ± 20	0.87
Enstatite (M_2 site)	0.2158	6500 ± 500	802 ± 20	0.88
Diopside (M_1 site)	0.2076	9200 ± 200	818 ± 20	0.90
Diopside (M_2 site)	0.2498	5800 ± 800	907 ± 20	1.0
Anorthite	0.250–0.254	5300 ± 500	825 ± 20	0.88
Calcite	0.235	7200 ± 200	790 ± 20	0.85

†These values were calculated from luminescence excitation spectra [6,25,62] using trees-corrected, weak-field matrices for cubic symmetry. B_O is the free-ion value of the Racah parameter B taken as 910 cm^{-1}.

Similar considerations can be extended to Fe^{3+} spectra; there is usually no confusion between Fe^{3+} and Mn^{2+} since, although octahedral Fe^{3+} ligand-field spectra look like those of Mn^{2+}, calculation of 10 *Dq* will show a much higher value than for Mn^{2+} and the emission will invariably be in the infra-red.

It is just possible that octahedrally-coordinated Mn^{2+} could produce emission in the same spectral region as tetrahedrally-coordinated Fe^{3+} and that 10 *Dq* could be similar in magnitude for the two cases. However, tetrahedral Fe^{3+} has a characteristically different ligand-field spectrum,[25] and its emission is often on the edge of the infra-red at somewhat longer wavelengths than is usual for Mn^{2+}.

The emission spectrum of Cr^{3+} is very characteristic usually consisting of one or two closely-spaced sharp lines in the red. If further evidence is needed then the luminescence excitation spectrum of this emission yields the characteristic d^3 ligand-field spectrum[7] (see Fig. 4.7).

Paradoxically, all the ions considered so far have a luminescence transition that is forbidden. It is therefore to be expected that the decay times are rather long; even tetrahedral Fe^{3+} luminescence usually has a decay time of one or two milliseconds.

What of other transition-metal ions? From time to time there have been reports of Ti^{3+} or Ti^{4+} (?) in silicates[37] and quartz[38] giving rise to a blue emission band, but it is not as yet clear whether the emission is from the Ti^{3+} ion or from a defect caused by the introduction of titanium.

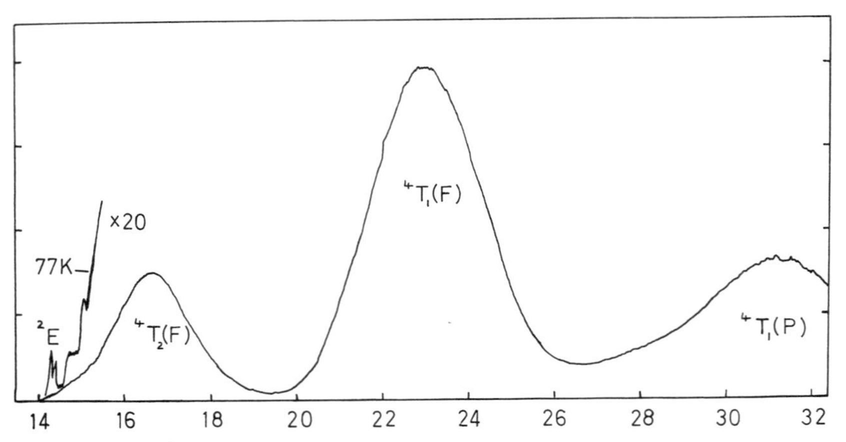

Fig. 4.7 Luminescence excitation spectrum of Cr^{3+} in grossularite (BM55491) at room temperature. At 77 K the characteristic R lines due to the ^2E state are apparent. The emission spectrum at room temperature (not shown) does not exhibit the R lines, probably because the $^4T_2(F)$ level is thermally populated, but at 77 K the spin-forbidden R lines of Cr^{3+} are clearly evident on the low energy side of the usual Mn^{2+} emission band. (Green and Walker, unpublished work.) The wavenumber scale is marked at 2000 cm^{-1} intervals.

It is common knowledge that an appreciable concentration of Fe^{2+} completely quenches any luminescence; Ni^{2+} and Co^{2+} are known to have a similar effect. The energy levels of these ions form a 'ladder' of states close in energy which facilitate non-radiative transitions. In particular, the energy gap between the first excited state and the ground state is small. However, emission from these ions has been observed in some materials in the infra-red but usually only at low temperatures. Recently, for example, Fe^{2+} emission was recorded at temperatures near absolute zero in gallium phosphide at a wavelength of about 3 μm;[39] Co^{2+} and Ni^{2+} emissions have also been observed in the infra-red at low temperatures.[40] However, to date, there are no reports of infra-red luminescence from these ions in naturally occurring minerals, although detection of weak luminescence beyond the spectral range of photomultipliers is difficult owing to the low sensitivity of infra-red detectors such as lead sulphide cells.

It may be thought that the higher the concentration of a luminescent ion such as Mn^{2+}, the more intense the emission. First of all it must be appreciated that concentrations as low as 100 ppm, or even less, can produce easily detectable emission; optimum concentrations are usually around 0.1–1%. Increasing the concentration further does not usually produce any significant increase in luminescence intensity; in fact, eventually it begins to reduce it (see Fig. 4.8). This behaviour is referred to as concentration quenching. Manganese minerals such as rhodochrosite and rhodonite are not usually luminescent at room temperature although Mn^{2+} luminescence is observed at low

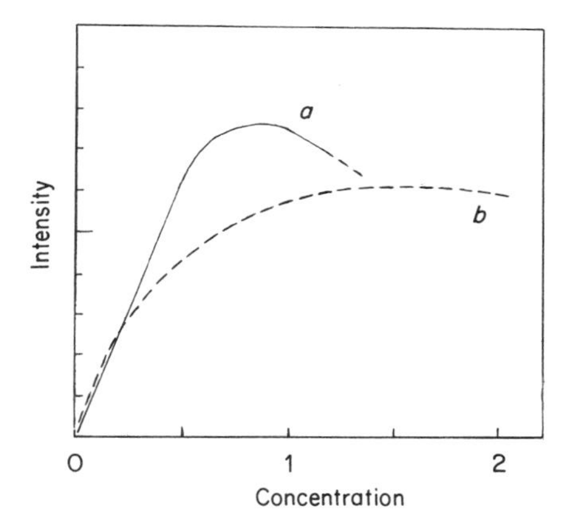

Fig. 4.8 The variation in luminescence intensity with concentration of luminescent impurities in synthetic anorthite. (a) Intensity of green emission with manganese concentration. (b) Intensity of infra-red emission with ferric ion concentrations. The intensity scales for (a) and (b) are unrelated but the concentration scale is the same and labelled in mol %.

temperatures.[41] Imbusch[7] has explained this behaviour in terms of a rapid transfer of excitation energy between neighbouring Mn^{2+} ions until it is trapped either by a shallow luminescent trap, which consists of an Mn^{2+} ion perturbed by an adjacent defect or impurity, or by a deep quenching trap such as an Fe^{2+} ion. At normal temperatures, the shallow luminescence trap is rapidly emptied by thermal activation before it can produce luminescence but at low temperatures luminescence decay becomes more probable.

(b) *Rare-earth centres*

The energy levels of rare-earth ions in solids can also be described by ligand-field theory.[42] However, here the interaction between the 4f electron states and the surrounding lattice is weak and the spectra of trivalent ions are dominated by sharp no-phonon lines. Moreover, luminescence transitions often occur between several different excited electronic levels and the ground state. Also because of the weak lattice interaction, the energy levels of a given ion do not vary much in position from one host structure to another, although the strength and symmetry of the ligand field may affect the degree of splitting of an energy level and the radiative transition probability. Nd^{3+} is, of course, well known as the centre responsible for infra-red laser emission in the synthetic yttrium aluminium garnet (YAG). However, reports of such lanthanide centres in minerals are sparse. The most common rare-earth centres appear to be trivalent and divalent europium. Mariano and Ring[31] detected several emission lines in scheelite which they ascribed to Eu^{3+}, Sm^{3+} and $Dy.^{3+}$ They also found that some apatites from carbonatites had a broad-blue luminescence band which was ascribed to Eu,$^{2+}$ although sometimes two or three sharp peaks at 590, 615 and 695 nm due to different electronic transitions of Eu^{3+} were also present, very often superimposed on the characteristic Mn^{2+} yellow band. Divalent rare-earth ions such as Eu^{2+} can produce broad-band emission rather than the familiar sharp line emission of trivalent ions. This is because transitions at energies in the visible region can involve an electron changing from a 4f to a 5d orbital (or vice versa in emission) instead of simply to another 4f orbital as in trivalent ions. Thus Eu^{2+} emission is more like that of Mn^{2+} except of course for spectral position; it is broad because of the larger lattice interaction with the 5d orbitals (i.e. the Huang–Rhys factor is much larger) and its spectral position is dependent on the ligand-field strength. It would appear that these lanthanide ions substitute for Ca^{2+}. Concentrations of europium in natural carbonatite apatites were measured and found to be between 100 and 500 ppm.[31] Mariano and Ring also studied other minerals from carbonatites and found that strontianite and fluorite also exhibited broad-blue Eu^{2+} emission.

Sharp line emission at room temperature is usually indicative of the presence of either trivalent rare-earth or Cr^{3+} centres. A luminescence excitation spectrum will distinguish between the two if there is any doubt (see Section 4.3.2 (b)).

(c) *Defect centres*

These are the most difficult types of centre to characterize. Luminescence involving such centres usually consists of a very broad emission band often without any discernible structure even at very low temperatures. Luminescence from F-centres in rocksalt for example is clearly due to intra-centre transitions with a very large Huang–Rhys factor.[43] However, at 4 K the F^+-centre in CaO does show a small no-phonon line with single phonon replicas in emission in addition to a large broad multiphonon band. (An F^+-centre is an electron trapped at an oxygen vacancy; since oxygen is divalent the centre has a residual positive charge.) The F^+-centre in periclase (MgO) on the other hand shows a very large Stokes shift and no structure.[44] These centres produce emission in the near ultra-violet centred around 350 nm. Defect centres in more covalent solids such as quartz are also often associated with oxygen vacancies; the E^1 centre in quartz is thought to be due to an electron in a 'dangling bond' of a silicon atom adjacent to an oxygen vacancy.[45] Whether this centre is responsible for one of the luminescence bands in quartz is, however, not yet proven and this problem will be discussed later (see Section 4.4.5).

The theoretical description of the electronic properties of vacancy defects is of course different in ionic and covalent crystals. In ionic crystals the starting point is similar to that of crystal field theory; an electron in an F- or F^+-centre for example can be considered in a potential due to point charges at the neighbouring cation positions. From the covalent point of view a vacancy produces spare valence electrons in 'dangling bonds' which previously participated in bonding to the missing atom. These electrons are most simply described by a 'defect molecule' model which constructs molecular orbitals from the dangling-bond orbitals.[46]

There are, of course, many types of defects in solids, some involving more than one vacancy, others not connected with vacancies at all; some are connected with impurities such as the so-called aluminium centre in quartz.[45] Many of these centres are paramagnetic having unpaired electron spins and have been characterized by electron spin resonance studies.

There are many luminescence bands in common minerals which are possibly due to point defects but very few, if any, have been unequivocably characterized. The broad blue emission band in quartz, and the similar and probably related emission which is observed in feldspars and other silicates, has been ascribed to various transitions, some involving structural defects, some involving impurity-related defects. Convincing proof of association with a known defect is however still awaited.

Defect centres can, of course, also be produced by radiation and some centres in minerals may be created in this way. For further details on defect centres in general the reader is referred to Stoneham[9] and parts of Crawford and Slifkin.[43,44] A recent review of luminescence centres in diamond is also of interest.[10]

4.3 EXPERIMENTAL TECHNIQUES

We shall now briefly review the techniques that have been used in the study of luminescence in minerals, including those involving spectral measurements, but omitting thermoluminescence techniques.

4.3.1 Cathodoluminescence techniques

Irradiation of a luminescence material with an electron beam has long been recognized as a very effective way of producing a high excitation density and therefore intense emission (as for example in a cathode-ray tube). The mechanisms involved in cathodoluminescence are complex and have been discussed at length by Garlick,[47] Meyer,[48] and others; the experimental problems and techniques have also been reviewed recently by the author.[49] Because effective luminescence centres are present in low concentrations and often absorb visible light or ultra-violet only very weakly, and because most minerals are large-bandgap insulators, photoluminescence is usually very weak and often barely visible even under intense excitation. Excitation of centres via the conduction band is often the most efficient method and electron excitation enables this to take place.

(a) *The electron microprobe*

Early observations of the luminescence of certain minerals under the highly-focused electron beam of a microprobe were probably responsible for the arousal of interest in cathodoluminescence in mineralogy and petrology. During the 1960s electron microprobes became one of the standard physical methods of analysis in geology. Using the technology of the electron microscope, the microprobe produces an electron beam focused down to a diameter of a few μm to 'probe' a small area of a rock section. One or more X-ray crystal spectrometers with appropriate geiger-tube detectors monitor the characteristic X-radiation (usually K_α) emitted from the irradiated area, thereby detecting the elements present. The electron beam energy must be greater than about 20 keV in order to produce such characteristic X-radiation. The beam itself can be rastered back and forth across the whole area of the section giving a picture of the distribution of any chosen element. Correlation of such pictures, displayed on an oscilloscope, with the distribution of luminescence of differing colours enables even very small areas of luminescence to be ascribed to a particular mineral phase. In some cases, where the concentration of luminescence centres such as Mn^{2+} in a mineral is not too low, it has been possible to correlate luminescence colour and brightness with the abundance of Mn or other dopant ion.[50] Anticorrelation of luminescence intensity with iron concentration has also been observed.[23] The sensitivity of the microprobe is, however, limited and very small concentrations of impurity centres which may produce efficient luminescence cannot always be measured. Moreover, measurements with the microprobe do not tell us the valence state of the element or give any information about its environment in the mineral structure.

(b) *Cathodoluminescence microscopes*

Although microprobes have proved very useful in the study of cathodoluminescence this was not the purpose for which they were designed. What was required was a much simpler, less-expensive instrument which would excite cathodoluminescence in a rock section preferably whilst the section was viewed under a petrological microscope. The electron beam needed only to be crudely focused so that all the field of view would be irradiated. Luminescence could then be observed and photographed under high magnification and compared with features observable in normal and polarized transmitted light. Ordinary uncovered thin sections could be used, although these must of course be mounted inside a vacuum chamber and yet be adjustable in position.

Sippel[51] designed a system which met these requirements and a few years later a commercial instrument was available called a luminoscope[52] (Nuclide Corp., Acton, USA). The source of electrons in these instruments is a cold-cathode discharge maintained at the optimum pressure for a good discharge by evacuating with a rotary pump and using a needle-valve air leak. Accelerating voltages of up to 18 kV can be used at a beam current of about 1 mA. High beam current densities must be avoided otherwise the specimen is damaged and the luminescence obliterated. There are basically two forms of damage; one is due to the heating of the specimen by the beam, the other is due to heavy negative ions from the discharge causing radiation damage.

Zinkernagel[20] has described a luminescence microscope device using a triode electron gun with a hot filament which overcomes some of the inherent disadvantages of the simpler cold-cathode luminoscope. There is no background blue light from the discharge and the risk of radiation damage due to negative-ion bombardment is reduced by being able to operate at a much better vacuum. The beam is also claimed to be much more uniform and stable since there is no air leak which is the usual cause of instability.

Using a similar instrument Ramseyer[53] has used the cathode-ray tube technique of covering the specimen with a thin layer of aluminium through which the electrons pass. Viewing the section through the other side he claims the luminescence yield is increased substantially by reflection of luminescence by the aluminium. The aluminium layer also has the beneficial effects of conducting away heat and charge build-up on the specimen as well as absorbing any heavy ions.

Zinkernagel has also used the instrument described for spectral measurements.

4.3.2 Spectral measurements

Although the cathodoluminescence colour and relative intensity can be observed visually or recorded on film, it is more precise to measure the spectral distribution of the luminescence; it is essential to do so if the nature of the centres is being sought. Moreover, it is difficult to analyse visually how many luminescence bands are contributing to the overall colour of luminescence. Clearly the overall colour will change if one emission band becomes stronger than another in a particular area of a mineral deposit. Subtle colour variations should therefore be investigated spectrally.

(a) *The measurement of emission spectra*

The essential requirements, in addition to a means of producing cathodoluminescence, are a high luminosity monochromator and a photomultiplier with a red-sensitive photocathode. Careful attention should be given to the method of light collection and illumination of the entrance slit of the monochromator; in particular the entrance cone of the instrument should be filled with light. It is clearly not sensible to have an expensive grating monochromator and then not to transfer the light to it efficiently. The spectrum can usually be scanned using a motor-driven sine drive to the grating mount. The photomultiplier output current is a linear function of the light intensity at any given wavelength and can be recorded using a high-impedance dc amplifier. However, there are advantages in chopping the signal in some way and using lock-in amplification techniques. Geake *et al.*[54] used pulsed 'square-wave' electron excitation which, when combined with lock-in amplification of the luminescence signal, eliminates any background light present and, by varying the frequency of excitation, enables time-resolved spectra to be measured and the decay times of luminescence bands to be estimated.

Whatever system is used, the emission spectrum is eventually plotted as luminescence intensity against wavelength. This is not, however, the 'true' spectrum since it has to be corrected for the variations in spectral response of the monochromator and detector. Such corrections may significantly alter the recorded spectrum causing relative intensities of bands to change and shifting band maxima. In spite of this fact, many published spectra are not corrected for instrument response and, in some cases, it is not even stated whether or not such a correction has been applied.

The luminescence 'intensity' measured by a photomultiplier is really a photon flux, i.e. it measures the relative number of photons incident per second. However, if a photomultiplier is calibrated against a device, such as a thermopile, which measures incident energy rather than the number of photons arriving per second, the 'intensity' will be in relative units of radiant power rather than in photons s^{-1}. A conversion to the latter units is, however, easily made; if E_λ is the radiant power per unit wavelength interval, and I_λ the number of photons per second per wavelength interval then clearly $I_\lambda = E_\lambda \lambda / hc$.

It is sometimes preferred to plot the spectrum as luminescence intensity against a linear scale of wavenumber (the reciprocal of wavelength), since then energy differences between spectral features can be read off directly. When such a conversion is done from a wavelength plot it is essential to remember that intensity ordinates should then be multiplied by the wavelength squared.[49]

In this chapter, all intensity scales of emission spectra are in units of photons per second per wavenumber (or, in some cases, wavelength) interval. Intensity scales of excitation spectra are in units of relative absorption coefficient.

The theory and practice of the measurement of luminescence spectra and methods of calibration have been discussed in detail by Hamilton *et al.*[49] to which the reader is referred.

(b) *The technique of luminescence excitation spectroscopy*

This technique enables the absorption spectrum of a particular luminescence emission centre to be measured. If, for example, a mineral emits a red luminescence band of unknown origin, it is possible to measure what absorption bands are connected with the centre causing this emission. These absorption bands may be particularly diagnostic; for example, in the case of Mn^{2+} centres, the ligand-field bands are immediately recognizable and the

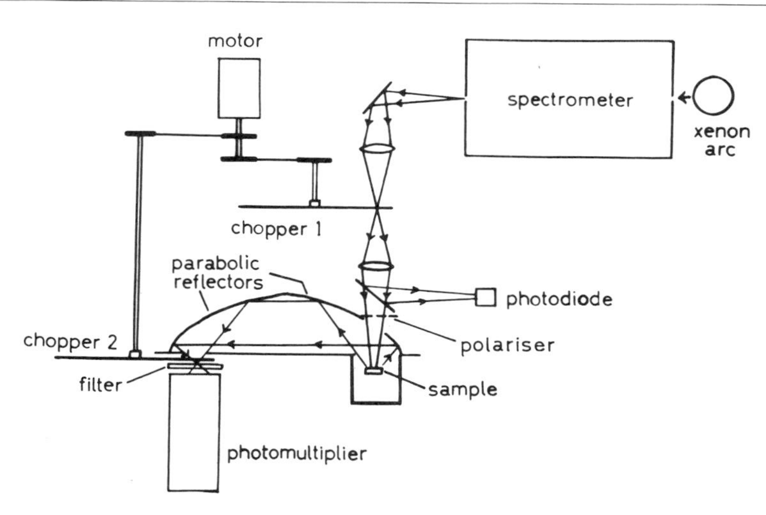

Fig. 4.9 Experimental arrangement for the determination of luminescence excitation spectra. The two chopper discs, which are in antiphase are driven from the same motor. The photodiode output is used to compensate for excitation intensity.

nature of the luminescence centre is instantly and unequivocally determined.

Figure 4.9 shows the experimental arrangement for this powerful technique. A high-intensity continuum light source as a xenon arc or tungsten halogen lamp illuminates the entrance slit of a high-luminosity monochromator, the output of which is scanned in wavelength and mechanically chopped before being focused onto the sample. Any luminescence emission band can be isolated by a filter and monitored by a photomultiplier. Immediately in front of the photomultiplier is a second mechanical sector-disc chopper which is driven in anti-phase with the first so that the photomultiplier can detect weak emission which may otherwise be swamped by scattered incident light. The chopping frequency used depends on the decay time of the luminescence. The integrated emission intensity is simply dependent on the strength of absorption by the centres at different wavelengths. In fact, in the limit of weak absorption, the luminescence excitation spectrum obtained as a plot of relative luminescence output against the wavelength of excitation is identical in profile with the absorption spectrum of the luminescence centre.[49] Since the photomultiplier is simply monitoring variations in intensity of the same spectral distribution the only correction to be applied is for the variation in intensity of the exciting light with wavelength.

The technique is valid when the luminescence transition and its intensity is independent of the original state of excitation. Whatever state is excited, the centre must relax to the emitting state, and this is true for many centres. It was first applied to minerals by Telfer and Walker[24] who were able to detect Fe^{3+} centres in plagioclase present in lunar soil; this measurement in particular shows the power of the method, since it was done on an ordinary soil sample without any form of mineral separation being necessary. In fact, the technique enables absorption profiles of luminescent ions to be easily measured in conditions where it would be impossible using conventional absorption spectroscopy. In so much as it is applicable to absorbing species present in very low concentrations, it is complementary to conventional absorption spectra. However, the luminescence excitation technique has the further advantage of measuring the absorption profile of one species in the presence of others without the problem of overlapping bands. It is even possible to measure separately the absorption profiles of the *same* ion in different sites (e.g. Mn^{2+} in diopsides, Fig. 4.6).

Although a great deal of information can be obtained from unpolarized spectra using powdered samples, polarized spectra can be obtained with correctly orientated single crystals just as in conventional absorption spectroscopy (see Fig. 4.10)

4.4 LUMINESCENCE CENTRES IN SOME COMMON MINERALS

No attempt will be made at encyclopaedic coverage of all known luminescence centres in all common minerals. We shall simply discuss commonly occurring centres in a few important minerals or mineral types.

4.4.1 Olivine

As expected only reasonably iron-free olivines show any appreciable luminescence. In cathodoluminescence, forsterites usually exhibit two broad emission bands, a strong red one centred around 650 nm (15 500 cm^{-1}) and a much weaker and even broader blue emission with a maximum in the range 460–470 nm (21 500 cm^{-1}). Sometimes, in natural samples, the red emission band is broader than in synthetic ones giving a more orange-red emission but this may be due to the influence of other impurities. The red emission is due to Mn^{2+}, as is evident from the excitation spectrum in Fig. 4.6, and has recently been studied in detail.[6,55] At room temperature the Mn^{2+} emission has a decay time of about 35 ms in pure synthetic samples. At low temperature the decay time increased slightly and a small no-phonon line is observed at 615 nm (16 250 cm^{-1}). Phonon replicas also occur at 180 cm^{-1} intervals and the Stokes shift is about 2000 cm^{-1} implying a Huang–Rhys factor of about 5.5. It appears that the Mn^{2+} ion mainly occupies only one of the two cation sites and this is in agreement with electron spin resonance studies.[56,57] The value of $10Dq$ calculated from the excitation spectrum (see Table 4.1) suggests that the Mn^{2+} ion is more likely to be in the slightly larger M_2 site. In fact, in some microcrystalline synthetic samples there is evidence that a very small fraction of the manganese ions occupy M_1 sites and these have a higher value of $10Dq$

(cf. diopside M_1 site). These ions in M_1 sites produce an extended tail on the long wavelength side of the emission band.

Polarized luminescence excitation spectra of a synthetic single-crystal forsterite are shown in Fig. 4.10. (This crystal was grown by C.B. Finch of Oak Ridge National Laboratories, Tennessee, USA and kindly loaned by R.A. Weeks of the same establishment.) These spectra were obtained by exciting the luminescence with linearly polarized light incident normally on a polished crystal surface. The plane of polarization was orientated parallel to one of the crystal axes which had been previously determined by X-ray methods.[6,55] The successful interpretation of such spectra depends on choosing the correct pseudosymmetry for the centre. The actual point symmetry of the distorted octahedral M_2 site in forsterite is C_S although there are higher symmetries to which the site approximates. It is clear from the spectra that the degeneracy has been removed from almost all states although the polarization dependence of band intensities clearly shows that the centre behaves as though the symmetry was higher than C_S. (In this symmetry there would be the same polarization dependence in two directions.)

For an electric dipole transition to be allowed in a particular direction the transition dipole moment must not change under a symmetry operation (Section 4.2). Using the terminology of group theory[3,4] this means that the direct product of symmetry representations of the initial and final states and the cartesian direction in which the electric vector lies should span the totally

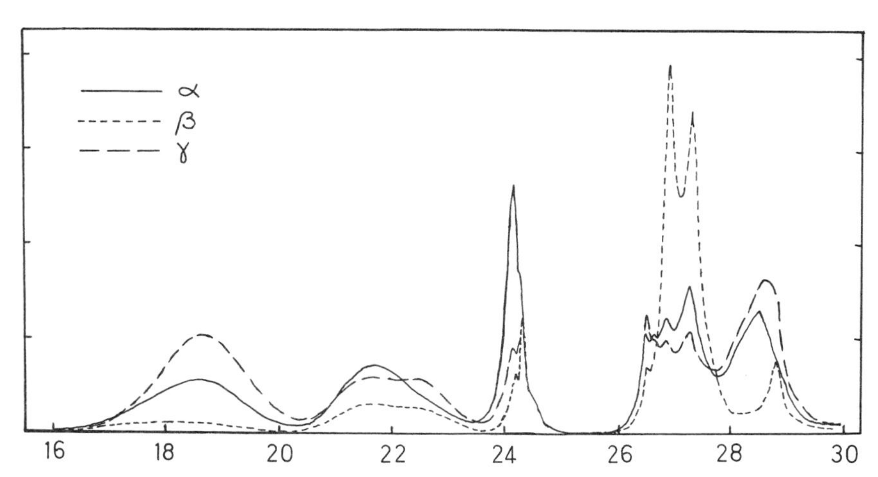

Fig. 4.10 Polarized luminescence excitation spectra of Mn^{2+} in synthetic single-crystal forsterite measured at room temperature (cf. unpolarized spectrum Fig. 4.6(a)). Low temperature spectra reveal even more detail but these will be given elsewhere with a complete analysis.[55] Spectra are corrected for variations in excitation intensity and the wavenumber scale is marked at 2000 cm^{-1} intervals. (Green and Walker, unpublished work.)

symmetric representation. For Mn^{2+} the ground state is totally symmetric and so for singly dimensioned representations this condition reduces simply to the fact that the final state (in absorption) must have the same symmetry representation as the cartesian direction of polarization. In a non-centrosymmetric site the final states will reduce to one of the non-degenerate symmetries A_1, A_2, B_1 or B_2. Consultation with the character table[3,4] of the particular symmetry group to which the site belongs reveals to which of these symmetry representations the cartesian coordinates, x, y and z belong. (The z-axis is usually chosen as the axis of highest symmetry.) If, for example, z has the same symmetry representation as A_1 then a transition to an A_1 state is allowed for polarization parallel to the z-axis for a Mn^{2+} ion in this site.

However, we have of course to choose the right symmetry group for the site and to determine how the degenerate octahedral states split in this symmetry. The latter is done using descent of symmetry tables[58] but the former is a matter of trial and error among likely candidates. C_{2v} pseudosymmetry has been used for the interpretation of Fe^{2+} absorption spectra in olivines[59] but there are in fact four possible C_{2v} pseudosymmetries depending on the choice of axis in the centre.[60] These axes are not in general the crystallographic axes for the crystal as a whole but their mutual orientation is known. Analysis has shown[55,6] that the polarized spectra of Mn^{2+} in forsterite are best explained by considering the M_2 site pseudosymmetry to be $C_{2v}(C_2, \sigma_d)$ rather than the $C_{2v}(C''_2)$ of Runciman *et al.*[59] Thus the symmetry as well as the position of almost every excited state of Mn^{2+} in the forsterite M_2 site has been determined. We have of course assumed that the differences in polarization are primarily due to the spatial symmetry of the ligand field but this can be justified.

Clearly a great deal is now known about the Mn^{2+} centre in forsterite; in contrast, relatively little is known about the blue emission centre. Excitation spectra reveal an absorption edge starting around 400 nm and still rising at 300 nm. The intensity of emission and the decay time increase substantially as the temperature decreases; at room temperature the decay time is of the order of 100 μs or less. The intensity of the blue emission is also less sensitive to the concentration of iron than the Mn^{2+} emission. In a synthetic olivine containing 1000 ppm Mn and 4 mol% iron the blue emission is reduced by a factor of about 20 in intensity at room temperature compared with an iron-free sample whereas the Mn^{2+} emission is reduced by a factor of about 90. Since the blue emission is more intense in very pure synthetic samples (with minimal manganese), it is likely to be due to a defect centre associated with the silica tetrahedra (cf. other silicates and quartz).

4.4.2 Pyroxenes and pyroxenoids

Pure iron-free enstatites seldom occur naturally and so most enstatites are not very luminescent. However, enstatite from stony meteorites is an exception and shows luminescence properties very similar to those of forsterite. The Mn^{2+} emission is, however, at slightly longer wavelengths (in spite of a smaller $10\,Dq$).[17,18]

Figure 4.6(b) shows an excitation spectrum of the Mn^{2+} emission of meteoritic orthoenstatite and comparison with the forsterite spectrum shows a smaller $10\,Dq$ but a large splitting of the $^4T_1$ level indicative of the large distortion of the M_2 site from octahedral symmetry. Differences between orthoenstatite and clinoenstatite are slight but nevertheless measurable. At low temperatures a no-phonon line is visible (see Fig. 4.2) but for clinoenstatite this is $140\,cm^{-1}$ lower in energy than for orthoenstatite. Sometimes both are present showing that the sample has ortho- and clinoenstatite components. The dominant phonon replicas are similarly spaced for the two phases ($165\,cm^{-1}$). There is no evidence at all of Mn^{2+} in the smaller M_1 site.

In meteoritic and synthetic samples it has been shown[18,50] that a very low manganese content results in the blue emission dominating although the Mn^{2+} emission is still apparent at concentrations of around 20 ppm in pure iron-free samples at room temperature (see Fig. 4.11). The effect of iron is similar to that in forsterite and therefore the blue emission can dominate at slightly higher manganese concentrations when a small amount of iron is present although the overall intensity is, of course, lower. Figure 4.11 shows the effect of manganese and iron concentrations on the emission spectrum at room temperature.

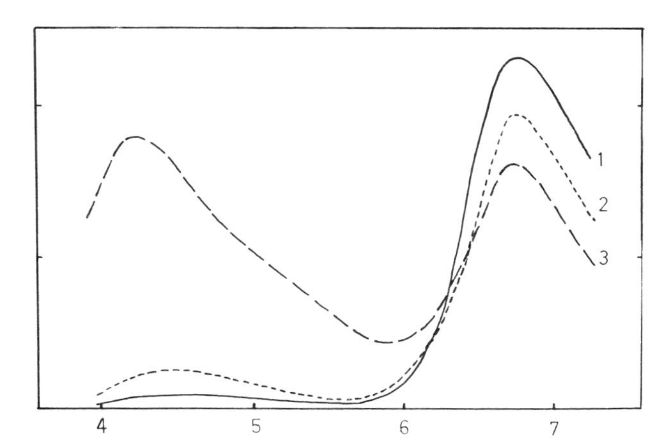

Fig. 4.11 Luminescence emission spectra of synthetic clinoenstatite. (1) Pure clinoenstatite with 320 ppm Mn added. (2) Clinoenstatite with 320 ppm Mn and 1 mol% Fe (3) Pure clinoenstatite with minimal manganese (20 ppm). The intensity scales for 1, 2 and 3 are unrelated. The wavelength scale is marked at 100 nm intervals. The red peak of curve 1 is actually about five times that of curve 2 and twenty times that of curve 3; the red peak intensity decreases very rapidly with increase in iron concentration above 1 mol%.

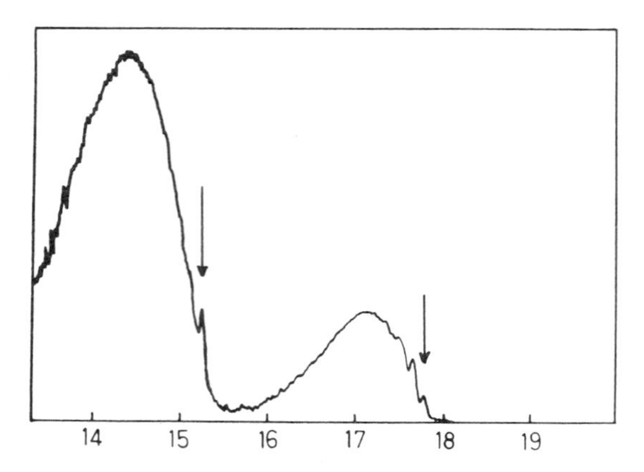

Fig. 4.12 Luminescence emission spectrum of a diopside (BM40397) measured at 5 K. Arrows indicate the no-phonon lines of the Mn^{2+} emission at 17 800 cm^{-1} (M_2 site) and 15 250 cm^{-1} (M_1 site). The wavenumber scale is labelled at 1000 cm^{-1} intervals. Corrections for instrumental response have been applied. See Fig. 4.6c for excitation spectra of these bands.

Figure 4.12 shows the cathodoluminescence emission spectrum of a diopside measured at 5 K. There are two strong luminescence bands showing phonon structure and separate excitation spectra of these two bands show that both are due to Mn^{2+} but in different environments (Fig. 4.6). The M_1 (Mg^{2+}) cation site in diopside is very similar to that in enstatite in size and symmetry but the M_2 (Ca^{2+}) site is now eight-fold coordinated, of irregular symmetry, and much larger than in enstatite. Consequently the ligand field in the M_2 site is expected to be considerably smaller than in the M_1 site. It is therefore readily apparent which spectrum is due to Mn^{2+} in the M_1 site and which in the M_2 site. Note also that the Racah parameter B is practically the same as the free-iron value in the large M_2 site where orbital overlap with the ligands is very small (see Table 4.1). The M_1 excitation spectrum also indicates where we would expect the ligand field bands for Mn^{2+} in an enstatite or forsterite M_1 site if such occupancy occurred (see Section 4.5.2).

Wollastonites often exhibit green or orange luminescence due to Mn^{2+} although as already mentioned (Section 4.2.2(a)) iron-free manganese pyroxenoids such as rhodonite are only luminescent at low temperatures.

4.4.3 Feldspars

In general, plagioclase feldspars exhibit three broad cathodoluminescence emission bands, one on the edge of the infra-red, one in the green, and a strong blue emission. Most plagioclase luminescence appears blue, bluish-green or even white. The colour is obviously dependent on the relative intensities of the three emission bands. In many plagioclases the infra-red band is dominant (see

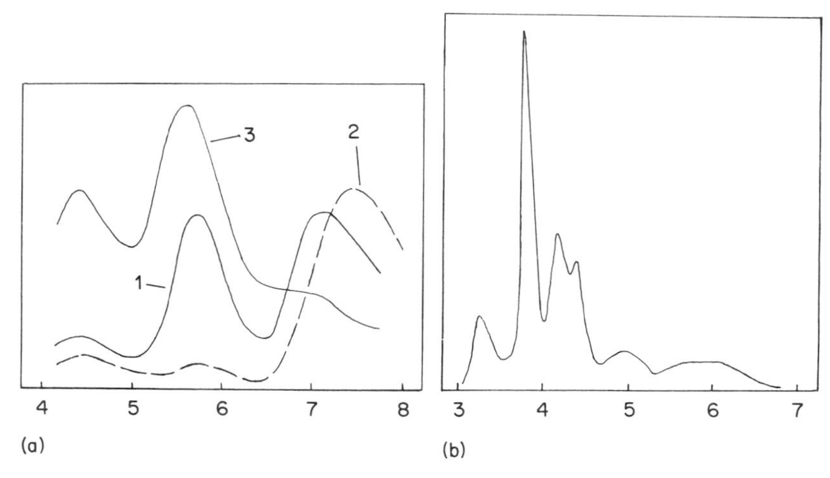

Fig. 4.13 (a) Luminescence emission spectra of (1) a typical plagioclase (Emmons Suite no 25), (2) an oligoclase (Emmons Suite no 4), (3) a lunar breccia 67455.3.1. Spectra are corrected for instrumental response. (b) Luminescence excitation spectrum of the infra-red Fe^{3+} band of a bytownite (Emmons Suite no 25). All spectra are measured at room temperature. The wavelength scales are marked at 100 nm intervals.

Fig. 4.13(a)) but since only the tail of this band is in the visible it often has little effect on the visual colour. Plagioclases from lunar or meteoritic sources, however, have a very much weaker infra-red band. The work of Geake *et al.*[23] and Telfer and Walker[24,25] demonstrated that it was due to Fe^{3+}. Figure 4.13(b) shows a luminescence excitation spectrum of this band which is characteristic of Fe^{3+} in a tetrahedral site. The ligand field parameter $10\,Dq$ is about $7500\,\mathrm{cm}^{-1}$ and the Racah parameter B is about $700\,\mathrm{cm}^{-1}$ which is considerably depressed from the free-iron value and lower than that for Fe^{3+} in an octahedral site.[61] It is expected that the Fe^{3+} occupies the larger Al^{3+} tetrahedral sites,[25] although it is not yet clear whether it is occupying one particular site or two or three similar sites. The appearance of no-phonon lines at low temperature would be helpful but so far none have been seen although the Stokes shift is not excessive. The decay time of the Fe^{3+} emission is around one millisecond at room temperature but increases on cooling.

The green emission centred around 560 nm is due to Mn^{2+} in metal cation sites,[22,25] most probably Ca^{2+} sites, and an excitation spectrum of this emission is shown in Fig. 4.5. This spectrum is characteristic of Mn^{2+} in a large site (cf. diopside M_2 site) and yields a low value of $10Dq$ ($5300\,\mathrm{cm}^{-1}$).[25,62] The calcium sites, of which there are four types in anorthite, are very irregular seven or six-fold coordinated sites and again there is, at present, no spectroscopic evidence to decide how many of these sites Mn^{2+} ions occupy.

Measurements on pure synthetic anorthites have shown[25] that the intensity of the green band depends on the amount of manganese added up to a critical concentration above which concentration quenching becomes apparent (see Section 4.2(a)). A similar behaviour is found for the dependence of infra-red emission intensity on Fe^{3+} concentration.

The Mn^{2+} emission is less evident in K-feldspars and it is not yet clear whether it is present at all; it seems unlikely that Mn^{2+} would substitute readily for potassium. The Fe^{3+} emission, however, appears to be present in all feldspars. Orthoclases sometimes appear red in cathodoluminescence suggesting that a considerable part of the Fe^{3+} emission band is in the visible.

The blue emission, which often appears quite strong in all feldspar cathodoluminescence as seen visually, has a much shorter decay time at room temperature than the other bands, once again indicating that it has a similar origin to that in quartz and other silicates.

4.4.4 Carbonates

The luminescence of calcite, dolomite and magnesite was investigated thoroughly by Medlin[15,16] twenty years ago and the Mn^{2+} emission interpreted in terms of ligand-field theory. He found, contrary to recent work,[63] that the efficiency of luminescence (in this case thermoluminescence) in synthetic calcite precipitates was dependent on manganese content and that above an optimum concentration, concentration quenching began to reduce the efficiency. Sommer[26] found a similar behaviour for Mn^{2+} in aragonite. Measurements were also carried out on calcite-magnesite solid solutions which showed that the emission band maximum shifted from 590 nm for pure calcite to 680 nm for pure magnesite.[26] However, dolomite is not a solid solution but has a definite crystal structure in which magnesium sites are considerably smaller (Mg–O distance 2.10 Å) than the calcium sites (Ca–O distance 2.39 Å).[64] It would therefore be expected that the emission would consist of two bands due to Mn^{2+} in both sites (cf. diopside). However, unlike diopside, the bands would probably overlap since the Ca^{2+} site in dolomite is octahedrally coordinated and smaller than in diopside. Sommer[26] reported that this indeed happens but unfortunately did not present any spectra. The emission spectra of Graves and Roberts,[27] and Medlin,[15,16] suggest that the peak emission wavelength for Mn^{2+} in dolomite is at longer wavelengths than for calcite. Natural calcites usually have a peak emission at about 610 nm[65] (see Fig. 4.5) rather than the 590 nm reported by Sommer. Electron spin resonance measurements by Wildeman[66] indicate that in dolomites the manganese surprisingly appears to have a fairly strong preference for magnesium sites, much more so than in diopside. The ratio of Mn^{2+} in magnesium sites to that in calcium sites appears to be about 4:1 in sedimentary dolomites but very much higher in hydrothermally altered dolomites. Clearly, dolomite is a prime candidate for luminescence excitation spectroscopy to corroborate such findings.

A luminescence excitation spectrum of Mn^{2+} in calcite is shown in Fig. 4.5 and the calculated ligand-field parameters are shown in Table 4.1. The Stokes shift is about 2700 cm^{-1} which is reasonably large and it remains to be seen whether careful low temperature measurements will reveal a no-phonon peak. The value of $10Dq$ is somewhat higher than might be expected for the large Ca^{2+} site.

As usual, the presence of Fe^{2+} leads to quenching. Pierson[63] has recently noted that dolomites containing more than 1% iron are virtually non-luminescent. He also found that there was *no* correlation between intensity of luminescence and manganese concentration across the various dolomites examined. This correlation is perhaps not surprising when it is noted that the samples were from different formations with different histories and that the iron concentration varies in a similar manner to the manganese concentration. Other workers[67] suggest that at concentrations of iron below 1% the intensity of luminescence of calcite cements appear to be controlled by the Fe/Mn ratio; this ratio is usually relatively constant in a particular zone of a crystal which luminesces uniformly.

Graves and Roberts[27] discovered that the thermoluminescence emission of calcites and dolomites at low temperatures consists of a broad emission band in the blue and that the characteristic Mn^{2+} orange-red emission did not become evident until normal temperatures were approached. The origin of this blue emission is not known.

4.4.5 Quartz

This mineral is one of the most important fundamental structures in mineralogy. The rather open structure with four-fold and two-fold co-ordinated atoms is more suggestive of a covalent rather than ionic solid and Harrison[68] has concluded that SiO_2 is best considered as a covalent solid with substantial polarity.

Luminescence centres in quartz have importance for all silicate structures, particularly those in which the silica tetrahedra are linked. It often appears that the simpler the structure the more complex the luminescence properties of the material (cf. diamond[10]). Certainly various forms of SiO_2 can commonly exhibit two or three luminescence bands in the visible not to mention at least one more in the ultra-violet (see, for example, Koyama[69]). What is more very few, if any, of the luminescence centres have been identified with any certainty, although a considerable amount of research has been done on vitreous and amorphous SiO_2 as well as crystalline quartz. It has been found that, in general, crystalline and non-crystalline quartz often show similar luminescence bands.

In the geological sphere, the work of Sippel[19] and the more recent spectral measurements of Zinkernagel[20] have shown that the visible cathodoluminescence of common natural α-quartz consists of two broad emission bands; a blue emission centred around 450 nm (2.7 eV) and a red emission at about 650 nm (1.9 eV).

Figure 4.14 shows the cathodoluminescence spectrum of a typical synthetic α-quartz at normal temperatures and at 77 K. The characteristic red emission has a reasonably short decay time at room temperature but the longer-lived blue emission increases both in intensity and decay time as the temperature is reduced and often completely swamps the red emission at low temperatures. The broad-band blue emission has a very long tail stretching into the red end of the spectrum. It is this latter emission alluded to in previous sections which is probably common to other silicates.

There are many well-known defect centres in quartz[45] which have been characterized by electron-spin resonance studies. The problem is which, if any, of these paramagnetic centres are responsible for luminescence transitions. There are, for example, various types of defect centres which are associated with the substitution of aluminium for silicon;[70] Durrani *et al.*[71] suggest that the alkali-compensated aluminium centre is responsible for the blue emission as seen in thermoluminescence. Other workers have correlated the intensity of the blue emission in quartz[38] and feldspars[37] with titanium content. However, it is found that neither the presence of aluminium nor titanium is necessary for the production of the characteristic blue cathodoluminescence since both ultra-pure fused quartz and crystalline quartz can exhibit it. In fact both the blue and red emission bands appear to be due to defects which are often present even when metal impurity levels are extremely low (i.e. less than 1 ppm).[20,72,73]

There is therefore considerable discussion about the nature of these defects,

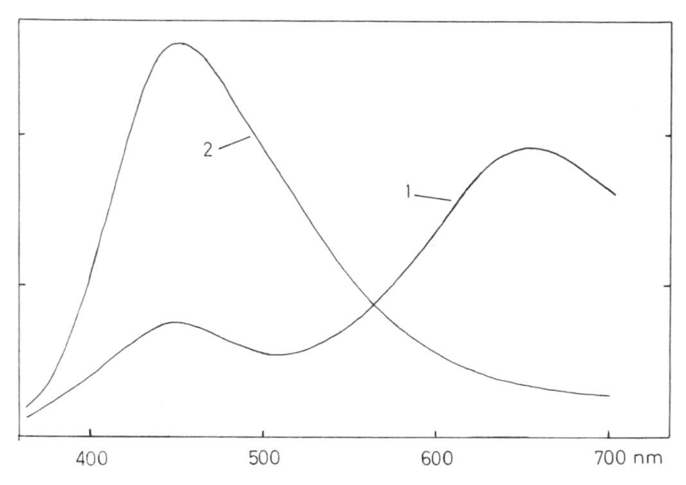

Fig. 4.14 Luminescence emission spectra of a synthetic α-quartz. (1) At room temperature, (2) at 77 K. Spectra have been corrected for instrumental response. Intensity scales are unrelated for 1 and 2.

about the transitions involved, and to what extent these defects are produced by irradiation. (The literature on the effects of neutrons and γ-irradiation on quartz is already extensive.)

Sigel[74] suggested that the blue emission was due to hole capture by an E′ centre, or, to put it another way, an electronic transition from a localized level of the E′ centre to the valence band. The generally accepted model for the E′ centre is the one proposed by Fiegl *et al.*[75] If an oxygen vacancy is formed then there will be a 'dangling' sp^3 orbital on each of the two silicon atoms to which the missing oxygen was bonded. It is thought that a hole is trapped by one 'dangling' orbital and that this silicon atom then relaxes into the plane of the remaining oxygens to which it is bonded, leaving the other silicon sp^3 orbital, containing an unpaired electron, 'dangling' into the vacancy. Sigel[74] also demonstrated that both the blue and the red emission bands could be excited directly by photons of energy well below the bandgap ($\sim$ 10 eV), in the same manner as, for example, Mn^{2+} emission can be excited in forsterite. The absorption is similarly weak probably on account of the low concentration of defect centres, and possibly because of the 'forbiddenness' of the transitions involved. At very low temperatures the blue emission can have a decay time exceeding a millisecond.

The ultra-violet emission band at 290 nm (4.3 eV), which is commonly seen in SiO_2 thin films, has also been associated with oxygen vacancies.[69,76,77]

Another intrinsic paramagnetic defect found in quartz is the oxygen-associated hole centre (OHC) and recently it has been suggested that the red emission is associated with such a centre.[78,79] An OHC is a 'non-bridging' oxygen atom which is bonded to one silicon atom only and has a trapped hole in a 2p orbital. Stapelbroek *et al.*[80] denoted this centre the 'wet' OHC since it was prevalent in silicas with a high OH content. Another 'dry' OHC has been detected by electron spin resonance (ESR) but its precise nature has not been determined.

There are many reports that the red emission, whatever its nature, can be induced by irradiation with neutrons or γ-rays.

Greaves[81] has proposed a model for point defects in silica which is based on the breaking of Si–O bonds to produce positively-charged trivalent silicons and negatively-charged monovalent oxygens. Such centres would be diamagnetic and therefore not detectable by ESR until the trivalent silicon trapped an electron and the oxygen trapped a hole. It is suggested that an oxygen vacancy is not necessarily an essential part of the E′ centre. According to Greaves the red emission is due to the recombination of a localized exciton at a paramagnetic silicon site. Skuja and Silin,[78] however, believe the red emission to be due to a transition between the $2p_z$ and $2p_y$ orbitals of a non-bridging oxygen. Clearly there is as yet no agreement or conclusive experimental evidence as to the transitions responsible for intrinsic defect luminescence in SiO_2.

Of course, it is unlikely that all luminescence from all forms of silica is due to intrinsic defects whether impurity-induced or not. For example, Fe^{3+} centres are known to occur and an emission at around 700 nm has been reported.[82] It may be that this emission is at least a component in the red emission of citrines.[38]

Finally, in view of the current technological interest in SiO_2, progress in the characterization of its luminescence centres can be expected in the near future.

4.5 SOME CONCLUSIONS

It is perhaps instructive to consider to what extent the questions posed in the introduction have been answered. At least some of the answers have become evident in the course of the text but in other cases it has been indicated that present knowledge is incomplete. However, there are two important topics which deserve further discussion.

4.5.1 Cathodoluminescence; intensity and colour variations

The intensity of emission has been shown to depend on the concentration of luminescence centres and the concentration of quenchers. We have seen that too high a concentration of luminescence centres can lead to a reduction in luminescence intensity. As for quenchers, it should not always be assumed that the only cause of quenching is the presence of iron, although this may be the dominant cause particularly at concentrations above about 1%. There is also the possibility that the spatial distribution of impurities may not be uniform, particularly on an atomic scale. If resonance transfer of excitation occurs between say an Mn^{2+} ion and Fe^{2+} ion then there will be a critical separation of the order of a few lattice spacings beyond which the probability of transfer will be significantly reduced. Other impurities and defects can also provide alternative pathways of recombinations which may be radiative or non-radiative depending on the relative values of k_f and k_i (see Section 4.2).

It is known from thermoluminescence studies[14] that electron-trap energy distributions in minerals depend critically on the history and conditions of formation and for this reason 'glow' curves can be used as a 'finger printing' technique. Similarly, two samples of the same mineral formed under differing conditions may not have identical luminescence properties even though they had similar impurity concentrations.

On a practical point, relative intensities should be measured using a linear photodetector and not estimated by eye or by photographic density since these methods, besides being inaccurate, have an approximately logarithmic rather than linear response to illumination.

Some minerals which are normally luminescent such as quartz[20] or feldspars[83] are often found to be non-luminescent when formed diagenetically at low temperatures. It has, however, been found that heat treatment can render authigenic quartz luminescent.[20] If the luminescence centres are mainly due to intrinsic defects then it appears to be possible to produce some of these by heat treatment.

Colour variations, as we have already mentioned, can be due to changes in the relative intensities of luminescence bands which are caused not only by changes in the relative concentrations of luminescence centres but also sometimes by changes in the concentration of quenchers (see Section 4.4.2).

Colour changes may also be due to emission bands shifting slightly in wavelength and broadening on account of lattice disorder or 'glassiness', or possibly because of solid-solution effects.

It has been claimed that the cathodoluminescence colour of quartz is indicative of the temperature of formation.[20] Quartz from igneous rocks shows blue, red or violet luminescence whereas lower temperature metamorphism produces the 'brown' luminescent variety.

4.5.2 Site occupancy and cation ordering in minerals

The distribution of cations among the various possible sites in minerals, particularly silicates, has been a topic of considerable interest for many years. Much experimental evidence has been accumulated from optical absorption, Mössbauer, electron spin resonance, and X-ray measurements. However, most of these techniques require good single crystals which are not necessary for luminescence excitation spectroscopy and the potential of the latter technique in this area has already been indicated.

Traditionally, site occupancy was predicted mainly on the grounds of ionic size and later modified for transition metals by consideration of the ligand or crystal-field stabilization energy (CFSE).[29] Of course the cation distribution among possible sites should be such as to minimize the lattice energy and the simple criteria used are based on this hypothesis. However, recently, precise calculations of lattice energies have been carried out for pyroxenes and pyroxenoids[84] which may prove very helpful. These calculations were made in an effort to predict which of several pyroxenoid structures would be the most stable for a particular choice of metal cations. It has since been suggested by the author that such calculations could be used to predict what cation ordering would be expected in different silicate structures.[85]

We have already seen that the distribution of Mn^{2+} ions among different cation sites is not always what we might expect and, although with hindsight an explanation in terms of ionic size may be possible, it would have been difficult to predict the actual Mn^{2+} distribution in diopside and dolomite. (d^5 ions have zero CFSE and so ligand-field effects cannot be invoked.) The M_1 site in diopside is almost identical with that in enstatite and actually slightly smaller than the M_1 site in forsterite yet there is virtually no Mn^{2+} in either of the latter two sites.[86] The Mn^{2+} ion, although often having a strong site preference, is however peculiarly tolerant of its environment. It is found in almost any type of site, from tetrahedral sites in willemite and spinels to the nine-fold coordinated Ca^{2+} site in aragonite, which makes it a particularly interesting ion for site occupancy and cation ordering predictions.

REFERENCES

1. Herzberg, G. (1944) *Atomic Spectra and Atomic Structure*, Dover, New York.
2. Curie, D. (1963) *Luminescence in Crystals*, ch. 1, Methuen, London.
3. Cotton, F.A. (1963) *Chemical Applications of Group Theory*, Interscience, New York.
4. Urch, D.S. (1970) *Orbitals and Symmetry*, Penguin, Harmondsworth.
5. Förster, Th. (1959) *Disc. Faraday Soc.*, **27**, 7.
6. Green, G.R. (1982) Ph.D. thesis, University of Manchester.
7. Imbusch, G.F. (1978) in *Luminescence Spectroscopy* (ed. M.D. Lumb) ch. 1, Academic Press, London.
8. Rebane, K.K. (1970) *Impurity Spectra of Solids*, Plenum Press, New York.
9. Stoneham, A.M. (1975) *Theory of Defects in Solids*, Oxford University Press, Oxford.
10. Walker, J. (1979) *Rep. Prog. Phys.*, **42**(10), 1605.
11. Smith, J.V. and Stenstrom, R.C. (1965) *J. Geol.*, **73**, 627.
12. Long, J.V.P. and Agrell, S.O. (1965) *Min. Mag.*, **34**, 318.
13. Sippel, R.F. and Glover, E.D. (1965) *Science*, **150**, 1283.
14. McDougall, D.J. (ed.) (1968) *Thermoluminescence of Geological Materials*, Academic Press, London.
15. Medlin, W.L. (1963) *J. Opt. Soc. Amer.*, **53**(11), 1276.
16. Medlin, W.L. (1968) in *Thermoluminescence of Geological Materials* (ed. D.J. McDougall), ch. 4.1, Academic Press, London.
17. Geake, J.E. and Walker, G. (1966) *Geochim. Cosmochim, Acta*, **30**, 929.
18. Geake, J.E. and Walker, G. (1967) *Proc. R. Soc. Lond.*, **A296**, 337.
19. Sippel, R.F. (1968) *J. Sediment, Petrol.*, **38**(2), 530.
20. Zinkernagel, U. (1978) *Contrib. Sedimentology*, **8**, 1.
21. Sippel, R.F., and Spencer, A.B. (1970) Proceedings Apollo 11 Lunar Science Conference. *Geochim. Cosmochim. Acta, Suppl. 1*, **3**, 2413.
22. Geake, J.E., Walker G., Mills, A.A. and Garlick, G.F.J. (1971) Proceedings 2nd Lunar Science Conference, *Geochim. Cosmochim. Acta, Suppl. 2*, **3**, 2265.
23. Geake, J.E., Walker, G., Telfer, D.Y. *et al.* (1973) Proceedings 4th Lunar Science Conference, *Geochim. Cosmochim. Acta, Suppl. 4*, **3**, 3181.
24. Telfer, D.J. and Walker, G. (1975) *Nature, Lond.*, **258**, 694.
25. Telfer, D.J. and Walker, G. (1978) *Mod. Geol.*, **6**, 199.
26. Sommer, S.E. (1972) *Chem. Geol.*, **9**, 257.
27. Graves, W.E. and Roberts, H.H. (1972) *Chem. Geol.*, **9**, 249.
28. Nickel, E. (1978) *Mineral Sci. Engng*, **10**, 73.
29. Burns, R.G. (1970) *Mineralogical Applications of Crystal Field Theory*, Cambridge University Press, Cambridge.

30. Orgel, L.E. (1955) *J. Chem. Phys.*, **23**, 1958 and 1824.
31. Mariano, A.M. and Ring, P.J. (1975) *Geochim. Cosmochim. Acta*, **39**, 649.
32. Marfunin, A.S. (1979) *Spectroscopy, Luminescence and Radiation Centres in Minerals* (Trans. V.V. Schiffer), Springer, Berlin.
33. Remond, G. (1977) *J. Luminescence*, **15**, 121.
34. Bell, P.M., Mao, H.K. and Ross, G.R. (ch. 1) and, Burns, R.G. and Vaughan, D.J. (ch. 2) (1975) *Infra-red and Raman Spectroscopy of Lunar and Terrestrial Minerals* (ed. C. Karr), Academic Press, New York.
35. Low, W. and Rosengarten, G. (1964) *J. molec. Spectr.*, **12**, 319; Orgel, L.E. (1955) *J. Chem. Phys.*, **23**, 1004.
36. Tanabe, Y. and Sugano, S. (1954) *J. phys. Soc. Jap.*, **9**, 753.
37. Mariano, A.N., Ito, J. and Ring, P.J. (1973) Geological Society of America annual meeting, Abstracts, 726.
38. Sprunt, E.S. (1981) *Scanning Electron Microscopy*, 525.
39. Clark, M.G. and Dean P.J. (1979) Proceedings 14th International Conference on Semi-Conductors, *Inst. Phys. Conf. Series*, **43**, 291.
40. Bishop, S.G., Dean, P.J. Porteous, P. and Robbins, D.J. (1980) *J. Phys. C*, **13**(2), 1331; Ralph, J.E. and Townsend, M.G. (1968) *J. Chem. Phys.*, **48**, 149.
41. Gorobets, B.S. Gaft, M.L. and Laverova, V.L. (1978) *J. appl. Spectr.* (USA) **28**(6), 750 (Translation from Russian).
42. Newman, D.J. (1971) *Adv. Phys.*, **20**, 197.
43. Klick, C.C. (1972) in *Point Defects in Solids*, (ed. J.H. Crawford Jr. and L.M. Slifkin), Vol 1, ch. 5, Plenum Press, New York.
44. Hughes, A.E. and Henderson, B. (1972) in *Point Defects in Solids* (ed. J.H. Crawford Jr. and L.M. Slifkin), Vol. 1, ch. 7, Plenum Press, New York.
45. Grimscom, D.L. (1978) in *The Physics of* SiO_2 *and its Interfaces* (ed. S.T. Pantelides), ch. 5, Pergamon Press, New York.
46. Coulson, C.A. and Larkins, F.P. (1971) *J. Chem. Phys. Solids*, **32**, 3345.
47. Garlick, G.F.J. (1966) in *Luminescence of Inorganic Solids* (ed. P. Goldberg), ch. 2, Academic Press, New York.
48. Meyer, V.D. (1970) *J. appl. Phys.*, **41**, 4059.
49. Hamilton, T.D.S., Munro, I.H. and Walker, G. (1978) in *Luminescence Spectroscopy* (ed. M.D. Lumb), ch. 3, Academic Press, London.
50. Grögler, N. and Liener, A. (1968) in *Thermoluminescence of Geological Materials* (ed. D.J. McDougall), ch. 11.2, Academic Press, London.
51. Sippel, R.F. (1965) *Rev. Sci. Instrum.*, **36**(2), 1556.
52. Herzog, L.F. Marshall, D.J. and Babione, R.F. (1970) Pennsylvania State University, MRL spec. publ. **70–101**, 79.
53. Ramseyer, K. (1982) Workshop-Conference on the Geological Applications of Cathodoluminescence, University of Manchester.
54. Geake, J.E., Walker, G., Mills, A.A. and Garlick, G.F.J. (1972) Proceedings 3rd Lunar Science Conference, *Geochim. Cosmochim. Acta, Suppl. 3*, **3**, 2971.
55. Green, G.R. and Walker, G. *Am. Mineral*, submitted for publication 1984.

56. Michoulier, J., Gaite, J.M. and Maffeo, B. (1969) *Compt. Rend. Acad. Sci. Paris*, **269B**, 535.
57. Chatelain, A. and Weeks, R.A. (1970) *J. Chem. Phys.* **52**, 5682.
58. Wilson, E.B., Decius, J.C. and Cross, P.C. (1955) *Molecular Vibrations*, McGraw-Hill, New York.
59. Runciman, W.A., Sengupta, D. and Gourley, J.T. (1973) *Am. Mineral*, **58**, 451.
60. Goldman, D.S. and Rossman, G.R. (1977) *Am. Mineral.*, **62**, 151.
61. Telfer, D.J. and Walker, G. (1976) *J. Luminescence*, **11**, 315.
62. Telfer, D.J. (1975) Ph.D. thesis, University of Manchester.
63. Pierson, B.J. (1981) *Sedimentology*, **28**, 601.
64. Wyckoff, R.W.G. (1964) *Crystal Structures*, Vol. 2, Interscience, New York, p. 588.
65. Aquilar, G.M. and Osendi, M.I. (1982) *J. Luminescence*, **27**, 365.
66. Wildeman, T.R. (1970) *Chem. Geol.*, **5**, 167.
67. Frank, J.R. Carpenter, A.B. and Oglesby, T.W. (1982) *J. Sediment. Petrol.*, **52**, 631.
68 Harrison, W.A. (1978) in *The Physics of Sio_2 and its Interfaces* (ed. S.T. Pantelides), Pergamon Press, Oxford, p. 105.
69. Koyama, H. (1980) *J. appl. Phys.*, **51**(4), 2228.
70. Weil, J.A. (1975) *Radiation Effects*, **26**, 261.
71. Durrani, S.A. Khazal, K.A.R., McKeever, S.W.S. and Riley, R.J. (1977) *Radiation Effects*, **33**, 237.
72. McKnight, S.W. and Palik, E.D. (1980) *J. Non-cryst. Solids*, **40**, 595.
73. Trukhin, A.N. and Plaudis, A.E. (1979) *Sov. Phys. Solid State*, **21**(4), 644.
74. Sigel, G.H. Jr. (1973) *J. Non-Cryst. Solids*, **13**, 372.
75. Fiegl, F.J., Fowler, W.B. and Yip, K.L. (1974) *Solid State Comm.*, **14**, 225.
76. Jones, C.E. and Embree, D. (1976) *J. appl. Phys.*, **47**, 5365; and (1978) in *The Physics of SiO_2 and its Interfaces* (ed. S.T. Pantelides), Pergamon Press, Oxford, p. 289.
77. Gee, C.M. and Kastner, M. (1980) *J. Non-cryst. Solids*, **40**, 577.
78. Skuja, L.N. and Silin, A.R. (1979) *Phys. stat. sol.* (a), **56**, K11; and (1982) *Phys. stat. sol.* (a), **70**, 43.
79. Sigel, G.H. Jr. and Marrone, M.J. (1981) *J. Non-cryst. Solids*, **45**, 235.
80. Stapelbroek, M., Grimscom, D.L., Friebele, E.J. and Sigel, G.H. Jr. (1979) *J. Non-cryst. Solids*, **32**, 313.
81. Greaves, G.N. (1979) *J. Non-cryst. Solids*, **32**, 295; and (1978) *Phil. Mag. B*, **37**(4), 447.
82. Pott, G.T. and McNicol, B.D. (1971) *Disc. Faraday Soc.*, **52**, 121.
83. Kastner, M. (1971) *Am. Mineral.*, **56**, 1403.
84. Catlow, C.R.A., Thomas, J.M., Parker, S.C. and Jefferson, D.A. (1982) *Nature, Lond.*, **295**, 658.
85. Walker, G. (1982) *Nature, Lond.*, **300**, 199.
86. Walker, G. (1983) *Chem. Britain*, **19**(10), 824.

5

Mössbauer Spectroscopy in Mineral Chemistry

A.G. Maddock

Mössbauer spectroscopy is a technique, applicable virtually only to solid materials, that yields information about the environment of particular nuclei in a solid. It is, unfortunately, limited in its application to a restricted number of elements but is especially favourable for iron where the less abundant ^{57}Fe gives rise to the spectra. Because of the widespread occurrence of iron in minerals Mössbauer spectroscopy has found a variety of applications in mineral chemistry. Although the nuclear properties of tin, iodine, tellurium, gold, iridium and several other elements permit Mössbauer spectroscopic studies, practically all the applications to mineral chemistry so far reported refer to studies of the environment of iron nuclei.

5.1 THE BASIS OF MÖSSBAUER SPECTROSCOPY

In this review only a rather brief account of the principles of Mössbauer spectroscopy will be given, just sufficient to make the chapter self-contained. The energy differences that are explored in these spectra arise from the hyperfine interactions of the nuclear charge, and electric and magnetic multipoles, with the electric and magnetic fields produced by the surroundings of the nucleus, especially the orbital electrons of the atom of which the nucleus forms part, and even the more distant environment.

A clear, but rigorous, account of these hyperfine interactions has been given by M.G. Clark[1] and briefer treatments will be found in most textbooks dealing with Mössbauer spectroscopy.[2–7]

Both absorption and emission Mössbauer spectra can be investigated but the applications in mineral chemistry are almost entirely of the former type. The primary radiation used arises from the decay of an excited state, $^{57m_1}Fe$, iron nucleus by gamma-ray emission. Only a proportion of the decay events lead to photon emission; the remainder suffer internal conversion involving the ejection of an orbital electron from the atom with kinetic energy equal to the difference between the alternative photon energy and the binding energy of the orbital electron.

The life-time of the excited state that gives the Mössbauer emission is invariably very short, generally between 10^{-9} and 10^{-6}s, so that the state must be fed by some longer-lived decay process. Figure 5.1 shows how this can be done with ^{57m_1}Fe using ^{57}Co, which suffers orbital electron capture with a half-life of 270 days. The first product of the decay is a still more excited state, ^{57m_2}Fe; some proportion of these nuclei emit 121.9 keV photons and yield the ^{57m_1}Fe, the remainder decay directly to the ground state emitting 136.3 keV photons. A source containing ^{57}Co will thus emit 136.3 and 121.9 keV photons and various iron X-ray photons, arising from orbital vacancies due to the orbital electron capture or internal conversion processes as well as the 14.4 keV Mössbauer emission.

No such decay process can yield a unique photon energy. Inevitably, the ideal line spectrum one would like suffers some broadening due to the short life of the excited state. This limits the uncertainty in the time and, since time and energy do not commute, some appreciable uncertainty in energy ensues.

In gaseous and liquid systems there is still more line broadening and disparity between the energy distribution of the emitted photons and that of the photons needed for excitation of an ^{57}Fe nucleus to ^{57m_1}Fe by resonant absorption. Since the new ^{57m_1}Fe will quickly de-excite, such resonant absorption and re-emission manifests itself as scattering of the photon beam. This discrepancy in energy distributions arises because photon emission leads to recoil of the emitting nucleus, in order to conserve momentum between photon and nucleus, so that some of the excitation energy must be used for this

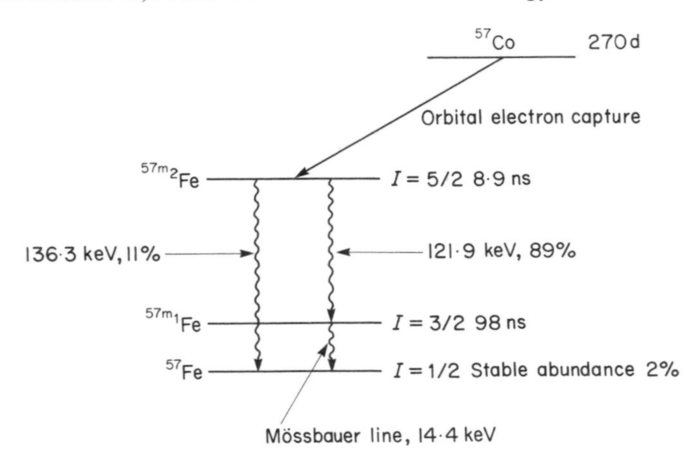

Fig. 5.1 ^{57}Co decay scheme.

recoil. A similar situation arises when the photon suffers resonant absorption by ^{57}Fe to give $^{57m_1}Fe$; some of the photon energy must appear in the $^{57m_1}Fe$ recoil. In addition to the recoil effect, the energy of the photon emitted on de-excitation of $^{57m_1}Fe$ will generally be modified by the Döppler effect due to the motion of the $^{57m_1}Fe$ atom. Mössbauer discovered that if the excited nuclei were present in atoms in a solid lattice and provided that the photon energy is not too large (in practice less than about 150 keV) a substantial proportion of the photon emission events avoid these energy-modifying processes and the line-width for this part of the photon emission is determined almost entirely by the uncertainty principle and thus the mean life of the excited state. This effect arises because energy can only be transferred to the vibrating $^{57m_1}Fe$ atom in the crystal lattice in discrete quanta, whose magnitudes are determined by the phonon spectrum of the solid. The Mössbauer emission corresponds to events where no phonon excitation occurs. The proportion of photon emission events of this kind is known as the Mössbauer fraction, f. Since the excited state lifetimes range from about 10^{-6} to 10^{-9} s, the width of the photon emission line at half maximum intensity will be from 7×10^{-10} to 7×10^{-7} eV–extremely sharp lines! The same effect is operative for the absorption of the photon by the stable ^{57}Fe, so that the collimated radiation from a solid source containing ^{57}Co will suffer resonant scattering, and attenuation of the beam, by an 'absorber' (more correctly scatterer) containing ^{57}Fe in the same chemical environment as the ^{57}Co in the source.

By deliberate use of the Döppler effect, moving the source in relation to the absorber, one can modify the energy of the photons by $\Delta E = v/c\, E_\gamma$, where E_γ is the unperturbed photon energy, v the velocity of the source in the direction defined by the line joining the source to the absorber, and c the velocity of light. For the emission from ^{57m}Fe displaying the Mössbauer effect $E_\gamma = 14.4$ keV, so that for a velocity of $10\,\text{cm}\,\text{s}^{-1}$ $\Delta E = \pm 4.82 \times 10^{-6}$eV, the sign being positive for movement towards the absorber and negative if away from it. Thus an extremely small range of energies in the vicinity of E_γ can be spanned in this way, but the width of the line spectrum emitted will be fairly small compared with this range. Because of the nuclear hyperfine interactions with electric and magnetic fields, if the source and absorber provide different environments for the ^{57}Fe nuclei, resonant scattering will not be observed for a static source but will occur at one or more velocities of the source corresponding to very slightly different photon energies. Since these energy differences are very small, much less than the probable error in our value of the primary photon energy, it is conventional to refer to them in terms of v, the velocity at which resonant scattering occurs.

5.2 THE HYPERFINE INTERACTIONS[1]

From our present point of view, the most important of these are interactions with the electric field at the nucleus. In principle, one should consider a multipolar expansion, first the monopolar interaction of the nuclear charge with the electric field in which it finds itself, then dipolar, quadrupolar, octupolar, etc. Fortunately, nuclei do not possess dipolar or octupolar moments and higher interactions are too small to be detected by this technique, so one is left with monopolar and quadrupolar effects.

5.2.1 The chemical isomer shift

Because of the Coulombic interaction of the positive nuclear charge (distributed over the small, but finite, volume of the nucleus) with the field due to the charge distribution represented by the orbital electrons, the energy levels of the nucleus are not the same when the nucleus is combined in an atom as they would be if the nucleus were a point charge or if one were dealing with a bare nucleus. If the ground state and the excited state of the Mössbauer nucleus were the same size, this would simply displace both levels by the same amount and the energy available for photon emission would be unchanged (Fig. 5.2). In fact, the two states of the nucleus are generally of different sizes, so that the energy available for the photon is no longer the same. Further, the difference, δ', depends on the charge density distribution of the orbital electrons in the vicinity of the nucleus and this quantity depends on the state of chemical combination of the atom concerned. The quantity δ' can be estimated with different levels of rigor[1] and the precise expression obtained depends on whether relativistic wavefunctions are to be used or not and other approximations chosen. Fortunately these details largely affect the constant terms in the expression:

$$\delta' = \frac{2\pi}{5} Ze^2(R_e^2 - R_g^2)|\psi(0)|^2 \tag{5.1}$$

where Z is the nuclear charge, e the electronic charge, R_e and R_g the mean radii of the excited and ground states of the nucleus and $|\psi(0)|^2$ measures the electron density at the nucleus (ψ is the wavefunction for the orbital electrons and $\psi(0)$ its value at $r = 0$). Noting that $R_e - R_g$ is very small, expressing δ'

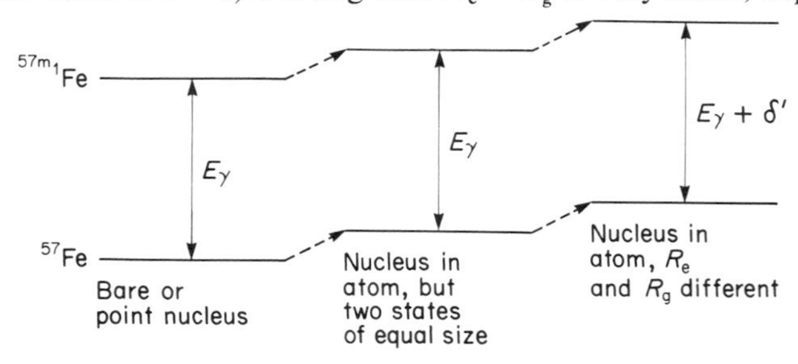

Fig. 5.2 Chemical isomer shifts.

as a velocity for the equivalent Döppler change of E_γ

$$v' = \frac{4\pi c}{5E_\gamma} Ze^2 R^2 \frac{\delta R}{R} |\psi(0)|^2 \tag{5.2}$$

where $\delta R = R_e - R_g$ and R is the mean of the two mean radii R_e and R_g. It should be noted that the sign of δR is not always positive and, indeed, in the particular case of iron, it is negative. Thus for iron, v' increases as $|\psi(0)|^2$ decreases. This expression implies that for a given Mössbauer nucleus

$$v' = C|\psi(0)|^2 \tag{5.3}$$

so that v' depends on the state of chemical combination of the atom containing the Mössbauer nucleus.

Now for Mössbauer scattering to take place it is necessary for the photon energy to equal the difference in energies of the excited and ground states of the ^{57}Fe in the absorber. For the source $v' = C|\psi_s(0)|^2$ and for the absorber $v'' = C|\psi_A(0)|^2$. Thus, $\delta = v'' - v' = C(|\psi_A(0)|^2 - |\psi_s(0)|^2$ where δ is the Döppler velocity required to produce resonant scattering for a source and absorber of different chemical composition. Since one cannot use a solid composed of bare nuclei, only relative measurements are possible. For a given source material we can set $C|\psi_s(0)|^2 = K$ so that

$$v = C|\psi_A(0)|^2 - K \tag{5.4}$$

It is normal to express the chemical shift δ in relation to a reproducible standard iron compound. Soft iron metal is commonly used, partly because it also serves to calibrate the velocity scan of the spectrometer, but a rather more reproducible substance, such as sodium nitroprusside, $Na_2Fe(CN)_5NO \cdot 2H_2O$ is often used as the reference standard. In addition to the shift arising from the monopolar interaction the measured shift will include a small temperature dependent contribution from the second order Döppler effect or thermal red shift, which arises from the very small mass difference between the excited and ground states of the nucleus.

Thus the chemical shift δ is a measure of the relative value of $|\psi_A(0)|^2$ compared to some arbitrary standard. Now $|\psi_A(0)|^2$, the electron density at the nucleus is determined predominantly by the occupation of the s orbitals in the atom containing the Mössbauer nucleus, since the wavefunctions for the non-spherically symmetrical orbitals fall to zero at the nucleus. However, changes in the occupation of the p and d orbitals of the atom can have a secondary effect in so far as they screen and modify the electron density at the nucleus due to the outer occupied s orbitals.

5.2.2 Electric quadrupole interaction; quadrupole splitting

If the nuclear spin quantum number I is greater than 1 for some nuclear state then, for that state, the nucleus will possess a quadrupole moment, eQ. In all

Mössbauer transitions, either or both the excited and ground states have a quadrupole moment. As seen in Fig. 5.1, for the $^{57m_1}Fe$ state $I = 3/2$ and for the ground state $I = \frac{1}{2}$, thus only the former state has a quadrupole moment. In the presence of a non-spherically symmetrical electric field the excited state can assume different orientations in relation to the field and the energy levels will split, the degeneracy of the original state $(2I + 1)$ being partly, or completely, removed. This splitting is known as the quadrupole splitting.

At this point some information about the electric field gradient or efg is needed.[7] The efg is the gradient of the electric field at a point, and the field is the negative of the gradient of the potential at the point. Thus efg $= \nabla E = -\nabla\nabla V$. It can thus be specified by a tensor

$$\text{efg} = -\begin{bmatrix} \dfrac{\partial^2 V}{\partial x^2}, & \dfrac{\partial^2 V}{\partial x \partial y}, & \dfrac{\partial^2 V}{\partial x \partial z} \\ \dfrac{\partial^2 V}{\partial y \partial x}, & \dfrac{\partial^2 V}{\partial y^2}, & \dfrac{\partial^2 V}{\partial y \partial z} \\ \dfrac{\partial^2 V}{\partial z \partial x}, & \dfrac{\partial^2 V}{\partial z \partial y}, & \dfrac{\partial^2 V}{\partial z^2} \end{bmatrix} \tag{5.5}$$

Now, $\partial^2 V/\partial x \partial y = \partial^2 V/\partial y \partial x$, etc., and by choice of appropriate axes the tensor can always be diagonalized so that

$$\text{efg} = -\begin{bmatrix} \dfrac{\partial^2 V}{\partial x^2} & 0 & 0 \\ 0 & \dfrac{\partial^2 V}{\partial y^2} & 0 \\ 0 & 0 & \dfrac{\partial^2 V}{\partial z^2} \end{bmatrix} \tag{5.6}$$

Further these axes will be labelled so that

$$\left|\frac{\partial^2 V}{\partial x^2}\right| < \left|\frac{\partial^2 V}{\partial y^2}\right| < \left|\frac{\partial^2 V}{\partial z^2}\right|$$

It should be noted however that the axes chosen in this way do not necessarily bear a simple relation to the axes used for crystallographic purposes. But it is only in single crystal studies and mixed electric quadrupole and magnetic dipolar interactions that this relation becomes important. Finally, for an electric field, the above tensor must be traceless, thus

$$\frac{\partial^2 V}{\partial x^2} + \frac{\partial^2 V}{\partial y^2} + \frac{\partial^2 V}{\partial z^2} = 0 \tag{5.7}$$

Using an obvious condensed notation this may be expressed as

$$V_{xx} + V_{yy} + V_{zz} = 0 \tag{5.8}$$

Hence, the efg can be specified in terms of V_{zz} and $\eta = (V_{yy} - V_{xx}/V_{zz})$, called the asymmetry parameter. It will be noted that $0 < \eta < 1$, and, importantly, that a three or more fold axis of symmetry passing through the Mössbauer nucleus implies that $V_{xx} = V_{yy}$ and $\eta = 0$. Two such axes, mutually perpendicular, lead to efg $= 0$, i.e. $V_{zz} = \eta = 0$. The electric field gradient arises from the orbital electrons of the atom containing the Mössbauer nucleus, from electron density transferred to orbitals of this atom by covalent interaction, and from all the more remote charged entities in the crystal.

Consider a point charge -e with polar coordinates r, θ, ϕ specified in relation to the nucleus as origin. The potential produced at the origin is $V = -e/r$ and thence

$$\begin{aligned} V_{zz} &= e(3\cos^2\theta - 1)/r^3 & V_{yy} &= e(3\sin^2\theta\sin^2\phi - 1)/r^3 \\ V_{xx} &= e(3\sin^2\theta\cos^2\phi - 1)/r^3 & V_{xy} &= e(3\sin^2\theta\sin\phi\cos\phi)/r^3 \\ V_{xz} &= e(3\sin\theta\cos\theta\cos\phi)/r^3 & V_{yz} &= e(3\sin\theta\cos\theta\sin\phi)/r^3 \end{aligned} \tag{5.9}$$

These quantities specify the electric field gradient tensor due to this charge distribution. Diagonalization will be equivalent to arranging that the charge lies on the Z axis, so that $\theta = \phi = 0$, and $V_{xy} = V_{xz} = V_{yz} = 0$, while $V_{zz} = 2e/r^3$, $V_{xx} = V_{yy} = -e/r^3$ so that $\eta = 0$.

This can clearly be extended to cover any distribution of charge. Cloud shells of orbital electrons, giving spherically symmetrical distributions of charge, make no contribution to the efg; nor will a partly filled degenerate subset of orbitals, such as the e_g^* or t_{2g} sets in truly octahedral symmetry. But any asymmetric occupation of p or d orbitals will lead to an efg.

The Hamiltonian operator for the quadrupolar interaction with an electric field gradient specified by V_{zz} and η is

$$\hat{H} = \frac{eQV_{zz}}{4I(2I-1)}\left\{3\hat{I}_z + \hat{I}^2 + \eta\frac{(\hat{I}_+^2 + \hat{I}_-^2)}{2}\right\} \tag{5.10}$$

In this equation $\hat{I}$ is the nuclear spin operator, so that $\hat{I}^2|Im> = I(I+1)|Im>$, where m is the quantum number for the z component of I. $\hat{I}_z$ is the operator for this component, so that $\hat{I}_z^2|Im> = m|Im>$ and $\hat{I}_+$ and $\hat{I}_-$ are spin raising and lowering operators, thus

$$I_{\pm}|Im\rangle = \sqrt{(I \pm m)(I \pm m + 1)}|Im \pm 1\rangle \tag{5.11}$$

Applying this $\langle Im|H|Im'\rangle$ to the various combinations of m and m', which assume values $I, I-1, \ldots, -I$, one obtains the matrix elements for the Hamiltonian which, after diagonalization if necessary, yield the different energy terms.

Applying this to the particular case of $^{57m_1}Fe$ and ^{57}Fe: for the latter $I = \frac{1}{2}$ so that $Q = 0$ and the level remains unsplit, but doubly degenerate; for the former $I = 3/2$ and the matrix is shown in Table 5.1, where $\alpha = eQV_{zz}/4$. Thus one

Table 5.1

$m \backslash m'$	$3/2$	$1/2$	$-1/2$	$-3/2$
$3/2$	α	0	$\frac{\sqrt{3}\eta\alpha}{12}$	0
$1/2$	0	$-\alpha$	0	$\frac{\sqrt{3}\eta\alpha}{12}$
$-1/2$	$\frac{\sqrt{3}\eta\alpha}{12}$	0	$-\alpha$	0
$-3/2$	0	$\frac{\sqrt{3}\eta\alpha}{12}$	0	α

finds two doubly degenerate levels with energies

$$\pm 3/2 \ldots \frac{eQq}{4}\sqrt{\left(1+\frac{\eta^2}{3}\right)}$$

$$\pm 1/2 \ldots \frac{-eQq}{4}\sqrt{\left(1+\frac{\eta^2}{3}\right)}$$

in relation to the original level (Fig. 5.3). Hence there will now be two transition energies between the excited and ground states of the ^{57}Fe with a difference of energies of $(eQV_{zz}/2)\sqrt{[1+(\eta/3)]}$. It will be noted that since $\eta \ngtr 1$ the maximum effect the asymmetry parameter can have only amounts to $\sim 15\%$ of this energy difference or quadrupole splitting.

For less simple combinations of spin values than with ^{57}Fe, more levels will be found and, if the spin quantum numbers I are integral, the efg can remove all the spin degeneracy so that $2I+1$ energy levels will arise. The selection rule for transitions between the two sets of states of the nucleus is $\Delta m = 0, \pm 1$.

The angular distribution of the emission or absorption events is not isotropic but depends on the direction the photons travel in relation to the efg

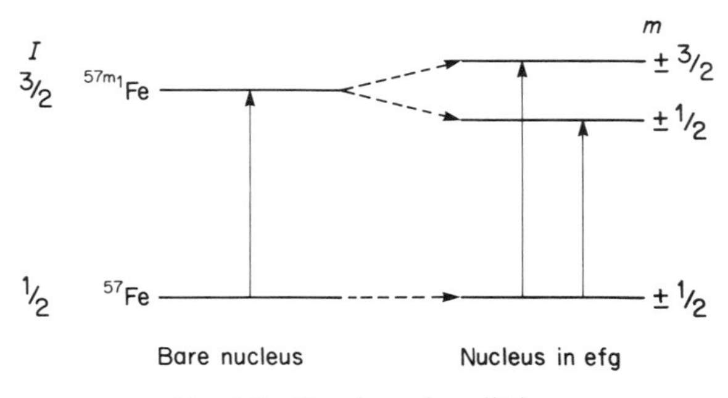

Fig. 5.3 Quadrupole splitting.

axes. In the simplest case, where $\eta = 0$, if θ is the angle between the direction of the incident or emitted photons and the z axis of the efg, the two kinds of event involving either the $\pm 3/2$ or the $\pm 1/2$ states of the $^{57m_1}Fe$ show different probabilities as a function of θ

$$\begin{array}{ll} \pm 3/2 \leftrightarrow \pm 1/2 & 1 + \cos^2\theta \\ \pm 1/2 \leftrightarrow \pm 1/2 & 2/3 + \sin^2\theta \\ {}^{57m_1}\mathrm{Fe} \quad {}^{57}\mathrm{Fe} & \end{array}$$

Thus for orientated absorbers, single crystals or specially arranged samples, the two lines corresponding to the two transitions vary in intensity from 2 to 1 for the $\pm 3/2 \leftrightarrow \pm 1/2$ transitions and 2/3 to 5/3 for the $\pm 1/2 \leftrightarrow \pm 1/2$ transitions as the absorber is rotated in relation to the incident photon beam. Fortunately for randomly orientated powder samples these variations average out to a 1:1 intensity ratio for the two lines.

The combined effect of these two electrical interactions is seen if the number of 14.4 keV photons penetrating an absorber is measured as a function of the velocity of the source in relation to the absorber, the mean distance between the vibrating source and detector being kept constant.

Figure 5.4 shows the spectrum obtained when there is no electric field gradient in the source environment, but the absorber, of different chemical composition to the source, provides an electric field gradient at the ^{57}Fe nuclei. The absorber consists of randomly orientated crystallites in a powder sample.

The quadrupole splitting, Δ, is given by $(eQV_{zz}/2)\sqrt{[1 + (\eta^2/3)]}$. The chemical or isomer shift, δ, appears as the displacement of the centre of the quadrupole split pair from zero velocity. This will give a value with reference to $|\psi(0)|^2$ for the source material. Attention must be drawn to two features of such spectra. Firstly, since the spectrum only yields Δ one cannot obtain both V_{zz} and η from the measurement. Secondly, one does not know which of the lines

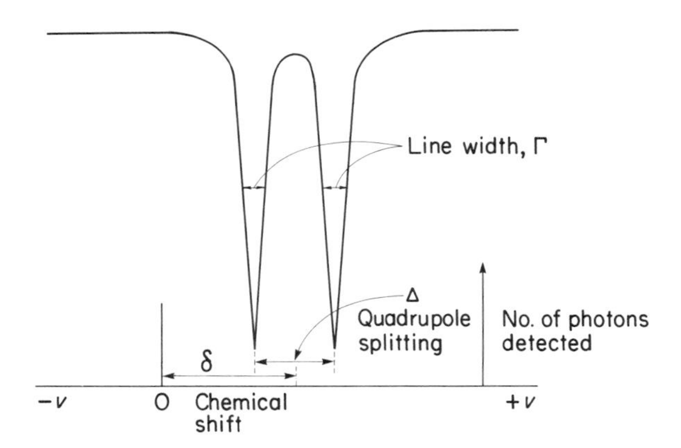

Fig. 5.4 Representation of a Mössbauer spectrum showing chemical isomer shift δ and quadrupole splitting Δ.

corresponds to the $\pm 3/2 \leftrightarrow \pm 1/2$ transitions and which to the $\pm 1/2 \leftrightarrow \pm 1/2$ transitions, because which of the $\pm 3/2$ or $\pm 1/2$ states, lies at the higher energy depends on the sign of the efg parameter V_{zz}; that is to say, it depends on whether the excess of negative charge over a spherically symmetrical distribution lies along the z axis or in the equatorial x–y plane.

5.2.3 Magnetic dipolar interaction

Magnetic monopoles probably do not exist, but nuclei with $I \neq 0$ possess a magnetic moment. In principle, octupolar effects should be considered but they are apparently too small to be revealed by this technique.

If the Mössbauer nucleus is subject to a steady unidirectional magnetic field of strength H, then all the spin degeneracy is removed and if the state possesses a spin I, $2I + 1$ separate energy levels will arise. The energy levels will be given by $E_{\mathrm{m}} = -g_{\mathrm{N}}\mu mH$, so the separation between levels will be given by $g_{\mathrm{N}}\mu H$. In these equations, μ is the nuclear magnetic moment and g_{N} the nuclear equivalent of Lande's fine splitting factor. Thus the excited state $^{57m_1}Fe$ yields $+3/2$, $+1/2$, $-1/2$ and $-3/2$ levels and the ground state ^{57}Fe, $+1/2$ and $-1/2$ levels. The selection rule governing the transitions between these states is $\Delta m = 0, \pm 1$. The sequence of levels is shown in Fig. 5.5. It should be noted that the separation of the levels is different for the excited and ground states and the sequence of m values is different. This is because μ is different for the two states and while it is positive for the ground state, it is negative for the excited state of ^{57}Fe.

As for the efg split transitions, the transition probability is not isotropic but depends on the angle θ between the direction the photon travels and the magnetic field (Table 5.2).

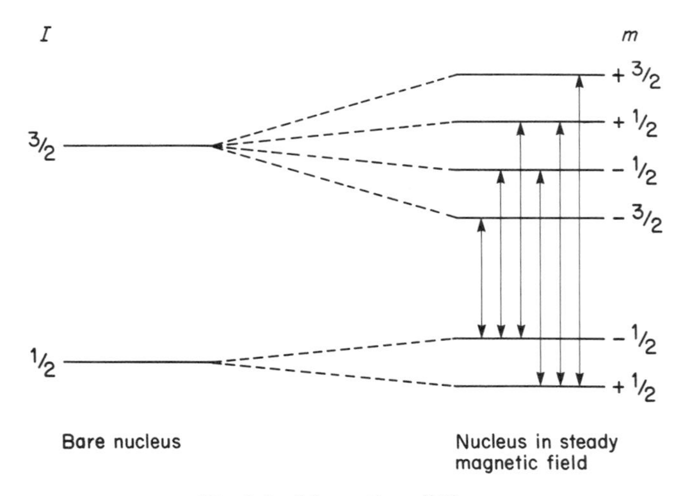

Fig. 5.5 Magnetic splitting.

Table 5.2

Transition	*Angular dependence*
$\pm 3/2 \leftrightarrow \pm 1/2$	$3/4(1+\cos^2\theta)$
$\pm 1/2 \leftrightarrow \pm 1/2$	$\sin^2\theta$
$\pm 1/2 \leftrightarrow \mp 1/2$	$1/4(1+\cos^2\theta)$

Upper or lower signs to be taken together.

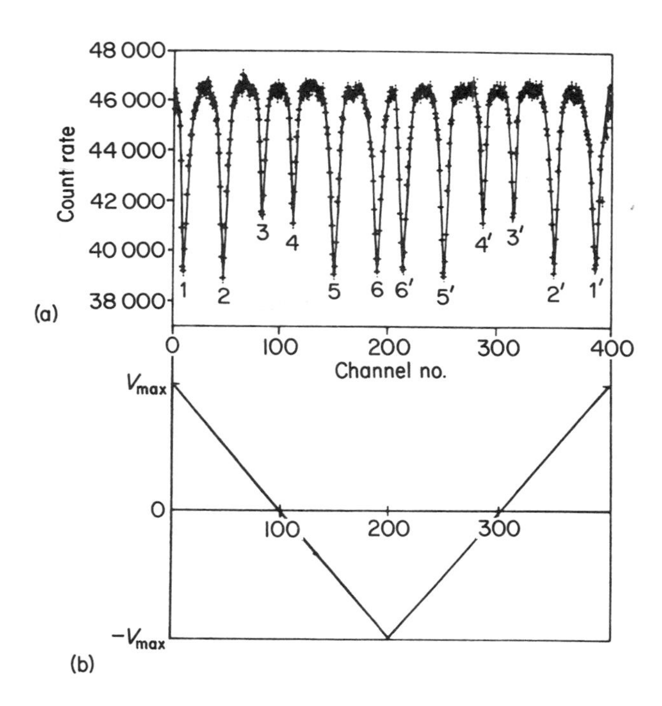

Fig. 5.6 (a) Magnetically split Mössbauer spectrum of metallic iron recorded with a spectrometer giving a mirror image of the spectrum in channels 200–400. The positions of the six lines for metallic iron are: lines 1 and 6 $\pm$ 5.328 mm s^{-1}, 2 and 5 $\pm$ 3.083 mm s^{-1}. 3 and 4 $\pm$ 0.839 mm s^{-1}. The lower part of the figure (b) illustrates the velocity as a function of the channel number and demonstrates the application to calibration of the symmetric sawtooth drive spectrometer. (Reproduced with permission from Gutlich, P., Link, R. and Trautwein, A. (1978) *Mössbauer Spectroscopy and Translation Metal Chemistry*, Springer Verlag, Berlin.

If the magnetic fields at the nuclei are randomly orientated in relation to the direction of the photon beam these average out to intensity ratios of 3:2:1:1:2:3 for the six permitted combinations. In these circumstances a source, emitting a single line spectrum, will give an absorption spectrum with an absorber displaying randomly orientated but constant magnetic fields at the ^{57}Fe nuclei, a spectrum such as is shown in Fig. 5.6.

Now the magnetic field at the nucleus can find its origin either in an externally applied magnetic field or in the field due to unpaired orbital electrons. The magnetic fields needed to produce appreciable splittings are in fact large and superconducting magnets are needed to give easily measurable splittings. Sufficiently large fields are produced at the nucleus by unpaired orbital electrons, but it is necessary that the field persists in a fixed direction for a time longer than the life-time of the excited state $^{57m_1}Fe$.

For paramagnetic iron compounds at room temperature spin-spin and spin-lattice relaxation processes usually lead to a rapid change of direction of the field, fast in comparison to the life-time of the $^{57m_1}Fe$, and the average magnetic field will be zero so that no magnetic splitting will be observed. But at a sufficiently low temperature, usually well below liquid nitrogen temperature, the relaxation may slow down sufficiently for a magnetic spectrum to develop. By contrast in ferromagnetic or antiferromagnetic materials the cooperative interactions keep the nuclear fields aligned and, below the Curie or Néel temperatures, a magnetically split spectrum is observed. Thus both soft iron (ferromagnetic) and iron(III) oxide (antiferromagnetic) yield magnetically split spectra at room temperature. When such a spectrum is obtained, if the magnetic moment of the ground state nucleus is known as is usually the case, the magnetic field at the nucleus can be calculated.

The situation when the field at the nucleus is relaxing neither very quickly nor very slowly in comparison with the life-time of the $^{57m_1}Fe$ state is complex, but two cases demand attention. The first effect that is noticed, as the relaxation time is slowed down from a very fast value on the $^{57m_1}Fe$ time scale, is that the absorption lines start to grow broader. This situation often appears in iron(III) complexes even at room temperature, because, if the iron atoms are rather widely separated as for example in a complex with large ligands, the spin-spin relaxation will be slow and the spin-lattice relaxation, which is temperature dependent, is still rather slow at room temperature. Hence some line broadening from this cause is often observed with iron(III) absorption lines.[8]

The second case provides us with a valuable opportunity for determining the sign of the electric field gradient. Figure 5.7(a) shows a typical quadrupole split spectrum. To determine the sign of the electric gradient one needs to know which of the two lines, $\pm 3/2 \leftrightarrow \pm 1/2$ and $\pm 1/2 \leftrightarrow \pm 1/2$, lies at the more positive velocity. Now the magnetic fields readily available from small superconducting magnets are not strong enough to give a clearly magnetically split pattern. However as the field removes the degeneracy of $\pm 3/2$ and $\pm 1/2$ states,

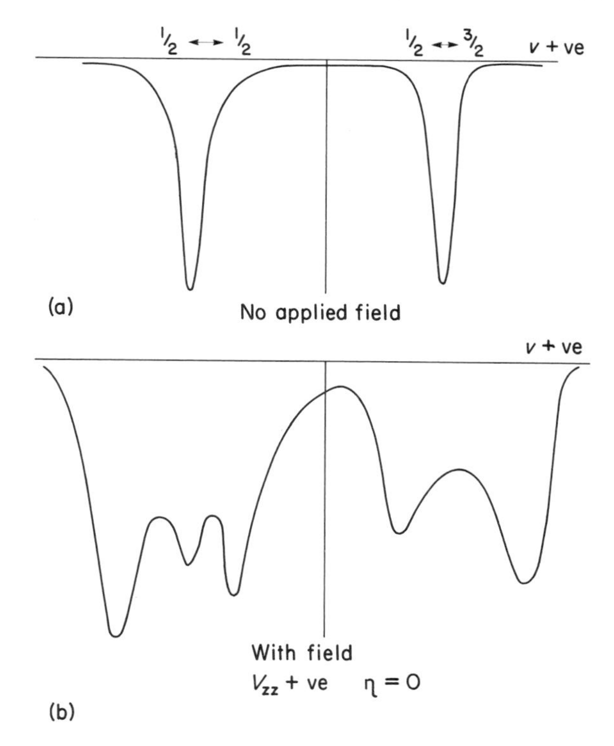

Fig. 5.7 Determination of the sign of V_{zz}.

after line broadening at low fields, the $\pm 1/2 \leftrightarrow \pm 1/2$ line begins to split, ultimately to yield four lines, because all the pairs of sub-states can combine with each other ($\Delta m = 0, \pm 1$). For the $\pm 3/2 \leftrightarrow \pm 1/2$ only two combinations are possible, because $+3/2$ cannot combine with $-1/2$, nor $-3/2$ with $+1/2$. Thus the spectrum begins to look like Fig. 5.7(b), supposing the $+1/2 \leftrightarrow \pm 1/2$ line lies to lower velocity in Fig. 5.7(a). The precise form of these spectra can be calculated and the effect provides a fairly straightforward means of determining the sign of V_{zz}.

5.3 THE MÖSSBAUER FACTOR, *f*, AND THE INTENSITY OF THE ABSORPTION LINES

Using the Debye model for a solid, in which the atoms display a small group of vibrational frequencies ranging up to some maximum value ν_D, the energy of the vibrating atoms is quantized and, using a simple harmonic oscillator approximation, its energy levels are given by $h\nu(n + 1/2)$ with similar expressions for other frequencies. Considering first only one frequency, ν_D, the separation between vibrational levels is $h\nu_D$ and kinetic energy can only be transferred to the atoms in units of $h\nu_D$. Typically, these quanta lie in the range 10^{-3}–10^{-1} eV. Now the energy the photon-emitting atom should take up to conserve momentum with the photon is $E_R = E_\gamma^2/2Mc^2$, where M is the mass of the emitting atom. Döppler modification of the energy of the emitted photon will also involve some acquisition or loss of energy by the atom. The recoil energies calculated in this way lie in the range 10^{-4}–10^{-2} eV, so that in some proportion of the events it is not possible for this energy to lead to excitation of the vibrational energy of the atom or phonon excitation. In these zero phonon events, momentum is effectively conserved between the emitted photon and all the atoms in the crystallite containing the emitting atom; these comprise a very large mass compared with the atom, so that negligible recoil occurs. Neither recoil nor Döppler modification of the photon energy takes place.

One can associate a characteristic temperature θ_D, the Debye temperature, with each solid, such that $h\nu_D = k\theta_D$ (k is Boltzmann's constant). Now classically the energy of the simple harmonic oscillator is given by $M\omega_D^2\langle x^2\rangle$, where $\langle x^2\rangle$ is the mean square vibrational amplitude of the atom and $\omega_D = 2\pi\nu_D$ is the angular velocity. Thus the phonon quanta have energy $\hbar^2/2M\langle x^2\rangle$. If we put $E_\gamma/c = \mathbf{k}\hbar$, so that $\mathbf{k}$ is the wave vector for the radiation, then the recoil energy becomes $\mathbf{k}^2\hbar^2/2M$ and it can be shown that the proportion of recoil free events, f, when phonon excitation does not occur, will be $f = \exp[-\mathbf{k}^2\langle x^2\rangle]$.

This important result shows that f will decrease rapidly as $\mathbf{k}$ increases, that is as the photon energy E_γ increases. In addition, f will decrease as $\langle x^2\rangle$ increases. Now $\langle x^2\rangle = \hbar^2/2M\theta_D$, so that the higher the Debye temperature of the solid the larger the f factor.

Taking into account that several phonon levels may be populated and that more than one ν will be involved, it can be shown that

$$f = \exp - \left\{\frac{3m\mathbf{k}^2\hbar^2}{4Mk\theta_D}\left[1 + 4\left(\frac{T}{\theta_D}\right)^2 \int_0^{\theta_D/T} \frac{x}{e^x - 1}\mathrm{d}x\right]\right\} \tag{5.12}$$

This gives

$$\text{(i) for } T \ll \theta_D \quad f \approx \exp - \left[\frac{3\mathbf{k}^2\hbar^2}{4Mk\theta_D}\right]$$

and

$$\text{(ii) for } T > \theta_D \quad f \approx \exp - \left[\frac{3\mathbf{k}^2\hbar^2 T}{Mk\theta_D^2}\right]$$

Thus f generally increases as the temperature is reduced.

The f factor is closely connected with the Debye–Waller factor in the analysis of X-ray diffraction patterns.

Since most minerals have high Debye temperatures, satisfactory f factors are usually found for room temperature measurements.

5.4 ^{57}Fe MÖSSBAUER PARAMETERS AND DEDUCTIONS FROM SUCH DATA

Having outlined the factors that determine the principal Mössbauer parameters, the particular case of ^{57}Fe will be considered in some detail,[3] because the main applications in mineralogy relate to this element.

5.4.1 The chemical shift, δ

In the case of ^{57}Fe the sign of $\delta R/R$ is negative, that is to say the excited state, ^{57m}Fe, is smaller than the ground state, ^{57}Fe. Thus the chemical shift δ grows larger as the charge density at the nucleus $|\psi(0)|^2$ decreases.

In principle, values of $|\psi(0)|^2$ for iron atoms in different compounds are susceptible to calculation. In practice, the calculations are lengthy, complicated and expensive in all but extremely simple systems. Further, although $\delta R/R$ can be obtained from other data, such as information on internal conversion, calculation of δ from $\delta R/R$ may not lead to very accurate values. The problem is aggravated by the fact that the difference in δ arises from small differences in the comparatively large values of $|\psi(0)|^2$. Although a more fundamental treatment of chemical shifts is a currently active area of study, for the present purpose it is better to choose a more empirical approach.

The main factors that will determine $|\psi(0)|^2$ in an iron compound will be:

(i) The occupation of the 4s orbital of the iron. In oxidation states two and three, the 4s electrons are formally lost, but more or less covalent transfer of electron density from the ligand species to this orbital will take place. Such transfer will be much more important for the low-spin derivatives than for the high-spin species. Increased 4s occupation will reduce δ.

(ii) Because of their screening effect the occupation of the 3d orbitals will also affect the chemical shift. Increased occupation of these orbitals will reduce $|\psi(0)|^2$ and enhance δ.

(iii) In addition to these more obvious effects, changes in the radial function for either of these orbitals will influence δ.These nephelauxetic effects further complicate a simple interpretation of δ.

In keeping with these predictions it is found:

(i) For high-spin compounds, the chemical shift for the Fe^{2+} compound is larger than the shift for the Fe^{3+} compound with the same ligands. This arises, in large measure, from the screening due to the extra 3d electron in Fe^{2+}. This difference is sufficient that even with change of ligand the chemical shift of Fe^{2+} is always greater than that of Fe^{3+} in high spin situations. In minerals, the iron is most frequently coordinated by oxygen and Fe^{3+} can be readily distinguished from Fe^{2+}.

(ii) For oxygen coordination of the iron, the chemical shift for tetrahedral, four-coordinate Fe^{3+} is less than for octahedral, six-coordinate Fe^{3+}.

These shifts are both less than the shifts for Fe^{2+}, where tetrahedral sites also show a lower shift than the octahedral sites, which in turn are less than for the rarer cases of eight coordination. The relation of δ to the oxidation state and the coordination number of the iron in minerals is shown in Fig. 5.8(a).

The above relations suggest that the increasing occupation of the d orbitals with increasing coordination number is reducing $|\psi(0)|^2$ and increasing the chemical shift δ. However, the situation is probably more complicated as the single result for Fe^{2+} in a planar four-coordinate environment gives a lower chemical shift than for tetrahedral sites.

(iii) For less common oxidation states of iron in high spin configurations, the chemical shift again seems to reflect the occupation of the 3d orbitals; Fe^{+}

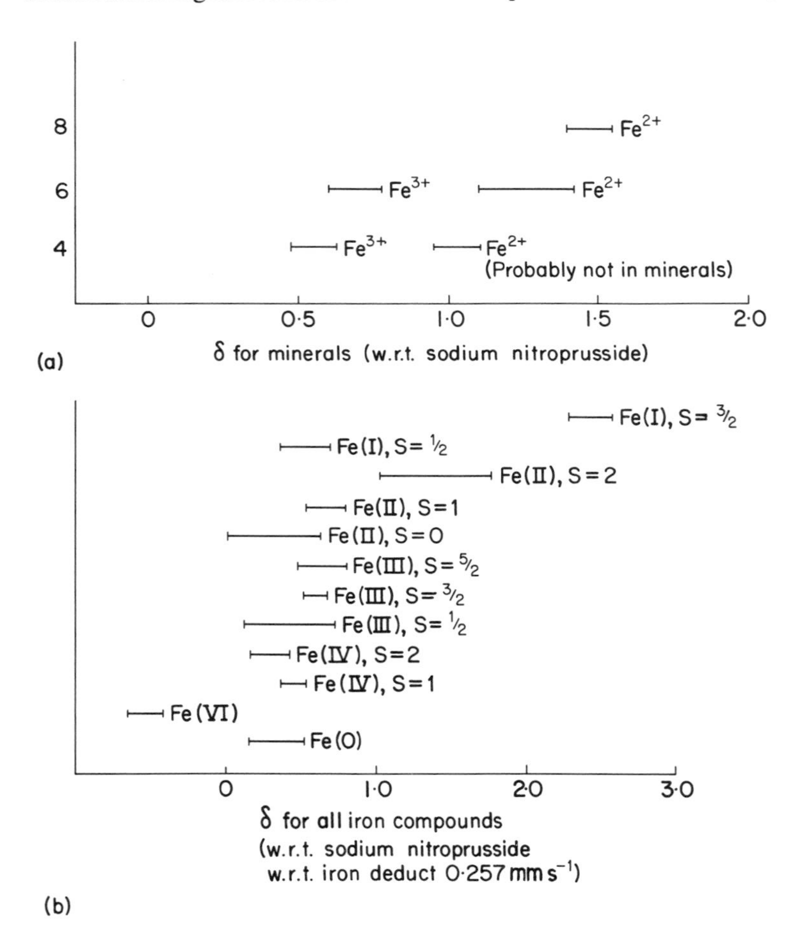

Fig. 5.8 Relationships between chemical isomer shifts and (a) co-ordination number; (b) spin states.

has been identified in iron-doped alkali metal halides and it shows a considerably larger chemical shift than Fe^{2+}. By contrast Fe^{6+}, in the tetrahedral environment in K_2FeO_4, shows one of the smallest recorded chemical shifts.

Thus the chemical shifts for high spin iron increase in the order

$$\delta Fe^{6+} < Fe^{3+} < Fe^{2+} < Fe^{+}$$

as does the number of 3d electrons. It may be noted here that the shifts for low-spin Fe^{IV} in the mixed oxides of the $M^{II}FeO_6$ type, a number of which have a perovskite structure, lie between the values for high-spin Fe^{6+} and Fe^{3+}.

(iv) Turning now to the variation of the chemical shift with the nature of the ligand for high-spin Fe^{2+} and Fe^{3+} in compounds of the same coordination number, it is soon found that the data for high-spin Fe^{2+} compounds are the more difficult to systematize. For these, there is some parallel with the nephelauxetic series; thus for octahedral Fe^{2+} in the FeX_2 halides, the fluoride has the largest chemical shift and the iodide the smallest. But for the smaller category of four-coordinate complexes the correlation is not as marked and for the more nephelauxetic ligands the changes in δ are small.

For high spin Fe^{3+} compounds, the pattern is rather simple and for both octahedral six-coordination and tetrahedral four-coordination the chemical shift decreases in parallel with the position of the ligand in the spectrochemical series. But for the four coordinate species the changes are rather small. Thus the bromo-complexes give the largest δ and the fluoro-complexes the lowest δ of the halide complexes.

(v) The low-spin complexes of iron are much less important from a mineralogical point of view and only a few details will be given. The chemical shifts found for low-spin complexes of Fe(II) and Fe(III) are smaller than for the high-spin compounds and not very different in magnitude. A distinction between low-spin Fe(II) and Fe(III) cannot be made on the basis of their chemical shifts. The covalent transfer of electron density from ligand to metal is greater for the low-spin compounds and the effective oxidation states of the iron in Fe(II) and Fe(III) are no longer so very different. The influence of the ligand characteristics will be greater, and parallel the position of the ligand in the spectrochemical series.

The transfer of electron density from the ligand to metal orbitals increases $|\psi(0)|^2$ and reduces δ. Supposing that the nephelauxetic effect, the change in the radial part of the iron atom wavefunction, is not too important it could be expected that $|\psi(0)|^2$ will increase with both the σ donor strength of the ligand, enhancing 4s occupation, and with π acceptance by the ligand, reducing the 3d occupation. Thus $|\psi(0)|^2$ might be expected to increase with $(\sigma + \pi)$, where the symbols measure the ligand ability to

donate and accept electron density in the sense mentioned above. This accounts for the parallel with the spectrochemical series.
It will be noted that the increased 4s occupation, with stronger σ donors, normally outweighs the accompanying increased screening due to increased 3d occupation, so that δ decreases as σ donor power increases.

(vi) The inclusion of low-spin complexes introduces compounds of still lower oxidation states. As would be expected from the above discussion the numerous Fe(O) complexes generally have lower chemical shifts than the low-spin Fe(III) or Fe(II) complexes. The formally negative oxidation state compounds e.g. $Fe(CO)_4^{2-}$ tend to still lower shifts and for chemically similar compounds, the isomer shift usually increases with coordination number.

A few general observations may be made. Chemical shifts are not usually sensitive to small changes in the stereochemistry of the complex, so that departures from idealized octahedral or tetrahedral forms are not very important. In general the magnitude of the shift is determined by the oxidation state of the iron and the identity of its ligands but changing the identity of the co-ion can lead to changes in δ much larger than the errors in the measurement; thus different salts of $FeCl_4^-$ and $FeBr_4^-$ give slightly different chemical shifts.

For the low spin Fe(II) complexes there is evidence that the chemical shift arising in complexes with more than one kind of ligand is at least approximately an additive property so that a system of partial chemical shift data can be developed. These tell one more about the ligand than the iron complex. Figure 5.8(b) summarizes the range of values of δ for these different situations.

5.4.2 The quadrupole splitting

Notionally, it is useful to divide the electric field gradient (efg) into three contributions:

$$V_{zz}(\text{total}) = V_{zz}(\text{cf}) + V_{zz}(\text{mo}) + V_{zz}(\text{lattice}) \tag{5.13}$$

(It should be noted that although this conceptual device is widely used, terminology for the components varies.)

In V_{zz}(cf) the cf stands for crystal field and this represents the part of the electric field gradient due to the valence electrons without taking into account electron density transferred by the ligands.

In V_{zz}(mo) the mo stands for molecular orbital and this takes account of the contribution to the efg from covalent transfer of electron density from ligand to metal or vice versa.

V_{zz} (lattice) allows for the contribution to the efg due to more remote charged species, the co-ions, for instance, in a polar solid.

Since the covalent transfer usually amounts to less electron density than that due to an asymmetrically distributed valence electron and because the efg due to a remote charged entity falls off as the inverse third power of the distance of the charge from the nucleus, in general.

$$V_{zz}(\mathrm{cf}) > V_{zz}(\mathrm{mo}) > V_{zz}(\mathrm{lattice})$$

unless the earlier terms assume zero values. However, the actual efg at the nucleus is modified by the screening effect of the closed shells of orbital electrons. These might be expected to reduce the efg but, because of the polarization of the cloud shells by the exterior charged species in the lattice, the V_{zz}(lattice) component is enhanced.

One may write

$$V_{zz} = (1-\gamma)V_{zz}(\mathrm{lattice}) + (1-R)V_{zz}(\mathrm{cf}+\mathrm{mo}) \tag{5.14}$$

where γ and R are the Sternheimer antishielding factors: γ is a negative number of the order -10, whereas R is of the order of ~ 0.2 for iron. Clearly the $(1-\gamma)$ factor magnifies the lattice contribution appreciably.

It should be noted that the value of Q, the quadrupole moment for the ^{57m_1}Fe, is not known with much accuracy so that a purely theoretical calculation of V_{zz} and η seldom gives good agreement with the experimental data:

(i) To consider some particular cases. The above considerations point to a few very simple situations:

(a) For high-spin Fe^{3+} the 3d orbitals are half-filled and for low-spin Fe(II), in an octahedral environment, the t_{2g} sub-set is fully occupied, so that in both these cases $V_{zz}(\mathrm{cf}) = 0$. Further, if all the ligands are the same $V_{zz}(\mathrm{mo}) = 0$ for octahedral and tetrahedral stereochemistries. Finally, $V_{zz}(\mathrm{lattice}) = 0$ for crystals in the cubic system.

(b) The very important case of Fe^{2+} in octahedral or near-octahedral environments justifies more detailed consideration.
As shown in Fig. 5.9 in the low-field, high-spin situation, the sixth d electron of the Fe^{2+} occupies the degenerate t_{2g} orbital and, since it is spread equally between all three orbitals, there will be no V_{zz}(cf).
However, the Jahn–Teller theorem shows that such situations are essentially metastable and should normally distort to remove the degeneracy. The effects of axial and rhombic distortion are also shown in the figure. If the splitting of the d orbitals introduced in this way is sufficiently great that at the prevailing temperature only the $|xy\rangle$ is doubly occupied, then V_{zz}(cf) will assume a limiting upper value 4/7 $e\langle r^{-3}\rangle_{3d}$, the factor 4/7 arising from the evaluation of the term $\langle 3\cos^2\theta - 1\rangle$ for the angular part of the $|xy\rangle$ wavefunction. It should be noted, however, that $\langle r^{-3}\rangle_{3d}$ is influenced by the covalency of the bonding, and that the free ion value probably sets an upper limit to

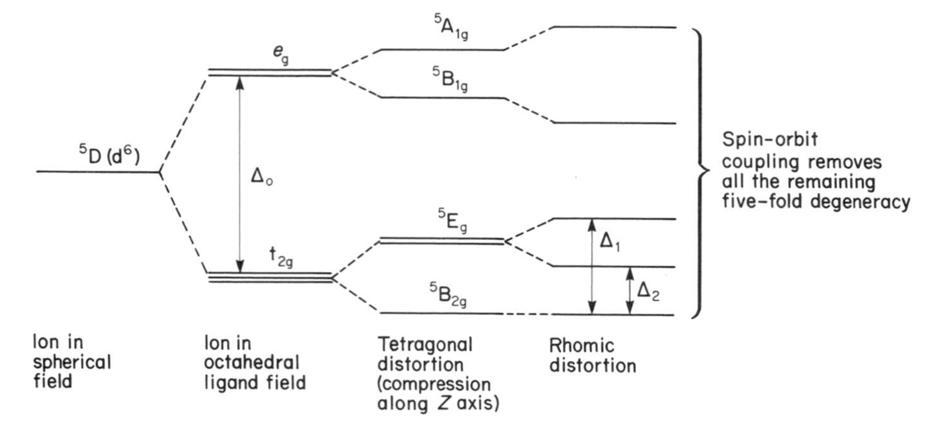

Fig. 5.9 Electronic levels for iron(II) in various environments.

V_{zz}(cf). If, however, the splittings Δ_1 and Δ_2 are such that some occupation of the upper orbitals will occur, then V_{zz}(cf) will be reduced to some value $F4/7e\langle r^{-3}\rangle_{3d}$ where $1 > F > 0.5$.

Even when Δ_1 and Δ_2 have values such that only the $|xy\rangle$ level is occupied at the ambient temperature, all these levels will be liable to spin-orbit coupling. The effect of this coupling is to yield mixed wavefunctions so that the states derived from the $|xy\rangle$ state can be expressed in terms of $|xy\rangle$, $|xz\rangle$ and $|yz\rangle$. The mixing coefficients will depend on the ratios Δ_1/λ and Δ_2/λ, where λ is the spin-orbit coupling constant. The overall effect will be similar to the effects of the stereochemical splitting of the d orbitals, but changes in the population of the substates will usually occur over a lower temperature range. The lowest energy spin-orbit coupled state has the largest admixture of $|yx\rangle$ and $|xz\rangle$ and thus the lowest value of F.

The lattice contribution also always has the opposite sign to V_{zz}(cf) and thus, so long as the regular octahedral environment is excluded, the more distorted the Fe^{2+} site, the smaller is the quadrupole splitting. The case of tetrahedral Fe^{2+} has been analysed in a similar way.

(c) For low-spin Fe(III) there is also a valence contribution to the efg. But in these complexes and for Fe(O) complexes the value of a partial V_{zz}(cf) and V_{zz}(mo) is doubtful.

(ii) Studies of low-spin Fe(II) complexes (for which $V_{zz}(\text{cf}) = 0$ and the efg arises mostly from the V_{zz}(mo) term, if more than one kind of ligand is present) suggest that in complexes of this kind the contributions of the different ligands to the efg are at least approximately additive and a system of partial quadrupole splitting contribution for the ligands can be constructed.[3,9] Since the σ donation by the ligand increases V_{zz}(mo) by covalent transfer to the 3d orbitals of the iron, and the π back donation decreases the occupation of 3d orbitals, it is found that the partial quadrupole splittings are a measure of σ–π for the ligands.[9]

5.4.3 Effect of temperature on δ and Δ

For the second order Döppler effect, although the mean velocity of the vibrating atom in a solid averages to zero the mean value $\langle v^2 \rangle$ does not. This leads to a temperature dependent contribution to the chemical shift, δ_{SOD}. For atoms vibrating as simple harmonic oscillators

$$\delta_{SOD} = \frac{-k}{2Mc^2}\langle r^2 \rangle$$

where k is the restoring force, M the mass of the atom and $\langle r^2 \rangle$ the mean square displacement of the atom. This term is related to the Mössbauer fraction, f, so that

$$\delta_{SOD} = \frac{3\hbar^2}{2ME_\gamma{}^2} k \ln f. \tag{5.15}$$

Hence

$$\delta = \frac{3\hbar^2 k}{2ME_\gamma^2} \ln f + \delta'' \tag{5.16}$$

is the true chemical shift. Thus, over some temperature interval ΔT, assuming k and δ'' have little temperature dependence

$$\left(\frac{\Delta\delta}{\Delta \ln f}\right)_{\Delta T} = \frac{3\hbar^2 k}{2\mathrm{ME}_\gamma{}^2}. \tag{5.17}$$

As a result, δ normally increases as T falls. Because of this effect one should be cautious of reading too much significance into small changes in δ between different compounds if only measurements at one temperature are available. From the above relations it is apparent that δ_{SOD} increases as the Debye temperature of the solid increases and also with the temperature of measurement.

The fact that δ changes with temperature and that the extent may be different for different iron environments, can sometimes be exploited in the investigation of complex overlapping Mössbauer mineral spectra. Changing the temperature of measurement may facilitate separation of two spectra.

For the reasons outlined in Section 5.4.2, the quadrupole splitting is usually temperature dependent, normally increasing at lower temperatures. This dependence can be utilized in two ways. It must be investigated if an attempt is to be made to extract ligand field parameters from the Mössbauer spectrum, but it is more frequently useful in mineralogical studies as a means of separation of spectra which overlap when measured at room temperature. A comparison of a room temperature and liquid nitrogen temperature measurement on a mineral is usally profitable.

5.5 EXPERIMENTAL DETAILS

Detailed accounts of Mössbauer spectrometers are to be found in several texts[2,3,4] and very useful information on technical improvement and instrumental devices can be found elsewhere.[10–23] Hence only a very brief account of the experimental details is given here.

5.5.1 General features of Mössbauer spectrometers

Like most spectrometric techniques, the Mössbauer spectrometer comprises a source, the sample called the absorber and a detector (Fig. 5.10). A cyclic movement is imposed on the source to give a small range of photon energies by deliberate use of the Döppler effect. The amplitude of movement of the source should be small compared with the distance between the source and the absorber. The output from the detector, which is chosen to respond only to the Mössbauer radiation, is fed to a recording device, synchronized with the movement of the source, so that during each of a number of small time intervals in the cycle of movement when the source is moving within a small band of velocities, the photons detected are recorded in a particular channel of the recorder. Thus, after a period of time, the recorder contains a spectrum of the number of photons reaching the detector, having passed through the absorber, as a function of the mean velocity of movement of the source in each of a large number of velocity intervals.

The detector should not subtend too large a solid angle at the source, so that

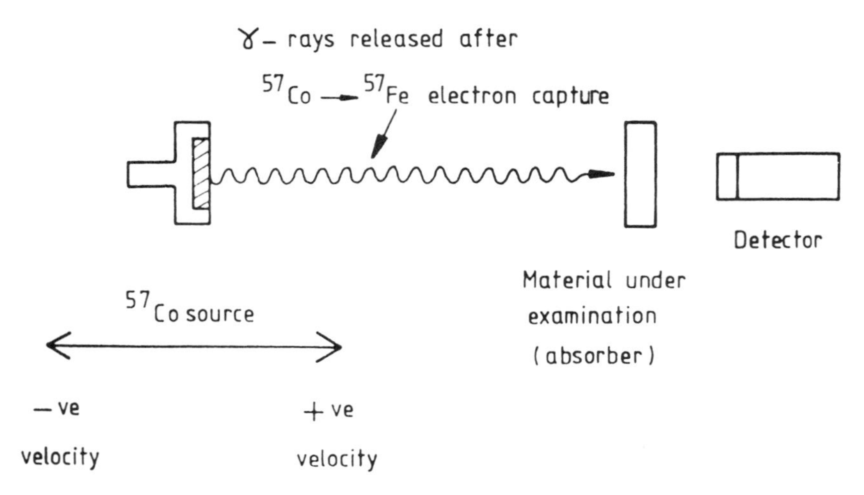

Fig. 5.10 Schematic representation of source, absorber and detector.

the cosine effect, modifying the change of photon energy due to the Döppler effect, is not too large for those photons reaching the detector from the source but not travelling parallel to the line joining the centre of the source to the centre of the detector. Thus the separation between the source and the detector should not be small.

The sample should be located fairly close to the source so that the probability of recording photons that have suffered a small angle Mössbauer scattering event is kept low.

Since it is frequently necessary to make measurements over a range of temperatures, a cryostat to hold the absorber, or source and absorber at a low temperature is desirable. For a lot of mineralogical work, it is really only necessary to be able to make measurements at room temperature, liquid nitrogen temperature and perhaps a few other values such as the solid CO_2–acetone temperature. For this purpose it is easy to arrange a simple absorber cryostat. If, however, it is intended to determine signs of the efg, for which a superconducting magnet is necessary, or if some study of magnetically split spectra or other studies demanding measurements at the temperature of liquid helium are envisaged, a much more elaborate helium cryostat with a built-in superconducting magnet will be essential.

5.5.2 The detector

The photon detector must be an energy-sensitive type, because it is essential to distinguish photons potentially liable to Mössbauer scattering from the other photon radiations emitted by the source. Three kinds of detector are in common use.

(a) *NaI/Tl scintillation detectors*

Most of the early work on Mössbauer spectroscopy was carried out using such detectors. For use with the 14.4 keV radiation from $^{57m_1}Fe$ some discrimination against more energetic photons can be obtained by the use of rather thin crystals; 1 mm thick is quite convenient. The detecting efficiency for the 14.4 keV radiation is high, as are the permissible counting rates. But the energy resolution possible with scintillation crystals is poor, perhaps 5 keV at 14.4 keV and often worse.

(b) *Proportional counters*

For work with ^{57}Fe these are especially suitable. The counter is a simple cylindrical ionization chamber with a thin wire collecting electrode and a beryllium window to admit the soft radiation. With a filling of xenon gas, the efficiency of a counter is acceptably high for the 14.4 keV radiation. The counting rate is restricted to about 60 kHz but the energy resolution is rather good, about 2 keV at 14.4 keV.

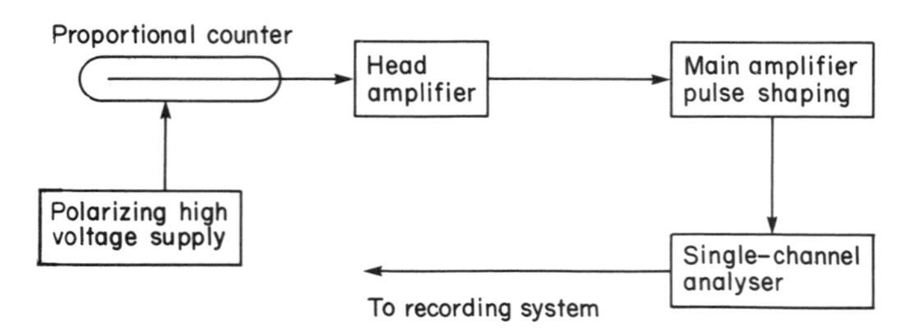

Fig. 5.11 Photon detection.

(c) *Solid state detectors*

For higher energy Mössbauer radiation than iron, and especially in cases where there is nearby X-radiation, it is best to use solid state detectors, silicon in the lower energy region and germanium at higher energies. With both of these an energy resolution of about 0.5 keV is possible.[11,12]

The electronic requirements of these detectors are shown in Fig. 5.11. The detector must be polarized by a well-stabilized high voltage supply. A head amplifier acts as an impedance changing device and feeds the pulses to a fast linear amplifier. The output from this amplifier consists of voltage pulses of sizes proportional to the energies of the photons that gave rise to the pulses. A single channel analyser then selects from these events only those of a size such that they must find their origin in the 14.4 keV photons from the $^{57m_1}Fe$.

5.5.3 The source drive and multichannel analyser

Although purely mechanical movement of the source has been used, practically all general purpose Mössbauer spectrometers now use an electromechanical drive.[13–15] The source is attached to the moving core of a vibrator comprising a coupled pair of solenoids. The low impedance winding, A, is the drive coil and the movement of the source is determined by the feed to this coil. A second higher impedance coil, B, in which there is another core attached to that moving in the drive coil and to the source, provides a signal that monitors the actual movement of the source (Fig. 5.12). The Mössbauer photon signals, selected by the single channel analyser, are fed to a pulse analyser used in the multiscaler mode. The incoming pulses are fed to the succession of recording channels in the analyser in phase with the movement of the source, the input sweeping the recording channels once during each cycle of motion of the source. Thus the photon signals recorded in the different channels correspond to events taking place when the source was in the same small interval during its cycle of movement. Hence each channel stores events taking place when the source was moving in a certain velocity interval v to $v + \delta v$. The number of channels used is set by the analyser, values of 256, 400, 512, 800 and 1024 being common. It should be noted that for all analysers there is a dead time while the analyser switches the input from one channel to the next. The analyser should be chosen

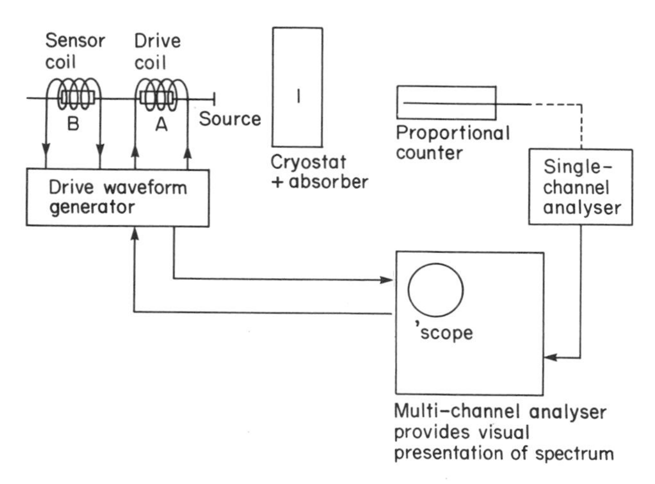

Fig. 5.12 Schematic representation of a Mössbauer spectrometer.

to keep this dead time, when the apparatus is not collecting spectral information, as small as possible compared to the dwell time, the period it feeds input pulses to a given channel during each cycle. The function of the multichannel analyser can also be fulfilled by some types of small computer.[22,23]

There are two ways of achieving the synchronization of the movement of the source with the sweeping of the recording channels. In one method the drive waveform unit also generates pulses which intiate the channel sweep and control the switching from channel to channel. The analysers are normally provided with circuits to permit such switching. The drive unit must then be instructed as to the number of channels the analyser provides, and switching pulses are then generated at intervals dividing the cycle of movement into the requisite number of intervals.

In another method the analyser controls the drive waveform. Most analysers operating in the multiscaler mode provide facilities for repetitive sweeping of the channels by the input with a pre-selected dwell time in each channel. In addition most analysers provide for splitting the channels into two equal sets. In doing this, the analyser usually provides a square wave with the polarity inverting each time the input moves from the first to the second set of channels, or at the end, returns to the first channel. This square wave can be used to generate a necessarily synchronized drive waveform. The frequency of the cycle of movement will be determined by the number of channels, the switching dead time and the dwell time. Convenient dwell times are characteristically from 20 to 100 μs. For either method of synchronization it is advisable that the frequency of the drive waveform be near, but not at, the natural mechanical resonance of the vibrator.

Various drive waveforms can be used. A sinusoidal motion is rather easily achieved and has the advantage of no discontinuities in the movement of the source. However, it has a serious disadvantage that if constant dwell times are used the mean velocity of movement of the source is not linearly related to the channel number. Thus some computer processing of the primary data is necessary to yield linear spectra. The visual presentation of the spectrum in the oscilloscope display, provided by the analyser, is hard to appreciate.

Using the square wave generated by the analyser itself enables a fairly easy conversion to a constant acceleration symmetric sawtooth drive which changes sign when the second set of recording channels is reached, and again at the end of the channels. Such a spectrometer will store two spectra, which should be mirror images of each other, one in each set of half the channels. Each spectrum provides a linear mean velocity–channel number relation and the computer programme eventually used to analyse the spectra can make use of all the data obtained. The weakness of this type of drive is due to the electromechanical discontinuity when the acceleration changes sign. The output high impedance sensor coil is used to provide feedback to the drive waveform generator to improve the adherence of the motion of the source to the imposed waveform. It is convenient to monitor the drive and sensor waveform motions on an oscilloscope. Under favourable conditions the change of velocity with channel number is closely linear except for a few channels after the change of sign of the acceleration, that is, at the beginning of the two sets of channels.

Finally, a single spectrum can be collected in the whole set of channels provided by the analyser by using an asymmetric sawtooth waveform drive with the channel switching in the analyser controlled from the drive generator. The method uses constant acceleration of the source, which covers a span of velocities, symmetrically from a negative to a positive value. This is the ascending ramp of the sawtooth and the input in the analyser is switched from channel to channel while the source follows this velocity ramp. At the end of the set of channels, instead of the acceleration simply changing sign, the source is returned to its initial state at the start of the velocity ramp without any attempt at linearity of change and during which no data is being collected. At the moment constant acceleration recommences, the drive instructs the analyser to start another sweep of the collection channels. A single spectrum is collected and the problem of non-linearity on change of acceleration of source is avoided. However, the apparatus is inactive during each cycle of motion for the time taken to return to the beginning of the velocity ramp. It is also possible to move the absorber rather than the source and this keeps the geometrical efficiency of the detection constant because of the constant source–detector separation, and larger amplitudes of movement are acceptable.

5.5.4 The source

The source of ^{57}Co should emit a single line of minimum width at half-maximum absorption, Γ, a value of about 0.13 mm s^{-1} being possible. This is close to the limit set by the uncertainty principle, 0.115 mm s^{-1}. But one must take account of the fact that a measured linewidth must include contributions from both the emission and absorption events. It is also desirable that the Mössbauer fraction, f, be as large as possible.

These requirements imply a solid of high Debye temperature and cubic crystal structure. For general purposes a source of ^{57}Co diffused into a small disc of one of the platinum group metals meets these demands. $^{57}Co/Pd$ sources are widely used. If work at liquid helium temperature is planned a $^{57}Co/Rh$ source is rather better.

5.5.5 Calibration

For many purposes, the velocity scan of the spectrometer is most conveniently calibrated by measuring the spectrum of a soft iron absorber. The positions of the six absorption lines are accurately known and serve as a check on the linearity of the velocity scan (Fig. 5.6).

^{57}Fe-enriched foils are useful for rapid checking of the spectrometer, but it should be remembered that multiple scattering events in such absorbers lead to considerable line broadening. Even with natural foils, the high intensity extreme lines 1 and 6 (Fig. 5.6) usually suffer some broadening and the 3 and 4 lines are better for estimating the Γ for the assembly.

If the total velocity scan is small, as is often desirable in mineral studies, then a thin sodium nitroprusside absorber which gives a quadrupole split pair of lines, is convenient for calibration. In this case, no check on linearity is possible.

It should be noted that the calibration with soft iron gives the reference zero for δ with respect to this compound. To convert δ with respect to soft iron to δ with respect to nitroprusside one must add 0.257 mm s^{-1}.

It is also possible to make an independent velocity calibration using interferometric methods.[17]

5.5.6 Extraction of parameters from the spectrum

Although the multichannel analyser gives one an immediate visual presentation of the accumulated spectrum in an oscilloscope, which is useful for qualitative purposes, quantitative processing of the data demands a modern computer treatment.[18–20] The data, being digital, are especially suited to such treatment. The analyser should be chosen to provide punched tape, magnetic tape cassette or other output suitable for easy introduction into the computer.

Because of the random nature of the decay process, the standard deviation on the number of events recorded in a channel, N, will be proportional to $\sqrt{N}$. Thus, for good statistics, the collection of data should continue until about 10^6 counts per channel have accumulated. The absorption line shapes are fairly closely Lorentzian. The base line, or level of counts in absence of Mössbauer scattering, may have a slightly sinusoidal shape if a moving source is used because of the change in geometrical efficiency between the extremes of motion of the source; however the fitting programme can allow for this.

Several programmes fitting the recorded spectrum with up to ten or twelve Lorentzian lines are available. That due to A.J. Stone has been widely used in mineralogical work.[19] This is a steered programme with the possibility of various kinds of constraints. A steered programme, fed with suggested values of several of the parameters to be computed, leads to efficient use of the computer. This is important since the calculations involve the inversion of large matrices and plenty of computer capacity is required.

It is frequently necessary to use constraints to obtain significant results. For instance, the linewidth may be known from some other data, so that the computer can be instructed to recognize wide lines as overlapping pairs. The intensities of the lines of a quadrupole split pair should be equal for a randomly orientated sample; using this restraint the computer may extract one member of the pair from another overlapping line.[3] In favourable cases up to ten or twelve lines may be identified in a spectrum and values of δ and Δ extracted. The analysis of the data should lead to a value of χ^2/n not very much greater than 1.0. In this expression χ^2 is Pearsons chi-squared fitting function and n is the number of degrees of freedom for the fitting, that is, the number of channels less the number of fitting parameters. A discussion of some of the problems of fitting mineral spectra can be found in Bancroft.[3]

5.6 MINERALOGICAL APPLICATIONS

A number of reviews of this subject have appeared.[24–35]

5.6.1 Analytical applications

(a) *Relation of the line intensity to the number of iron atoms per unit area of the absorber*[36–50]

Unfortunately, in the most general case, the relationship between the line intensity, or area under the absorption line expressed as a fraction of the number of photons in the absence of absorption, and the number of iron atoms per unit area of the absorber, is rather complex. Unless the absorber is extremely thin, the Lorentzian-shaped emission line from the source and the similarly-shaped distribution of cross-sections for the resonance scattering process, combine to give an absorption line which is no longer exactly Lorentzian. However, for moderate thicknesses of absorber, it can be closely represented by a Lorentzian by changing the half-width. Thus, except for very thin absorbers, the width at half-maximum of the absorption line is markedly greater than twice the value set by the life-time of the $^{57m_1}Fe$.

For this discussion, it is convenient to define the thickness of the absorber, t_a, by the dimensionless expression $t_a = f_a n \sigma_O$, where f_a is the Mössbauer fraction for the iron in the absorber, n is the number of ^{57}Fe atoms per cm^2 of the absorber and σ_O is the cross-section for the scattering process at the resonance maximum. For $t_a \leqslant 4$ it has been shown that $\Gamma_{ex} \simeq (2 + 0.027t_a) \cdot \Gamma_o$, where the subscripts ex and o distinguish the observed and life-time determined widths of the absorption line at half maximum. With suitable experimental data, this equation can be used[41,45] to obtain a value for f_a.

For non-overlapping lines, the absorption area

$$A = f_s \int_{-\infty}^{+\infty} (-\mathrm{e}^{-nf_a\sigma(E)})\,\mathrm{d}E \tag{5.18}$$

where $\sigma(E)$ is the scattering cross-section which is a function of the photon energy. This function is given by the Breit–Wigner expression

$$\sigma(E) = \frac{\sigma_o(\Gamma/2)^2}{(E - E_o)^2 - (\Gamma/2)^2} = \sigma_o K(E) \tag{5.19}$$

where σ_o is the scattering cross-section at the resonance maximum energy, E_o, and Γ the linewidth.

If there are i non-overlapping lines from our iron compound then

$$A_i = f_s \int_{-\infty}^{+\infty} (1 - \mathrm{e}^{-t_a B_i k_i(E)})\,\mathrm{d}E \tag{5.20}$$

The B_i are normalizing factors so that $\sum_i B_i = 1$. If $B_i t_a \ll 1$,

$$A_i \simeq \frac{\pi}{2} f_s t_a B_i \Gamma_i$$

or

$$A_i \simeq \frac{\pi}{2} f_a f_s \sigma_o \Gamma_i B_i n \tag{5.21}$$

Now this approximation is only acceptable up to $t_a \simeq 0.2$. For iron $\sigma_O = 2.56 \times 10^{-18}\,cm^2$. Hence for an absorber with $f_a \simeq 0.6$, $n \leqslant 1.3 \times 10^{17}$ atoms cm^{-2}. Allowing for the natural abundance of ^{57}Fe, 0.02, this implies that the absorber should not contain more than $0.6\,mg\,cm^{-2}$ of natural iron, which will correspond to, perhaps, $6\,mg\,cm^{-2}$ of a mineral.[47,48,50]

The above analysis assumes that the absorption has been corrected for non-Mössbauer photons counted in the absorption line. These include events leading to recoil and other photons accepted by the single channel analyser, e.g. X-rays and scattered radiation. This correction is not easily made.

The distortion of the line intensity arising from saturation effects due to thick absorbers makes it necessary to take special precautions with sample preparation if one component of a mixture has a much higher iron content than the others, for instance meteoritic material containing particles of iron metal. In such cases the sample must be very finely ground to avoid intensity distortion of the spectrum due to the high iron content component. This will occur because the t_a value for particles of this component will be much greater than for the rest of the material and the sample will behave like a transparent matrix containing some black spots in optical spectroscopy.

Such effects of absorber preparation, such as the need to ensure random orientation of crystallites, are usually referred to as texture effects.[21,46,49,51] It is clear from the above discussion that direct use of Mössbauer spectroscopy for the determination of iron is hardly practical. These difficulties of intensity correction for sample thickness have been discussed in detail by Stone.[50]

The situation is more favourable for comparisons of the amounts of two iron-bearing species or of iron in two or more sites in one mineral.

For *thin* samples one will have

$$A_1 = \frac{\pi}{2} f_s \sigma \Gamma_1 f_{a_1} n_1$$

$$A_2 = \frac{\pi}{2} f_s \sigma \Gamma_2 f_{a_2} n_2 \qquad (5.22)$$

so that

$$\frac{A_1}{A_2} = \frac{\Gamma_1 f_{a_1} n_1}{\Gamma_2 f_{a_2} n_2}$$

Now for many silicate minerals Γ is almost constant. Further, for measurements at low temperatures, the f values tend towards a fairly constant value, so that

$$\frac{A_1}{A_2} \simeq C \frac{n_1}{n_2} \qquad (5.23)$$

Even if some thickness corrections are necessary, one will obtain

$$\frac{A_1}{A_2} = \frac{\Gamma_1 f_{a_1} G(n_1, f_{a_1}) n_1}{\Gamma_2 f_{a_2} G(n_2, f_{a_2}) n_2} \qquad (5.24)$$

and the two correction factors G will not differ much if n_1 and n_2 are not widely different and will approach[36,37,40] the value 1 as $t_a \rightarrow O$.

Theoretically the treatment of partly overlapping spectra is much more complex, but experience has shown that for reasonably thin samples $A_1/A_2 \simeq f_{a_1} n / f_{a_2} n_2$. Thus one finds that for relative measurements of amounts of iron in different environments in an absorber Mössbauer spectroscopy certainly gives reproducible results, and when it has been possible to compare the absolute values with other procedures, agreement is rather good. As the theory would suggest, the discrepancies become larger when the ratio of n_1/n_2 assumes very large or very small values. But the overall sensitivity of Mössbauer spectroscopy is limited since, in general, an iron environment present in less than about 2 or 3% will not be distinguishable.

(b) *Fe^{2+}/Fe^{3+} ratios and the detection of Fe^{2+} and Fe^{3+}*

One of the simplest applications of the above principles is to obtain an independent check on the reliability of chemical analysis for the determination of Fe^{2+}/Fe^{3+} with some minerals.[52–66] Ferrous iron is fairly easily oxidized and the opening up of some minerals for chemical analysis is necessarily a rather energetic process; one might anticipate that the chemical value for Fe^{2+}/Fe^{3+} sometimes underestimates the true value. A number of investigators have reported[53,54,63,64] on the correlation between the chemical analysis and Mössbauer spectroscopic values for this ratio. When appropriate precautions are taken with the chemical analysis, the agreement between the two methods is good except, as expected, at rather large or small values.[52–67]

Besides these somewhat straightforward cases there are more interesting applications of this kind. For various minerals the oxidation state of the iron cannot be readily concluded from chemical analysis, for example, in sulphidic materials. In addition, there are minerals where, for instance, both Ti^{3+} and Fe^{3+} may be present and since these react immediately upon solution, the state of the iron in the mineral cannot be ascertained by analysis. Thus, chalcopyrite can be shown by Mössbauer spectroscopy to contain ferric iron, as also does the natural iron antimonate tripuhyite.[55] But neptunite does not appear to contain ferric iron.[52] It may be remarked that Mössbauer spectroscopy is generally more sensitive for detecting Fe^{2+} in the presence of much Fe^{3+} rather than the reverse situation.

(c) *Fingerprinting and petrological analysis*[68–168]

In principle, every mineral and every different iron site in the mineral yields a characteristic Mössbauer spectrum. The parameters δ and Δ characterizing the spectra from Fe^{2+} ions in different sites in the silicate lattice are, fortunately, rather sensitive to the precise geometry of the environment of the iron ion, so that the spectra from the different sites can often be resolved. For Fe^{3+} the situation is not quite so favourable and, although the quadrupole splittings vary appreciably, they are all rather small and the changes in the isomer shift are more limited so that overlapping spectra are rather common. In some cases measurements over a range of temperatures (down to 80 K) will facilitate resolution.

There is another factor which complicates the mineral spectra. The nearest-neighbour cations to the iron may not always be the same for a given type of cationic site. Varying proportions of different divalent cations may be involved in the next-nearest-neighbour environment. Each such environment will lead to a particular quadrupole splitting but since the varying species are rather remote from the iron nucleus, the differences will usually be rather small. In most cases, all that will be observed is a line-broadening, but in favourable cases the spectra yield information about the varying environment.

The first case in which it was shown that a single type of cationic site could,

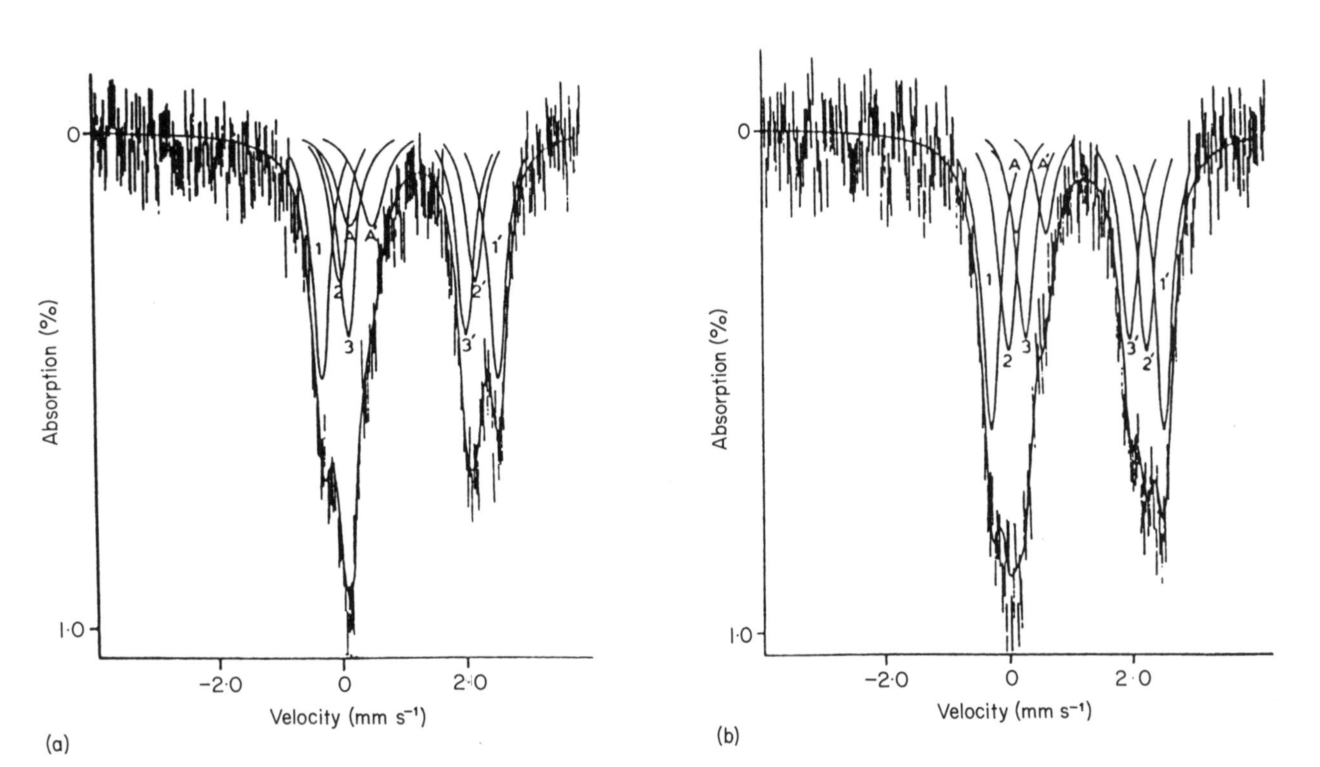

Fig. 5.13 Mössbauer spectra recorded from omphacites. (a) P2/n form; (b) C2/c form. (Reproduced with permission from Bancroft, G.M. (1979) *J. Physique*, Colloq. C2, **40**, C2–464.)

because of different combinations of next-nearest-neighbours, give rise to more than one quadrupole split pair was due to Dowty and Lindsley,[121] but an even more spectacular range of quadrupole splittings differing for this reason has been established for the omphacites.[169] In a certain composition region omphacites can be obtained belonging to the space group P2/n for which there are four cationic sites (M1, M1(1), M2 and M2(1)). On heating to 1000 °C at a pressure of 15–18 kbar, these transform to a C2/c structure for which there are only M1 and M2 cationic sites. Figure 5.13 shows the Mössbauer spectra of such a mineral before (a) and after (b) the structural change. Ignoring the pair AA′ due to Fe^{3+} iron, it is clear that in both cases three Fe^{2+} doublets are needed to fit the spectrum. In principle, this could represent Fe^{2+} in three of the four cationic sites in the P2/n form. However, in the C2/c form there are only two sites. Further, the quadrupole splittings of each pair vary considerably with temperature, which would not be expected if the Fe^{2+} occupied M2 or M2(1) sites, which are more distorted and should lead to a much less temperature-dependent splitting. X-ray diffraction data support this assignment of all the Fe^{2+} to the M1 sites. Hence Fe^{2+} in a simple cationic site can give rise to three quadrupole split pair spectra. These can be associated with M1 sites with $3Ca^{2+}$, $2Ca^{2+}$ and $1Na^{+}$ and $1Ca^{2+}$ and $2Na^{+}$ on the nearest M2 sites. This conclusion can be supported by consideration of the way the intensities vary with the Ca/(Ca + Na) ratio and also by the small changes in intensity on changing from the P2/n to the C2/c structures. The noteworthy feature is that a decrease in Δ of about 1 mm s^{-1} can be produced by the change of next-nearest-neighbours, probably effected, at least in part, by changes in the Fe–O distances in the essentially similar sites (see also refs. 112, 125).

Thus Mössbauer spectroscopy has been used extensively in petrological analysis of rocks. A data base of spectra of the principal rock forming minerals was accumulated in the 1960s and has been progressively enlarged.[68–168] The nature of the materials which have been examined is large, ranging from iron-oxides, hydroxides and carbonates to amphiboles, pyroxenes, tourmalines, olivines and sulphides. The nature of the minerals to which the references refer[68–168] is indicated in brackets after the author's name.

(i) *Lunar rocks and regoliths*

The non-destructive character of Mössbauer spectroscopy was an obvious attraction for the study of these precious materials.[170–200]

Figure 5.14, from a paper[172] by Gay, Bancroft and Bown illustrates the complexity of the spectra of the soil samples from Tranquillity Base. Nonetheless, the presence of ilmenite (lines 1, 1′) pyroxene (2, 2′ and 3, 3′) olivine (4, 4′) and the components iron (6, 6′ and 6″), troilite (7, 7′) and a spinel (8, 8′, 8″), which give magnetically split spectra, could be demonstrated. Estimates of the proportions of each from such overlapping spectra, with materials of uncertain *f* values, must be considered very approximate. The

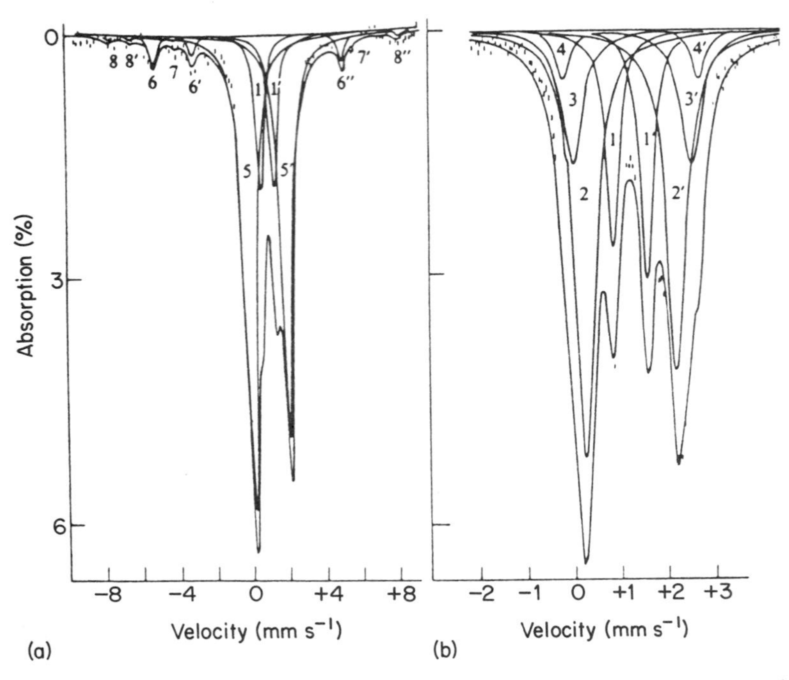

Fig. 5.14 Lunar regolith. (a) High velocity scan: (b) low velocity scan. (Reproduced with permission from Gay, P., Bancroft, G.M. and Brown, M.G. (1970) *Proc. Apollo Lunar Sci. Conf., Geochim. Cosmochim. Acta*, Suppl., **1**, 481.)

problem has been discussed by Herzenberg and Riley[178] and more recently by Gibb, Greatrex and Greenwood.[190] Quite naturally, one of the problems attracting attention has been the incidence of Fe^{3+} in such materials. Although the limit to the amount present that can be set by Mössbauer spectroscopy is rather high, it could be concluded that less than 3% of Fe^{3+} could be present in the majority of the samples.[174,175,177,179] After some kind of proximate analysis, magnetic separation, hand picking under the microscope, etc., the spectra of the different fractions are still more informative. Thus, an examination of the plagioclase fraction from a lunar basalt gave evidence for a small (< 1%) incorporation of Fe^{3+}. This conclusion has been confirmed by electron spin resonance measurements.[181,200]

It has also been shown by conversion electron Mössbauer spectroscopy (Chapter 8) that the composition of the superficial material of the rock is not measurably different from their interior and is indicative of the absence of any atmospheric weathering effects.[192] However, some evidence for an enhancement of the superparamagnetic Fe^0 fraction in the surface layers may indicate some reduction of Fe^{2+} to Fe^0 by the solar wind bombardment.[189]

(ii) *Meteoritic materials*[201–225]

The study of these extraterrestrial materials naturally predates the lunar studies. Again the non-destructive nature of the technique is valuable, but the study of meteorites also demonstrates simple problems that are difficult to solve satisfactorily by chemical analysis.

A long time ago it was noted by Prior that the Ni/Fe ratio in the metallic phase increased with increase in the Fe/Mg ratio in the stony phase of a meteorite. To explore this relation, suggesting transfer of iron from the metallic to the stony phase, it is necessary to have data on the iron content of the two phases.[201,203] Chemically this is difficult without phase separation, but, in principle, it should be possible using Mössbauer spectroscopy to establish the ratio of the iron in the two phases. However there are formidable difficulties in obtaining reliable results. The iron–nickel alloy phase has a large t_a compared to the stony phase and it is very difficult to avoid the texture effects in the absorbers already described.

Other applications have been more successful. The primary meteoritic material, like the lunar samples, has very little Fe^{3+}. As it falls to Earth, considerable superficial oxidation occurs and this is followed by still further atmospheric oxidation on, or in, the ground. Thus measurements of the Fe^{3+}/Fe^{2+} ratio in meteoritic material serve to distinguish falls, or fresh meteoritic material associated with a known fall, from finds, or meteoritic material from some earlier unidentified fall.[208,210] For similar reasons, the same ratio is useful in distinguishing tektites, the glassy meteoritic material from impactites, similar glassy material formed by meteorites on impact with the earth.[214,227] Unfortunately, this is not possible with the lunar materials.

Mössbauer spectroscopy has also proved fruitful in distinguishing the different iron–nickel alloy phases occurring in meteorites, taenite, josephinite etc., and has been responsible for the identification of new phases. The mineral composition of the stony phase can be explored in the same way and rather more precise values obtained[212,213,215,218,220,223] for the different proportions present.

(iii) *Glassy materials*[226–236]

Mössbauer spectroscopy is especially valuable in the study of these naturally occurring solids which are not sufficiently well crystallized for study by X-ray diffraction methods. Glasses are prominent members of the group. They are sufficiently solid to give reasonable Mössbauer fractions but since the iron will be present in a variety of environments the linewidths tend to be larger than for crystalline minerals because of a range of similar quadrupole splittings. However, distinctions between Fe^{2+} and Fe^{3+} and even the different coordination numbers of these ions is usually possible. The occurrence of four coordinate, approximately tetrahedral Fe^{3+} is common in these materials.[226–230]

Several papers have appeared on both synthetic silicates, aluminosilicate, borate and phosphate glasses as well as natural materials, like obsidians and other vitreous minerals.[230–236] Work on European and Mexican obsidian shows some possibility of identifying the source of obsidian from its Mössbauer spectrum.[232–234] The differences often arise from the properties of a magnetically split component in the Mössbauer spectrum due to the inclusion of dispersed particles of magnetite.

(iv) *Clays, clay minerals and micas*

Another group of minerals for which X-ray diffraction is rather unsatisfactory includes the clays and clay minerals. There is an extensive literature in this area.[51,237–298] In fact, Mössbauer spectroscopy is not at its best with these materials because most of them contain some or all of their iron as Fe^{3+} and the changes in the isomer shift and quadrupole splittings with environment are small so that the 'fingerprinting' application is rather hazardous.

A large proportion of the numerous publications have been devoted to identifying and characterizing the sites occupied by the Fe^{2+} and Fe^{3+} in these minerals. In some cases the Fe^{3+} occupies tetrahedral as well as octahedral sites as is the case for nontronite,[261,273] phlogopite[245,249,277,279] and biotites,[277,279] but not with glauconites.[268,285] Occupation of both cis and trans octahedral sites with two OH ligands can be distinguished for both oxidation states of iron.[285] In glauconites, the Fe^{3+} prefers the cis octahedral site and Fe^{2+} the trans but occupation is variable even for glauconites of the same age.[268] The glauconite studies are a good example of the more satisfactory analysis of the data possible with measurements at low temperature; two Fe^{2+} and two Fe^{3+} doublets can be identified.

It is important to recognize that clay minerals are of variable composition and in some cases their spectra change considerably with composition, for example the spectra of illites vary quite a lot with their composition whereas the chlorites give a more or less constant spectrum.[298]

The assignment of spectra to X-ray identified cationic sites is not always straightforward. The elementary principles have been outlined in Section 5.4. But other observations can be useful. For example, for the phyllosilicates the δ and Δ parameters for the cis sites seem to vary only slightly from mineral to mineral. Such observations facilitate assignment but must be used with caution.[286]

A question of industrial significance in relation to these materials concerns the state of the iron in kaolinites. Some admixture of dispersed haematite, or hydroxy compounds, such as goethite, is usually present. This material might be removed by leaching with acid, but cationic iron in the silicate lattice is unlikely to be removed in this way. Mössbauer spectroscopy reveals that some iron indeed may be present as goethite or dispersed haematite but that it is also possible for some iron to be present in octahedral sites as Fe^{3+}. With synthetic ^{57}Fe doped kaolinites the iron has been found as Fe^{2+} in octahedral sites.[255,293]

(v) *Sediments and nodules*[299–332]

A numerous collection of sedimentary materials are too poorly crystallized for X-ray diffraction study, but some useful information has been obtained from Mössbauer spectroscopy. They include iron–manganese deep-sea nodules as well as sea, river and lake sediments.[324–332] Very many of these sediments give spectra showing the presence of Fe^{3+} and Fe^{2+}. The quadrupole split pair, due to Fe^{3+}, has very much the same parameters as are found for iron(III) hydroxide gels,[318] while the Fe^{2+} component could arise from clay minerals, such as chlorite, illite or in some cases biotite.[309,317,319] A further characteristic of these materials is that they often yield magnetically split spectra at, or below, room temperature.[312,314] Over a range of temperatures the ratio of the intensities of the magnetically split to the quadrupole split spectra increases as the temperature decreases.[302] Such behaviour is characteristic of superparamagnetic particles.

Ferro- and antiferromagnetic properties are not molecular properties but depend on cooperative effects in an assembly of ions or molecules. Whether the alignment of the nuclear fields in a particular orientation persists for a time long compared to the Larmor frequency of the $^{57m_1}Fe$ nucleus, a condition that must be fulfilled to obtain a magnetically split spectrum, depends on the volume of the crystallites and on the temperature. The relaxation time for the nuclear field, $\tau = ae^{Cv/kT}$, where T is the temperature, v the volume of the crystallites and a and C are constants is determined by the composition of the crystallites. If τ is much larger than τ_o, the Larmor precession time for $^{57m_1}Fe$ ($\tau_o = 2.5 \times 10^{-8}$ s), a magnetically split spectrum is found, if it is much less then the quadrupole split pair appears. In the intervening range $\tau \approx \tau_o$, the spectra will be complex[248] but since the sizes of the crystallites are usually broadly distributed the proportion for which $\tau \sim \tau_o$ is usually small.

The product Cv can be regarded as an energy of activation for reorientating the nuclear field, so that if we suppose that there is a sharp transition from quadrupole split to magnetically split at $\tau = \tau_o$, the ratio of the intensities of the magnetic to the quadrupole split spectra, $R(T) = f(E)$ the fraction of crystallites with energy less than E, where $E = kT \ln \tau_0/a$. Now $f(E) = \int_0^E P(E)\,dE$ with $\int_0^\infty P(E)dE = 1$, $P(E)$ being the probability density for crystallites with energy E. But we can put

$$\int_0^E P(E)\,dE = \int_0^T g(T)\,dT \qquad (5.25)$$

where

$$g(T) = \ln \frac{\tau_0}{a} P\left(kT \ln \frac{\tau_0}{a} \right) \qquad (5.26)$$

Thus $g(T) = dR/dT$ and if C is known the distribution and mean value of v can be deduced.

A good description of the application of this method to some clays has been

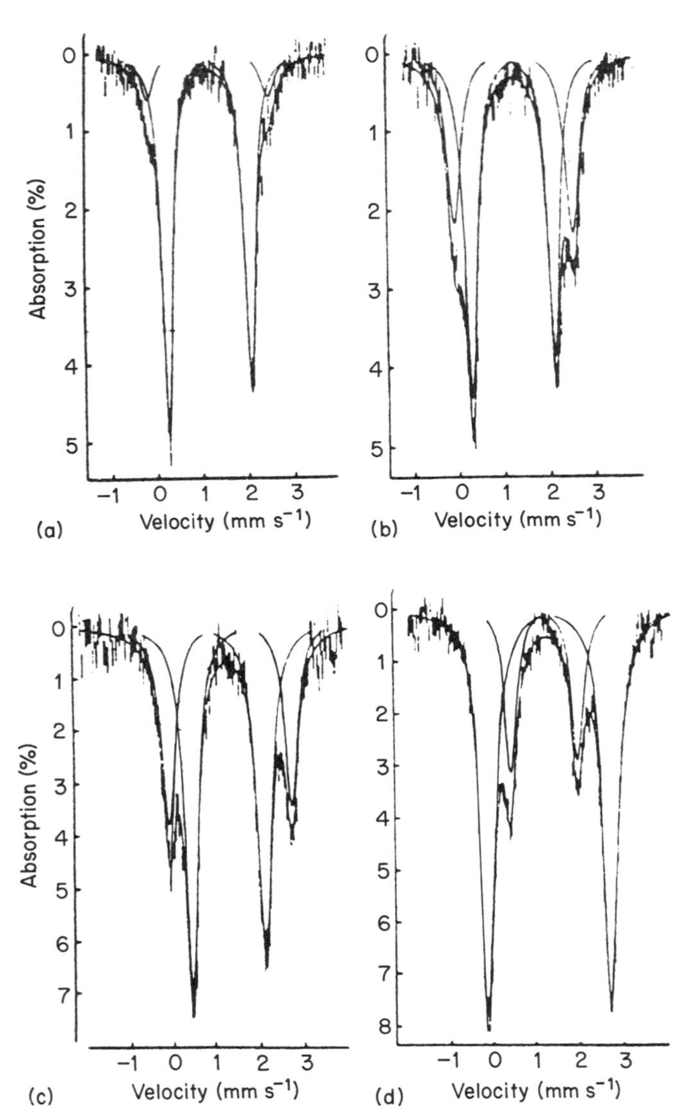

Fig. 5.15 Mössbauer spectra of series of anthophyllites of increasing iron content. (a) Anthophyllite, 23% Fe(II); (b) anthophyllite, 31.6% Fe(II); (c) cummingtonite, 35.4% Fe(II); (d) grunerite, 95.3% Fe(II). (% cationic sites occupied by Fe(II).) (Reproduced with permission from Bancroft, G.M., Maddock, A.G., Burns, R.G. and Strens, R. (1966) *Nature, Lond.*, **212**, 914.)

reported in the literature.[248] Goethite is an example of a mineral where the Mössbauer spectrum can vary from a simple quadrupole split pair through to a clean magnetically split spectrum according to the size of the crystallites.[102]

5.6.2 Cation ordering[333–375]

Perhaps one of the most fruitful of the applications of Mössbauer spectroscopy to mineralogy is its use in exploring how far the iron ions in a silicate lattice prefer one cationic site rather than another when more than one such site is available. This problem has indeed been studied using X-ray diffraction methods but this technique is rather tedious and not very accurate in this application. It is of course necessary that iron ions situated at the different sites yield resolvably different Mössbauer spectra. The alternative occupants of the sites are normally Fe^{2+} and Mg^{2+} ions. Other ions, such as Ca^{2+}, if present are not labile and are restricted to one site. It is usually assumed that the f factors are sufficiently nearly the same that one can write $Fe(1)/Fe(2) = I(1)/I(2)$, where Fe(1) and Fe(2) are the proportions of iron in the two sites and $I(1)$ and $I(2)$ the areas under the Mössbauer absorptions corresponding to the respective sites. In fact, by making measurements at a number of different thicknesses of sample this assumption could easily be verified or appropriate corrections made.

An early illustration of such ordering is shown in Fig. 5.15, where it is seen that in a series of anthophyllites of increasing iron content the iron ions prefer the more distorted cationic site that yields the inner doublet before they make much use of the other sites.[333,334] In these materials there are four octahedrally coordinated cationic sites M1, M2, M3 and M4, present in the proportions 2:2:1:2. The inner doublet arises from the occupation of the M4 sites; the doublets due to M1, M2 and M3 are not resolved under these conditions. Such overlap must, in fact, limit the accuracy of the relation assumed between iron content and line intensity referred to above. The ordered occupation continues into the chemically related, but crystallographically different, cummingtonites and grunerites. These differences in occupation reflect the equilibrium, or departure from equilibrium, for the reaction Fe^{2+} (site 1) + Mg^{2+}(site 2) $\rightleftharpoons$ Fe^{2+} (site 2) + Mg^{2+}(site 1). For this reaction $K = x_2(1 - x_1)/x_1(1 - x_2)$ where x_1 and x_2 are the mole fractions of iron in sites 1 and 2 ($x_1 = Fe^{2+}$ (site $1/Fe^{2+}$ (site 1) + Mg^{2+} (site 1)) etc. Thus if there are equal numbers of the two kinds of cationic site and $Fe^{2+}(\text{total})/(Fe^{2+}(\text{total}) + Mg^{2+}(\text{total})) = x$, then $K = (1 + y - 2xy)/(y^2 + y - 2xy)$ where $y =$ Fe(site 1)/Fe(site 2), which is directly available from the experimental data, and x is available from chemical analysis. For a random distribution of the iron $y = 1$. Equilibrium distribution isotherms[353] for orthopyroxenes are shown in Fig. 5.16.

One might expect that exchanging an Fe^{2+} ion on a site(1) with a magnesium on a site(2) would lead to a moderate enthalpy change and a small

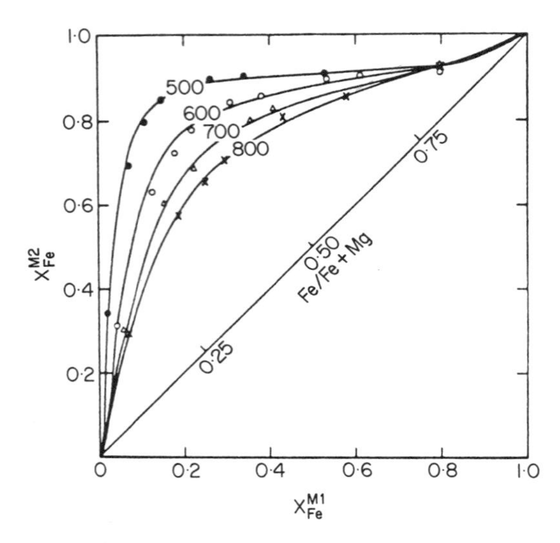

Fig. 5.16 Equilibrium distribution isotherms for orthopyroxenes. (Reproduced with permission from Virgo, D. and Hafner, S.S. (1970) *Am. Mineral.*, **55**, 201.)

entropy change, both quantities being nearly temperature independent. If we identify the sites so that the iron prefers site(1) and $K < 1$, then ΔH° for the above reaction can be expected to be positive and K increases as the temperature increases and more disordered systems ensue.[341]

The above treatment is oversimplified in so far as ideal behaviour is assumed; no activity coefficients are introduced. Indeed the experimental data suggest that this approximation is not too serious. A rather more sophisticated treatment, which seems very appropriate to these systems is to consider the equilibrium as a mixing process and to treat the activity coefficients on the basis of regular solutions. This implies that the entropy term is given by a purely statistical calculation and a modest enthalpy term arises. In such systems the activity coefficients will be given by the expressions

$$\ln \gamma_{\mathrm{Fe}}^{\mathrm{M1}} = \frac{W^{\mathrm{M1}}}{RT}(1 - x_1)^2, \quad \ln \gamma_{\mathrm{Mg}}^{\mathrm{M1}} = \frac{W^{\mathrm{M1}}}{RT} x_1^2 \tag{5.29}$$

and correspondingly for site 2 (it will be noted that these coefficients are rather different from the usual ionic mean activity coefficients).[341,352,362]

A non-equilibrium distribution of the iron between the two sites in a mineral, measured at room temperature, is therefore an indication that at some stage the mineral has been equilibrated at some higher temperature and that the precise distribution found will depend on this higher temperature and the subsequent cooling history of the mineral. Thus, dynamic as well as equilibrium factors are involved. Fortunately, it is possible to explore the kinetics of the disordering and ordering processes, over some temperature range, using the same sample as was used for the initial measurement of the distribution. Hence the energies of activation of these processes can be established.

Measurements on pyroxenes and clinopyroxenes show that at temperatures above 1000 K the equilibrium is established rather rapidly compared with the usual rates of cooling of the mineral. One can therefore only expect to find disorder corresponding to a temperature above 1000 K for samples that have been rapidly cooled or quenched. This may have happened with some lunar samples because of the impact of meteorites.[187,192]

At sufficiently low temperatures, the re-equilibration takes times of geological significance and if the kinetic data is obtained it is possible to deduce something of the cooling history of the mineral from the initial iron distribution.[353,359,367,372,375]

5.6.3 Kinetics and mechanism of mineral solid state reactions[376–385]

Mössbauer spectroscopy is one of a very small number of techniques permitting quantitative in situ studies of reactions in solids.

(a) *Reactions in various minerals*

The possibility of distinguishing changes in oxidation state of the iron atoms at specific cationic sites has encouraged numerous studies of the aerial oxidation of minerals.

In the case of vivianite,[376,377] an almost colourless iron(II) phosphate, the mineral rapidly turns blue on exposure to air. The mineral has two kinds of cationic site. There are single octahedra with 4 O^{2-} and 2 trans H_2O ligands and there are paired sites composed of two octahedra sharing a common edge and two O^{2-}, each octahedron having, in addition, a trans orientated pair of O^{2-} and water molecules at the two remaining positions. The number of pairs of sites is equal to the number of single sites.

On grinding the mineral in air, the amount of Fe^{2+} on the isolated octahedral sites decreases and a new spectrum due to Fe^{3+} on these sites appears. The Fe^{2+} on the paired sites shows much less change. The reaction appears to involve deprotonation of a water ligand at the single sites, leading to OH^- formation and Fe^{3+}. The sharp line spectra obtained suggest that the OH^- remains in the Fe^{3+} coordination octahedron. On prolonged treatment, one of the Fe^{2+} on the paired sites can be oxidized, but not apparently two. It is very likely that these Fe^{2+}–Fe^{3+} pairs are responsible for the unusual colour and oxidized vivianite is probably one of the minerals of the class discussed in Section 5.6.4.

In a similar way it has been shown that in the aerial oxidation of amosite, crocidolite and cummingtonite, the M1 and M3 sites first suffer oxidation and the M4 site is the least easily oxidized. This suggests that a deprotonation mechanism is again involved.[378–381]

(b) *Micas and clay minerals*

(i) *Reactions*

Numerous studies of weathering, thermal decomposition and oxidation–reduction reactions have been made and excellently reviewed.[298] The weathering of biotites to yield vermiculite or even kaolinite has been studied in some detail on natural sites. A question of interest is how far the growth of the proportion of Fe^{3+} arises from leaching out of Fe^{2+} and how far from the deprotonation in situ oxidation process referred to above. Both processes appear to be important. These transformations often yield other phases and the various forms of FeOOH and haematite in crystallites of various sizes may form. To distinguish these components may require a lengthy study measuring Mössbauer spectra at different temperatures down to 4 K.

(ii) *Deferrugination techniques*

The presence of admixed FeOOH, haematite and related substances is a common complication in the study of clay minerals and considerable attention has been given to the possibility of a simple chemical separation of the silicate phases – deferrugination.[298] Most of these treatments depend on the use of aklali carbonate solutions of reducing agents, oxalate, dithionite, etc., or a controlled treatment with hydrochloric acid. Mössbauer spectroscopy has provided a means of monitoring these processes. In fact, it has been shown that some of the treatments with reducing agents can alter the oxidation state of the iron in the silicate phase. There is even some specificity of different reducing agents as to whether they will reduce Fe^{3+} at tetrahedral sites or only at octahedral sites. There is evidence again for a mechanism involving protonation of OH^- groups. It has also been shown that even very mild treatments, such as washing with water, can decrease the amount of Fe^{2+} in montmorillonite. Studies of this kind should establish how reliable these treatments are.

(iii) *Firing of pottery*

Numerous papers have appeared on the firing of clays and the reactions involved in the manufacture of pottery. Heating biotite to 600 °C in an inert atmosphere leads to oxidation of the Fe^{2+} by deprotonation of OH^-, but with vermiculite or montmorillonite, partial reduction of Fe^{3+} occurs at 300 °C. The fully oxidized nontronite heated in hydrogen only suffers reduction after dehydroxylation. Clearly, the thermal decompositions of these materials are complex and much more detailed studies are needed.

Possible archeometric applications have attracted much attention. The Mössbauer spectra of pottery artefacts do not appear very useful as far as provenance is concerned but they may tell one something about the firing techinques that were used. The Fe^{2+}/Fe^{3+} ratios provide some indication of the redox condition prevailing during firing and it is possible that the Mössbauer parameters of the pottery are indicators of the temperature used in firing.[298]

(c) *Mining and metallurgical applications*

Mössbauer spectroscopy has found some applications in the area of industrial mineralogy.

Exploratory studies have been reported for its use in controlling the feed material for iron and steel furnaces[386,387] and rather more interesting results have been obtained in studies of the mineral content, especially pyrites, of coal.[388–392]

(d) *Dating*

Several possibilities of using Mössbauer spectroscopy for the dating of mineral materials have been suggested. As yet none of these can be said to have been fully established, although in some cases at least the principles of the methods seem sound.

A suggestion that the quadrupole splitting in biotites, glauconites and phlogopites increases with the age of the mineral is difficult to understand if the doublet incurred arises from a single iron site.[393,394]

Two proposals refer to minerals in which the iron occurs in octahedra containing two hydroxyl ligands. It has been noted above that such minerals often undergo oxidation of the iron on heating, as do, for example, biotites and phlogopites. It has been shown that the same oxidation can be brought about by irradiation with ionizing radiation. Thus if the local radiation dose rate is known the amount of Fe^{3+} produced will be a measure of the age of the mineral. It is necessary to know the initial Fe^{3+} content, if any, of the mineral. The method also requires that the mineral has not been subject to any thermal treatment producing a similar effect.[395–397]

A related proposal concerns the ratio of cis to trans Fe^{2+} sites. A transformation of one to the other may occur by proton migration and such changes may occur in biotites, phlogopites as well as in tourmalines.[398] A more extensive study of this process is needed.

Another possibility refers to the age of glauconites. In these minerals the quadrupole splittings and linewidths may change with time due to a slow ordering process. It is not clear at present how far such a change is influenced by heat, local radiation dose and other factors.[298]

A related method suitable for some sediments and materials containing superparamagnetic iron oxide or hydroxide, depends on the slow growth of the crystallites in these systems and the consequent change in the proportions of the magnetically split and quadrupole split spectra.

For pottery artefacts, however, there is evidence of the reverse change with age; the superparamagnetic fraction increases with the time elapsed since firing of the clays.[298,399,400]

5.6.4 Electron delocalization in minerals

One of the most interesting and active areas of research in mineral chemistry during recent years has been the investigation of the interaction between iron atoms in different oxidation states in minerals. It was recognized in the early part of this century that minerals containing both iron(II) and iron(III) frequently showed unusual optical and other physical properties. Their absorption spectra often displayed new features not found for iron(II) or iron(III) separately and they were often very dichroic. There was speculation already at that time that the origin of these effects might be similar to that leading to the behaviour of Prussian blue and compounds such as $Cs_4Sb^{III}Cl_6Sb^{V}Cl_6$.

(a) *Mixed valence minerals*[401–435]

Mixed valence compounds have attracted very much attention from inorganic chemists in recent years and the proceedings of a recent NATO Advanced Study Institute Conference provides much of the necessary chemical background to the study of these systems.[401,402]

Several years ago, Robin and Day showed that if we consider two iron ions, Fe^{2+} and Fe^{3+} in adjacent cationic sites, A and B, that are not very different stereochemically, the wavefunction for the pair can be represented by

$$\psi_g = \sqrt{(1-\alpha^2)}\psi_A^{2+}\psi_B^{3+} + \alpha\psi_A^{3+}\psi_B^{2+} \tag{5.28}$$

Associated with this will be an excited state

$$\psi_{ex} = \sqrt{(1-\beta^2)}\psi_A^{3+}\psi_B^{2+} + \beta\psi_A^{2+}\psi_B^{3+} \tag{5.29}$$

The value of α depends on E_t, given by

$$E_t = \langle\psi_A^{3+}\psi_B^{2+}|\hat{H}|\psi_A^{3+}\psi_B^{2+}\rangle - \langle\psi_A^{2+}\psi_B^{3+}|\hat{H}|\psi_A^{2+}\psi_B^{3+}\rangle \tag{5.30}$$

and V, which is defined by $V = \langle\psi_A^{2+}\psi_B^{3+}|\hat{H}|\psi_A^{3+}\psi_B^{2+}\rangle$

$$\alpha = \left\{1 - \left[1 + \frac{2E_t^2 - 2E_t(E_t^2+4V^2) + 4V^2}{4V^2}\right]^{-1}\right\} \tag{5.31}$$

It can be seen that α ranges from 0 for very large E_t to $1/\sqrt{2}$ for $E_t = 0$. This mixing coefficient is a measure of the degree of delocalization of the odd electron over the two iron atoms.

A useful classification of such systems can be based on the values of α. If the stereochemistries of the sites A and B are very different, E_t, which measures the energy of transferring the electron, will be large and $\alpha \simeq 0$. For such class I mixed valence compounds no new features arise, the Fe^{2+} and Fe^{3+} remain localized on the A and B sites and preserve their individual spectroscopic characteristics.

When the two sites become rather similar, having perhaps the same point group symmetry, E_t falls to a much lower value, perhaps a few eV, so that α may assume a value of about 0.005. The Fe^{2+} and Fe^{3+} retain their identities and their normal spectroscopic characteristics, but in addition a new strong absorption corresponding to $\psi_{ex} \leftarrow \psi_g$ will appear. Since such interactions may be confined to particular directions in the crystals, dichroic properties may also arise. The Mössbauer spectra of the Fe^{2+} and Fe^{3+} in such a compound will be normal because the iron atoms persist in a given oxidation state for a time that is long on the Mössbauer time scale, which is set by the half-life of $^{57m_1}Fe$. Undoubtedly several mineral species belong in this category.

If, however, the two sites are very similar indeed E_t may become very small and α will become very appreciable, 0.1–0.7, and the life-time of a given state of iron on either site will decrease until at $\alpha = 1/\sqrt{2}$ the iron atoms are indistinguishable. This class of compounds, III, can usefully be divided into two sections, (a) and (b). In class III(a) the electron delocalization only affects discrete groups of iron atoms in the compound. For example the similar atoms in a cluster or a polynuclear complex. In class III(b) the delocalization extends over the whole crystallite perhaps in one, two or three dimensions. This will be associated with electrical conductivity of the compound.

Since some minerals can have a range of compositions and iron contents, the delocalization may correspond to class IIIa, affecting pairs, or small groups of iron atoms. Alternatively there may be chains or sheets that can delocalize throughout the crystallite. Hence a given mineral may behave as class IIIa or class IIIb according to its precise composition. The delocalization will also be affected by defects and impurity or foreign cations.

So far in this account there would appear to be no reason for the delocalization to be temperature dependent. This is because a static assembly of atoms has been assumed and vibrational effects ignored. Some idea of the vibronic effects can be obtained from Fig. 5.17 where a one-dimensional simplification is portrayed. In this figure the x coordinate relates to the Fe–L distance in the two sites. All other ligands and distances are assumed to be the same. Thus the vibrational excitation, E_v, is needed to equalize the two Fe–L distances. Now since transitions from one state to the other are possible, through the intermediary of the ligand, they cannot be orthogonal so that the two curves will not cross but will follow the dashed or dotted lines. The minimum separation of the curves will be $2V$ and the dotted curve will arise where $E_v \gg V$ and the dashed curve in the opposite situation. The latter

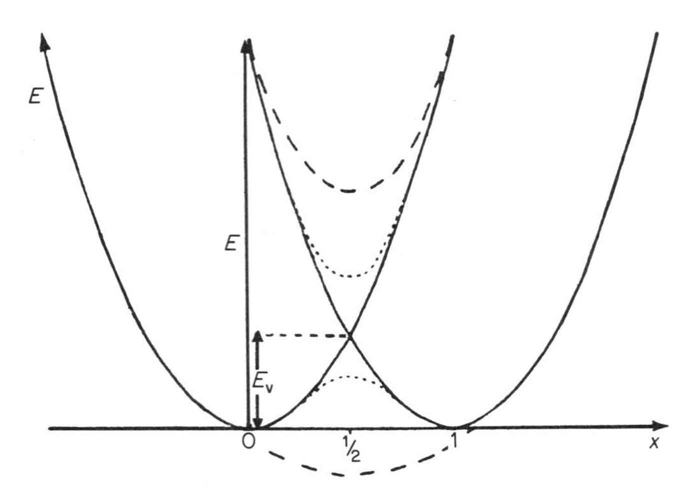

Fig. 5.17 A one-dimensional simplification of the vibronic effects on electron delocalization. Dashed curve, α large; dotted curve, α small. (Reproduced with permission from Mayoh, B. and Day, P. (1972) *J. Am. chem. Soc.*, **94**, 2885.)

corresponds to complete delocalization and the former to the appearance of localized Fe^{2+} and Fe^{3+} at a low enough temperature.

In this way E_v and V and therefore α become temperature dependent. One can foresee three situations.

(i) At low temperatures α may be sufficiently small that Mössbauer spectroscopy will distinguish the Fe^{2+} on A sites and Fe^{3+} on B sites, their life-times being long compared with that of $^{57m_1}Fe$.
(ii) At high temperatures α may be large enough that the life-times of Fe^{2+} on A and Fe^{3+} on B will be short compared with that of $^{57m_1}Fe$. The Mössbauer spectrum will show abnormal parameters but sharp lines. The chemical shift will correspond to the mean of those for Fe^{2+} and Fe^{3+}, weighted for the proportions of Fe^{2+} and Fe^{3+} delocalizing, if this is not 1:1. The quadrupole splitting will be a similar weighted mean, but care will need to be taken to allow for the signs of the two efg values.
(iii) Situations where the two life-times are comparable. Now since α is temperature dependent it may, in favourable cases, be possible to move from $(\alpha) \rightarrow (\gamma) \rightarrow (\beta)$ as the temperature is raised. The changes in the Mössbauer spectrum will resemble the changes in an nmr spectrum when increased temperature leads to an acceleration of the rate of exchange of two non-equivalent atoms in the molecule. The lines will at first grow broader, pass through collapsed, ill-defined spectra, and finally sharpen up as the averaged spectrum.

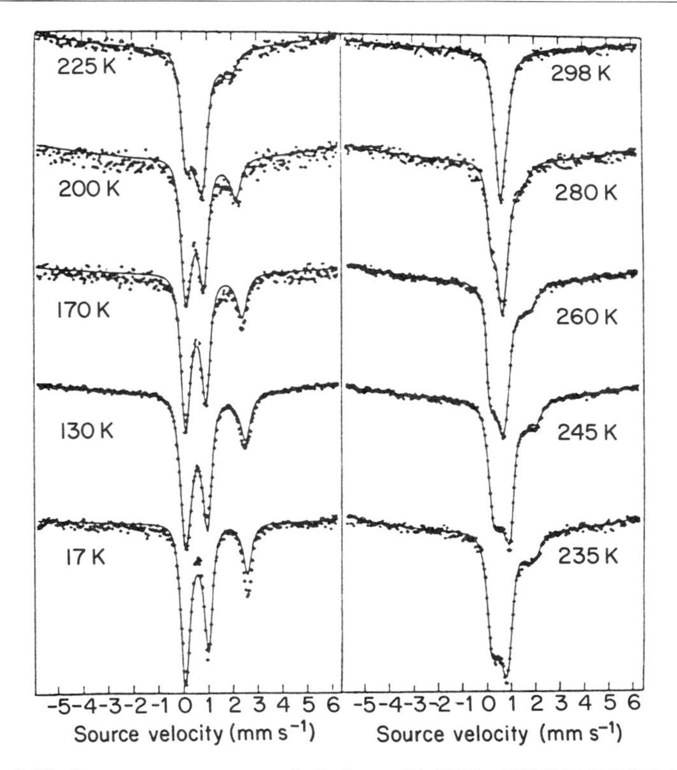

Fig. 5.18 Mössbauer spectra recorded from $Fe(II)Fe_2(III)O(OOCCH_3)_6(H_2O)_3$. (Reproduced with permission from Dziobkowski, C.T., Wrobleski, J.T., and Brown, D.B. (1980) *Inorg. Chem.*, **20**, 679.

At present, the mineral data are so complex that no such sequence of behaviour has been established, although the situations (i) and (ii) are undoubtedly identified in a number of cases. However, Fig. 5.18 shows the transition fairly clearly for[435] the case of $FeII(FeIII)_2\ O(OOC\cdot CH_3)_6 3H_2O$.

(b) *Electron delocalization in minerals*

A large number of minerals belong to classes II and III of the above classification. The most interesting cases are those in category III, where, at a temperature still feasible for Mössbauer spectroscopy, the delocalization leads to abnormal Mössbauer parameters corresponding neither to Fe^{2+} nor to Fe^{3+}. Unfortunately, most of the minerals concerned give very complex spectra and rather sophisticated analyses of the data are needed. In several cases, notably ilvaite, there is no general agreement as yet as to the best fitting of the data.

Two characteristics serve to identify these extensively delocalized systems:

(1) the isomer shifts are quite unusual, generally in the 0.7–0.85 mm s^{-1} region, too high for Fe^{3+} but too low for Fe^{2+}. In the past such a shift was attributed to a tetrahedral Fe^{2+}. (2) Because α changes comparatively rapidly with temperature, the Mössbauer spectrum is unusually temperature dependent.

Amongst the minerals that seem to give these abnormal spectra one finds ilvaite, where this unusual behaviour was first noted by Gerard and Grandjean.[403,407,432] Another is magnetite where the situation is further complicated by the Verwey phase transition at 119 K.[410] Other include tourmaline, deerite, howeite, vesuvianite and cordierite.[415,419,423,425,426,431,433,434] The minerals where the delocalization is too slight to exclude the appearance of individual Fe^{2+} and Fe^{3+} spectra include vivianite (Section 5.6.3a) glaucophane, biotite, crocidolite, riebeckite and augite.[406,411,415,416,419,421]

Some spectra for ilvaite, taken over a range of temperatures, are shown in Fig. 5.19. Since there is still some disagreement over the deconvolution of

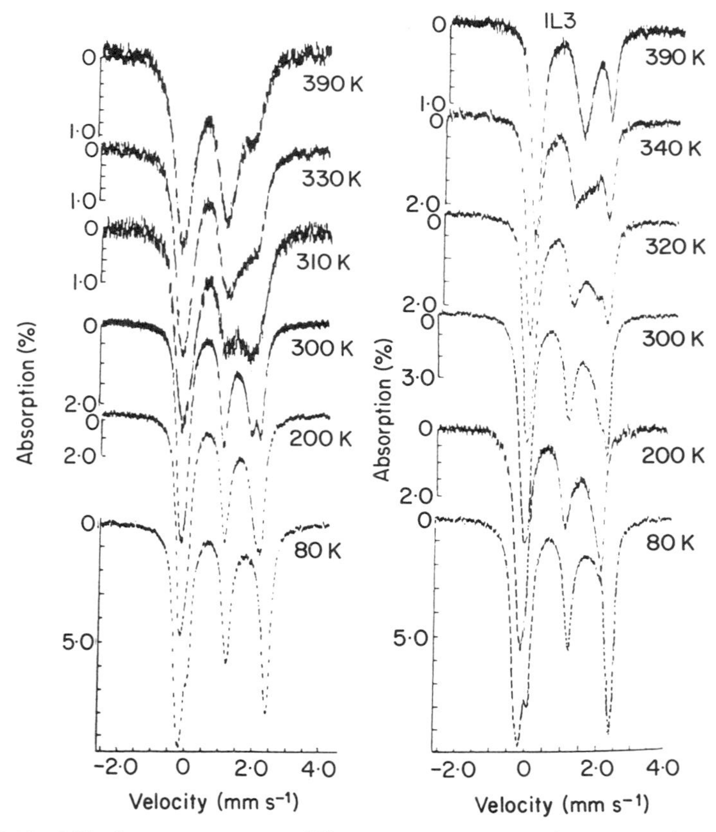

Fig. 5.19 Mössbauer spectra at different temperatures of two samples of ilvaite. (Reproduced with permission from Nolet, D.A. and Burns, R.G. (1979) *Phys. Chem. Min.*, **4**, 221.)

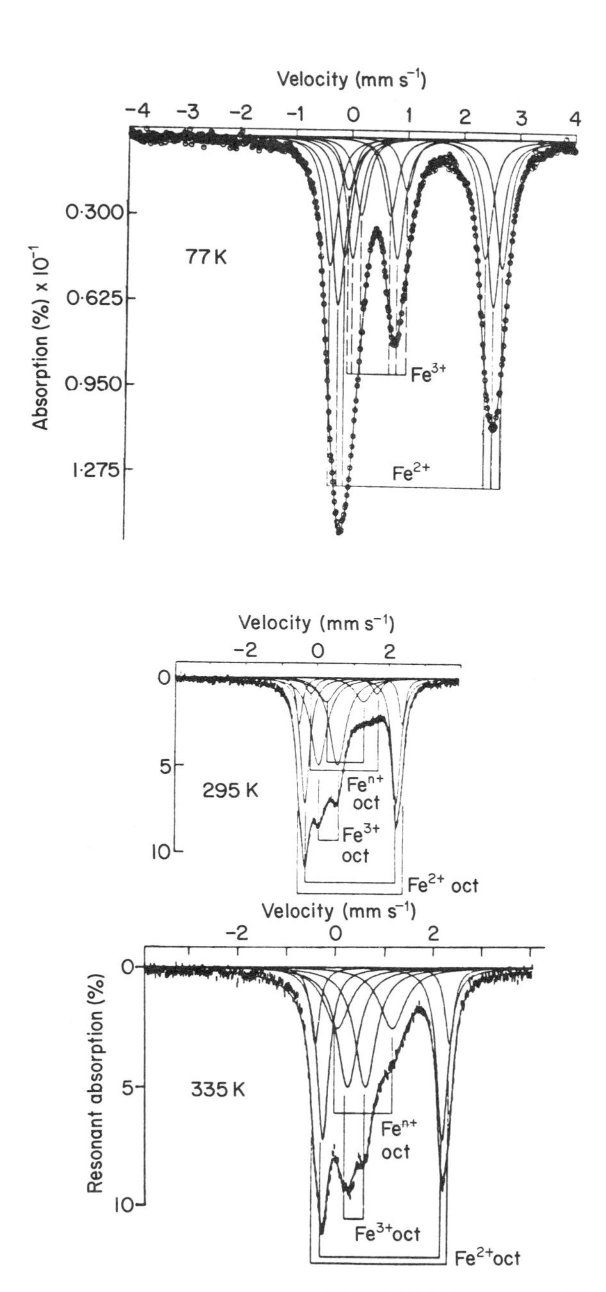

Fig. 5.20 Mössbauer spectra of deerite recorded at 77 K, 295 K and 335 K. (Reproduced with permission from Amthauer, G., Langer, K. and Schleistedt, M. (1980) *Phys. Chem. Min.*, **6**, 19.)

the spectra into constituent doublets, attention is drawn only to the rapid change of spectrum with temperature in the region between +1 and $+2\,mm\,s^{-1}$.[420,422,428]

One of the comparatively few examples where part of the anomalous doublet is clearly resolved can be seen in Fig. 5.20 where the spectrum of deerite is shown for various temperatures. The doublets identified as Fe^{2+}oct are the spectra due to the delocalized Fe^{2+}–Fe^{3+} on octahedral sites.[426] It will be noted that the lines are rather broad at the highest temperature used, but they disappear completely at the lowest temperature. The mineral is crystallographically complex and has nine cationic sites. This study has shown that the delocalization is affected by substitution of iron by other cations and probably by other kinds of defects.

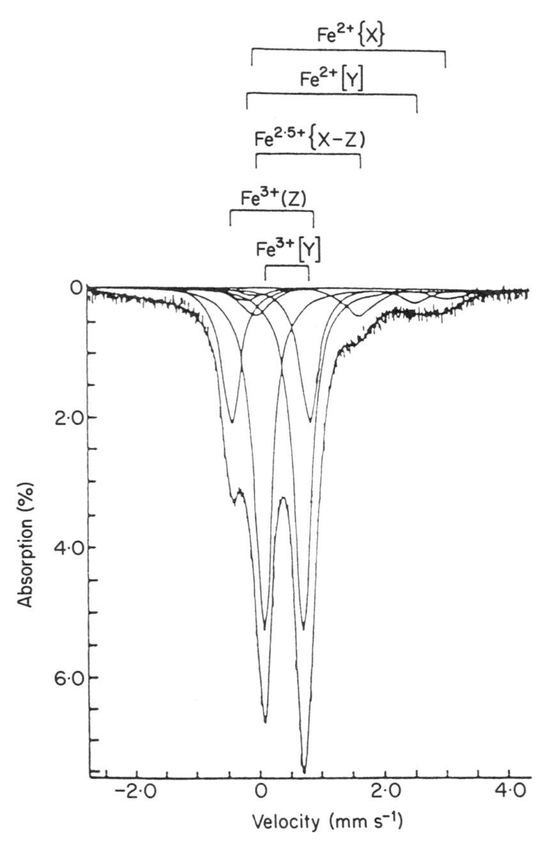

Fig. 5.21 Mössbauer spectrum of schorlomite fitted to five doublets assigned to octahedral Fe^{3+}[Fe^{3+}(Y)], tetrahedral Fe^{3+}[Fe^{3+}(Z)], 8-fold Fe^{2+}[Fe^{2+}{X}], octahedral Fe^{2+}[Y] and a doublet due to charge transfer between Fe^{2+}{X} and Fe^{3+}[Z] represented $Fe^{2.5+}${X–Z). (Reproduced with permission from Schwartz, K.B., Nolet, D.A. and Burns, R.G. (1980) *Am. Mineral.*, **65**, 142.)

Another example which seems clearly established is rather more surprising. It is the case of the titanium garnet schorlomite.[430] The doublet identified as $Fe^{2.5}$ (X – Z) in Fig. 5.21 has the anomalous chemical shift and has been attributed to delocalization between Fe^{2+} on an eight coordinate dodecahedral site sharing an edge with Fe^{3+} on a tetrahedral site. In the previous examples, the Fe^{2+} and Fe^{3+} were located on octahedra sharing an edge.

Minerals indeed provide a particularly useful group of solids for the systematic investigation of these delocalization processes.

5.7 ANTIMONY

The only other element for which there is an appreciable amount of information is antimony.[93]

5.8 OTHER PHYSICAL STUDIES

Several interesting areas of research on minerals using Mössbauer spectroscopy have been excluded from this review. They include single crystal work permitting the axes of the efg to be related to the crystallographic axes and establishing the sign of the efg, and magnetic studies at low temperatures, revealing magnetic ordering of the iron ions and other properties of the mineral.

REFERENCES

1. Clark, M.G. (1975) in *International Review of Science, Physical Chemistry Series 2* Vol. 2 (ed. A.D. Buckingham), Butterworths, London, p. 239.
2. Greenwood, N.N. and Gibb, T.C. (1971) *Mössbauer Spectroscopy*, Chapman and Hall, London.
3. Bancroft, G.M. (1973) *Mössbauer Spectroscopy*, McGraw-Hill, New York.
4. Bhide, V.G. (1973) *Mössbauer Effect*, Tata McGraw-Hill.
5. Gonser, U. (ed.) (1975) *Mössbauer Spectroscopy*, Springer-Verlag Berlin.
6. Gütlich, P., Link, R. and Trautwein, A. (1978) *Mössbauer Spectroscopy and Transition Metal Chemistry*, Springer-Verlag, Berlin.
7. Collins, R.L. and Travis, J.C. (1967) *Möss. Effect Meth.*, **3**, 123 (annual pub. Plenum Press ed., I.J. Gruverman).
8. Blume, M. (1965) *Phys. Rev, Lett.*, **14**, 96.
9. Clark, M.G., Maddock, A.G. and Platt, R.H. (1972) *J. chem. Soc. Dalton*, 281.
10. Kalvius, G.M. and Kankeleit, E. (1972) IAEA Panel on Mössbauer Spectroscopy Applications, Vienna.
11. Wurtinger, W. (1976) *J. Physique*, Colloq. C6, **37**, C6–697.

12. Davis, G.J., Maddock, A.G. and Williams, A.F. (1975) *Chem. Commun.*, 264.
13. Kankeleit, E. (1964) *Rev. Sci. Instrum.*, **35**, 194.
14. Bancroft, G.M., Maddock, A.G. and Ward, J. (1965) *Chemy Ind.*, 423.
15. Cohen, R.L. and Wertheim, G.K. (1974) *Methods Exp. Phys.*, **11**, 307.
16. Forster, A., Halder, N., Kalvius, G.M. *et al.* (1976) *J. Physique*, Colloq. C6, **37**, C6–725.
17. Cranshaw, T.E. (1973) *J. Phys. E.*, **6**, 1.
18. Reubenbauer, K. and Birchall, T. (1979) *Hyperfine Interact.*, **7**, 125.
19. Stone, A.J. (1967) *J. chem. Sec.*, **1967A**, 1966.
20. Muir, A.H. (1968) *Möss Effect Meth.*, **4**, 75 (annual pub. Plenum Press, ed., I.J. Gruverman).
21. Greaves, C. and Burns, R.G. (1971) *Am. Mineral.*, **56**, 2010.
22. Holbourn, P.E., Player, M.A. and Woodhams, F.W.D. (1979) *Nucl. Instrum. Meth.*, **165**, 119.
23. Goodman, R.H. (1967) *Möss. Effect Meth.*, **3**, 163 (annual Pub. Plenum Press, ed. I.J. Gruverman).
24. Sprenkel-Segel, E.L. and Hanna, S.S. (1966) *Möss Effect Method*, **2**, 89, (1966) (annual pub. Plenum Press, Ed., I.J. Gruverman).
25. Flinn, P.A. (1967) *Experimental Methods of Material Research*, (ed. H. Herman), Interscience, New York.
26. Bancroft, G.M. and Burns, R.G. (1968) Proc. *5th. Gen. Meeting Int. Mineral Ass., 1966*, Mineralogical Society, London, p. 36.
27. Hafner, S.S. (1968) *Application of Nuclear and Electronic Resonance Spectroscopy in Mineralogy*, American Geological Institute.
28. Herzenberg, C.L. (1969) *Möss Effect Method.*, **5**, 209 (annual pub. Plenum Press, ed. I.J. Gruverman).
29. Johnson, C.E. (1969) *App. Mod. Phys. Earth Planet. Interiors Conf*. (ed. S.K. Runcorn), *1967*, Wiley-Interscience, New York, p. 485.
30. Maddock, A.G. (1972) IAEA Panel on Mössbauer Spectroscopy Applications, Vienna.
31. Raclavsky, K. and Egiazerov, B.G. (1975) *Proc. 5th Int. Conf. Möss Spec. 1973*, Nuclear Information Centre, Prague.
32. Hafner, S.S. (1975) *Top. app. Phys.*, **5**, 167.
33. Bancroft, G.M. (1979) *J. Physique*, Colloq. C2, **40**, C2–464.
34. Govaert, A. and de Grave, E. (1979) *Phys. Mag.*, **1**, 15.
35. Jonas, K., Solymar, K. and Zoldi, J. (1980) *J. molec. Struct.*, **60**, 449.
36. Havens, W.W. and Rainwater, L.J. (1951) *Phys. Rev.*, **83**, 1123.
37. Preston, R.S., Hanna, S.S. and Herberle, J. (1962) *Phys. Rev.*, **128**, 2207.
38. Shirley, D.A., Kaplan, M. and Axel, P. (1961) *Phys. Rev.*, **123**, 816.
39. Margules, S. and Ehrman, J.R. (1961) *Nucl. Instrum. Meth.*, **12**, 131.
40. Lang, G. (1963) *Nucl. Instrum. Meth.*, **24**, 425.
41. Housley, R.M., Erickson, N.E. and Dash, T.G. (1964) *Nucl. Instrum. Meth.*, **27**, 29.

42. Hafemeister, D.W. and Brooks–Shera, E. (1966) *Nucl. Instrum. Meth.*, **41**, 133.
43. Margules, S., Debrunner, P. and Frauenfelder, H. (1963) *Nucl. Instrum. Meth.*, **21**, 217.
44. O'Connor, D.A. (1963) *Nucl. Instrum. Meth.*, **21**, 318.
45. Housley, R.M. (1965) *Nucl. Instrum. Meth.*, **35**, 77.
46. Bowman, J.D., Kankeleit, E., Kaufmann, E.N. and Pearson, B. (1967) *Nucl. Instrum. Meth.*, **50**, 13.
47. Gibb, T.C., Greatrex, G. and Greenwood, N.N. (1968) *J. Chem. Soc. A.*, 890.
48. Sprengel-Segel, E.L. and Hanna, S.S. (1964) *Geochim. Cosmochim. Acta*, **28**, 1913.
49. Ericsson, T. and Wäppling, R. (1976) *J. Physique*, Colloq. C6, **37**, C6–719.
50. Stone, A.J. (1973) *Nucl. Instrum. Meth.*, **107**, 285.
51. Hogg, C.S. and Meads, R.E. (Micas) (1970) *Mineral Mag.*, **27**, 606.
52. Bancroft, G.M., Burns, R.G. and Maddock,. A.G. (1967) *Acta Cryst.*, **22**, 934.
53. Lerman, A., Stiller, M. and Hermon, E. (1967) *Earth Planet. Sci. Lett.*, **3**, 409.
54. Alimarin, I.P. and Yakoulev, Y.V. (1969) (review) *At. Energ.*, **26**, 127.
55. Gakiel, U. and Malamud, M. (1969) *Am. Mineral.*, **54**, 299.
56. Anon. (1970) *Nature, Lond.*, **215**, 1002.
57. Burns, R.G. (1972) *Can. J. Spectrosc.*, **17**, 51.
58. Borg, R.J., Lai, D.Y.F. and Borg, I.Y. (1973) *Nature, Lond.*, **246**, 46.
59. Hermon, E., Simkin, D.J., Donnay, G. and Muir, W.B. (1973) *Tschermaks Mineral. Petrogr. Mitt.*, **19**, 124.
60. Ericsson, T. and Wäppling, R. (1975) *Bull. Geol. Soc. Finland*, **47**, 171.
61. Fatseas, G.A., Dormann, J.L. and Blanchard, H. (1976) *J. Physique*, Colloq. C6, **37**, 787.
62. Mitra, Sachinath (1978) *Indian Mineral.*, **19**, 1.
63. Bruyneel, W., Pollak, H. and Tack, L. (1980) *Int. J. appl. Radiat. Isotopes*, **31**, 639.
64. Stroink, G., Blaauw, C., White, C.G. and Leiper, W. (1980) *Can. Mineral.*, **18**, 285.
65. Meisel, W. (1975) *Proc. 5th. Int. Conf. Möss Spec., 1973*, Nuclear Information Centre, Prague, p. 200.
66. Loveland, P.J. (1980) *Min. Mag.*, **L13**, 682.
67. Perfilev, Y.D., Gorelikova, N.V. and Babeshkin, A.M. (1975) *Vestn. Mosk. Univ. Khim.*, **16**, 117.
68. De Coster, M., Pollak, H. and Amelinckx, S. (Numerous) (1963) *Phys. Stat. Solidi*, **3**, 283.
69. Goldanskii, V.I., Egizarov, B.G., Zaporozhets, V.M. *et al.* (Numerous) (1965) *Priklad. Geofiz.*, **44**, 202.

70. Bancroft, G.M., Burns, R.G. and Maddock, A.G. (Numerous) (1967) *Geochim. Cosmochim. Acta*, **31**, 2219.
71. Clark, M.G., Bancroft, G.M. and Stone, A.J. (Gillespite) (1967) *J. chem. Phys.*, **47**, 4250.
72. Marfunin, A.S., Mineeva, R.M., Mkrtchyan, A.R. *et al.* (Numerous) (1967) *Izv. Akad. Nauk S.S.S.R. Ser. Geol.* (10), 86.
73. Bancroft, G.M., Burns, R.C. and Stone, A.J. (Howeite, Deerite and Sapphirines) (1968) *Geochim. Cosmochim. Acta*, **32**, 547.
74. Brown, F.F. and Pritchard, H.M. (Orthoclase) (1968) *Earth Planet Sci. Lett.*, **5**, 259.
75. Takashima, Y. and Ohashi, S. (Numerous) (1968) *Bull. Chem. Soc. Japan*, **41**, 88.
76. Herzenberg, C.L., Lamoreaux, R.D. and Riley, D.L. (Ferberite, Wolframite) (1969) *Z. Krist.*, **128**, 414.
77. Banks, E., Deluca, J.A. and Berkooz, O. (Pyrochlores) (1973) *J. Solid State Chem.*, **6**, 569.
78. Manning, P.G. and Tricker, M.J. (Vivianite) (1975) *Can. Mineral.*, **13**, 259.
79. Helgason, Ö., Steinthorsson, S., Mørup, S. *et al.* (Icelandic Lavas) (1976) *J. Physique*, Colloq. C6, **37**, C6–829.
80. Hermon, E., Haddad, R., Simkin, D. *et al.* (Voltaite) (1976) *Can. J. Phys.*, **54**, 1149.
81. Regnard, J.R. (Staurolite) (1976) *J. Physique*, Colloq. C6, **37**, C6–797.
82. Calage, Y. and Pannetier, J. (Pyrochlores) (1977) *J. Phys. Chem. Solids.* **38**, 711.
83. Manning, P.G. and Tricker, M.J. (Grossulars) (1977) *Can. Mineral.*, **15**, 81.
84. Dorfman, M.D. Platonov, A.N. and Pol'shin, E. (Fenaksite) (1978) *Tr. Mineral Muz. Akad. Nauk S.S.S.R.*, **26**, 198.
85. Halenius, U. (Sillimanite) (1979) *Neues Jahrb. Mineral Monatsh.*, 163.
86. Higgins, J.B., Ribbe, P.H. and Herd, R.K. (Sapphirines) (1979) *Contrib. Mineral. Petrol.*, **68**, 349.
87. Johnston, J.H. and Knedler, K.E. (Hypersthene) (1979) *Min. Mag.*, **43**, 279.
88. Alves, K.M.B., Garg, R. and Garg, V.K. (Childrenite) (1980) *Radiochem. Radioanal. Lett.*, **45**, 129.
89. Emery, J., Cereze, A. and Varret, F. (Hematophanite) (1980) *J. Phys. Chem. Solids*, **41**, 1035.
90. Leclerc, A. (Jarosites) (1980) *Phys. Chem. Miner.*, **6**, 327.
91. Kunrath, J.I., Mueller, C.S. and Vasquez, A. (Wolframite) (1981) *Hyperfine Interact.*, **10**, 1013.
92. Kostiner, E.S. (Phosphate Minerals) (1972) *Am. Mineral.*, **57**, 1109.
93. Stevens, J.G. (Antimony Minerals) (1976) *J. Physique*, Colloq. C6, **37**, C6–877.
94. Rossiter, M.J. and Hodgson, A.E.M. ($\alpha, \beta, \gamma, \delta$, FeOOH) (1965) *J. inorg. nucl. Chem.*, **27**, 63.
95. Shinjo, T. (Goethite) (1966) *J. phys. Soc. Japan*, **21**, 917.

96. Herzenberg, C.L. and Toms, D. (Numerous) (1966) *J. geophys. Res.*, **71**, 2661.
97. Forsyth, J.B., Hedley, I.G. and Johnson, C.E. (Goethite) (1968) *J. Phys. C*, **1**, 179.
98. Goldanskii, V.I., Makarov. E.F., Suzdalev, I.P. and Vinogradov, I.A. (1968) **27**, 44.
99. Housley, R.M., Grant, R.W. and Gonser, V. (Siderite) (1969) *Phys. Rev.*, **178**, 514.
100. Kulgawczuk, D. (1969) (Review) *Nukleonika*, **14**, 778.
101. Eissa, N.A., Sallam, H.A., Gomaa, S.S. *et al.* (Goethite) (1974) *J. Phys. D.*, **7**, 2121.
102. Govaert, A., Dauwe, C., Plinke, P. *et al.* (1976) *J. Physique*, Colloq. C6-825.
103. Eymery, J.P., Moine, P. and Le Roy, A. (Haematite) (1978) *Bull. Mineral.*, **101**, 393.
104. Singh, A.K., Jain, B.K. and Chandra, K. (Ochres) (1978), *J. Phys. D.*, **11**, 55.
105. Johnston, J.H. and Logan, N.E. (Akaganite) (1979) *J. chem. Soc. Dalton*, 13.
106. Murad, E. (Akaganite) (1979) *Clay Minerals*, **14**, 273.
107. Murad, E. (Goethite) *Min. Mag.*, **43**, 353.
108. Borg, R.J. and Borg, I.Y. (1974) *J. Physique*, Colloq. C6, **35**, C6-553.
109. Mitra, S. and Bansal, C. (Hornblende) (1975) *Chem. Phys. Lett.*, **30**, 403.
110. Seifert, F. (Anthophyllites) (1978) *Am. J. Sci.*, **278**, 1323.
111. Tripathi, R.P. and Lokanathan, S. (1978) *Indian J. pure appl. Phys.*, **16**, 888.
112. Goldman, D.S. *Am. Mineral.*, **64**, 109.
113. Borg, R.J. and Borg, I.Y. (Riebeckite) (1980) *Phys. Chem. Min.*, **5**, 219.
114. Grozdanov, L., Tomov, T. and Asenov, S. (1980) *Geokim. Mineral. Petrol.*, **13**, 45.
115. Whitehead, H.J. and Freeman, A.G. (Anthophyllites) (1967) *J. inorg. nucl. chem.*, **29**, 903.
116. Ershova, P.Z., Babeshkin, A.M. and Perfilev, Y.D. (Cummingtonite) (1970) *Geokhimiya*, **1970**, 252.
117. Popp, R.K., Gilbert, M.C. and Craig, J.R. (1977) *Am. Mineral.*, **62**, 1
118. Shenoy, G.K., Kalvius, G.M. and Hafner, S.S. (1969) *J. appl. Phys.*, **40**, 1314.
119. Matsui, Yoshito, Syono, Yasuhiko and Maeda, Yonezo (1972) *Mineral. J.*, **7**, 88.
120. Ekimov, S.P., Krizhanskii, L.M., Nikitina, L.P. and Kristoforov, K.K. (1973) *Geokhimiya*, 761.
121. Dowty, E. and Lindsley, D.H. (1973) *Am. Mineral.*, **58**, 850.
122. Hafner, S.S. (1975) *Proc. 5th Int. Conf. Möss. Spectr., 1973*, Nuclear Information Centre, Prague, p. 372.

123. Kinrade, J., Skippen, G.B. and Wiles, D.R. (1975) *Geochim. Cosmochim. Acta*, **39**, 1325.
124. Mitra, S. (1976) *Neues Jahrb. Mineral. Monatsh.*, 169.
125. Ksenozov, S.Y., Kanzin, A.S., Suslov, G.I. and Cherepanov, V.A. (1977) *Fiz. Tverd Tela.*, **19**, 25.
126. Tripathi, R.P., Chanora, U., Krishnamurthy, A. and Lokanathan, S. (1977) *Proc. nucl. Phys. Solid State Phys. Symp.*, Department of Atomic Energy, India, vol. 20C, p. 521.
127. Annersten, H., Olesch, M. and Siefort, F.A. (1978) *Lithos*, **11**, 301.
128. Helgason, O. and Oskarsson, N. (1979) *J. Physique,* Colloq. C2, **40**, C2-452.
129. Siefert, F. (1979) *Mineral Mag.*, **43**, 313.
130. Amigo, J.M., Velasco, F. and Arriortua, M.I. (1980) *Neues Jahrb. Mineral. Monatsh.*, 337.
131. Gupta, R.G. and Mendiratta (1980) *Mineral. Mag.*, **43**, 815.
132. Marfunin, A.S., Mkrtchyan, A.R., Nadzharyan, G.N. and Platonov, Y.M. (1970) *Izv. Akad. Nauk S.S.S.R. Ser. Geol.*, 146.
133. Alvarez, M.A., Tornero, J. and Vara, J.M. (1975) *Ann Quim.*, **71**, 498.
134. Belov, A.F., Korneev, E.V., Khimich, T.A. *et al.* (1975) *Zh. Fiz. Kh.*, **49**, 1683.
135. Bhandari, S.S. and Varma, J. (1975) *Proc. nucl. Phys. Solid State Phys. Symp.*, vol. 18C Department of Atomic Energy, India, p. 552.
136. Dambly, M., Pollak, H., Quartier, R. and Bruyneel, W. (1976) *J. Physique*, Colloq. C6, **37**, C6-807.
137. Gorelikova, N.V., Perfil'ev, Y.D. and Babeshkin, A.M. (1976) *Zap. Vses. Mineral. O-va.*, **105**, 418.
138. Korovushkin, V.V., Kuzmin, V.I. and Belov, V.F. (1979) *Phys. Chem. Min.*, **4**, 209.
139. Pollak, H., Danon, J., Quartier, R. and Dauwe, C. (1979) *J. Physique*, Colloq. C2, **40**, C2-480.
140. Saegusa, N., Price, D.C. and Smith, G. (1979). *J. Physique*, Colloq. C2, **40**, C2-456.
141. Duncan, J.F. and Johnston, J.H. (1973) *Aust. J. Chem.*, **26**, 231.
142. Hrynkiewicz, A.Z., Kulgawczuk, D.S., Mazanek, E.S. *et al.* (1972) *Inst. Nucl. Phys., Cracow, Rep.* 814/PL.
143. Edel'shtein, I.I., Kirichok, P.P., Pilipenko, A.A. *et al.* (1979) *Izv. Vuz. Geol. Razvedka*, **22**, 35.
144. Takeda, M., Matsuo, M. and Tominaga, T. (1979) *Radiochem. Radioanal. Lett.*, **41**, 1.
145. Shinno, I. (1981) *Phys. Chem. Min.*, **7**, 91.
146. Ablesimov, N.E. and Bekhtol'd, A.F. (1978) *Dokl. Akad. Nauk S.S.S.R.*, **239**, 694.
147. Hafner, S. and Kalvius, M. (Troilite, Pyrrhotite) (1966) *Z. Krist.*, **123**, 443.

148. Hafner, S., Evans, B.J. and Kalvius, G.M. (Troilite, Pyrrhotite) (1967) *Solid State Comm.*, **5**, 17.
149. Greenwood, N.N. and Whitfield, H.J. (Cubanite) (1968) *J. chem. Soc.*, **1968A**, 1967.
150. Imbert, P. and Wintenberg, M. (Cubanite, Sternbergite) (1967) *Bull. Soc. Fr. Min. Crist.*, **90**, 299 .
151. Marfunin, A.S. and Mkrtchyan, A.R. (8 sulphidic minerals) (1967) *Geokhimiya*, **1967**, 1094.
152. Vainshtein, E.E., Valov, P.M., Barsanov, G.P. and Yakovleva, M.E. (Pyrites, Marcasite) (1967) *Geokhimiya*, **1967**, 876.
153. Borshagovskii, B.V., Marfunin, A.S., Mkrtchyan, A.R. and Stukan, R.A. (7 Sulphidic Minerals) (1968) *Izv. Akad. Nauk S.S.S.R., Ser Khim. Fiz.*, **1968**, 1267.
154. Raj, D., Chanora, K. and Puri, S.P. (Chalcopyrite) (1968) *J. phys. Soc. Japan*, **24**, 35.
155. Zhetbaev, A.K. and Kaipov, D.K. (Pyrites, Chalcopyrite, Bornite) (1968) *Izv. Akad. Nauk S.S.S.R., Ser. Fiz Mat.*, **6**, 78.
156. Morice, J., Rees, L.V.S. and Rickard, D.T. (Pyrites, Marcasite, Pyrrhotite, Greigite, Mackinawite) (1969) *J. inorg. nucl. Chem.*, **31**, 3797.
157. Goncharov, G.N., Ostanevich, Y.M. and Tomilov, S.B. (8 Sulphidic Minerals) (1970) *Izv. Akad. Nauk. S.S.S.R., Ser. Geol.*, **1970**, 79.
158. Vaughan, D.J. (Pyrrhotite) (1970) *Solid State Comm.*, **8**, 2165.
159. Vaughan, D.J. and Ridout, M.S. (Mackinawite, Greigite, Pentlandite) (1971) *J. inorg. nucl. Chem.*, **33**, 741.
160. Imbert, P., Varret, F. and Wintenberger, M. (1973) *J. Phys. Chem. Solids*, **34**, 1675.
161. Ioffe, P.A., Aleksandrova, L.S., Kazakov, M.I. *et al.* (Pyrrhotites) (1975) *Zh. Fiz. Khim.*, **49**, 7.
162. Pietsch, C., Fritzsch, E., Fritzsche, K. and Schneider, H.A. (Stannite) (1975) *Proc. 5th Conf. Möss. Spec., 1973*, Nuclear Information Centre Prague, pp. 1–3, p. 379.
163. Eissa, N.A., Sallam. H.A., El–Ockr, M.M. *et al.* (Chalcopyrite) (1976) *J. Physique*, Colloq. C6, **37**, C6-793.
164. Novikov, G.V., Egorov, V.K., Popov, V.I. and Bezmen, N.I. (Pyrrhotite) (1976) *Geokhimiya*, **1976**, 124.
165. Townsend, M.G., Goselin, J.R., Horwood, J.L. *et al.* (Violarite) (1977) *Phys. Stat. Solids*, **A40**, K25.
166. Blaauw, C., White, C.G., Leiper, W. and Clarke, D.B. (Djerfisherite, Pentlandite) (1979) *Mineral. Mag.*, **43**, 552.
167. Liu, Y.S. (Pyrites) (1979) *J. Physique*, Colloq., C2, **40**, C2-400.
168. Jagadeesh, M.S., Nagarathna, H.M., Montano, P.A. and Seehra, M.S. (Bornite) (1981) *Phys. Rev. B. (Cond. Mat.)*, **23**, 2350.
169. Bancroft, G.M. (1979) *J. Physique Colloq.*, C2, **40**, C2-464.

170. Fernandez–Moran, H., Hafner, S.S., Ohtsuki, M. and Virgo, D. (1970) *Science*, **167**, 686.
171. Gay, P., Bancroft, G.M. and Bown, M.G. (1970) *Science*, **167**, 626.
172. Gay, P., Bancroft, G.M. and Bown, M.G. (1970) *Proc. Apollo II Lunar Sci. Conf., Geochim. Cosmochim. Acta*, Suppl. 1, **1**, 481.
173. Greenwood, N.N. and Howe, A.T. (1970) *Proc. Apollo II Lunar Sci. Conf., Geochim. Cosmochim. Acta.* Suppl. 1, **3**, p. 2163.
174. Hafner, S.S. and Virgo, D. (1970) *Proc. Apollo II Lunar Sci. Conf.*, **3**, 2183.
175. Hafner, S.S., Janik, B. and Virgo, D. (1970) *Möss. Effect Method*, **6**, 193 (Annual Pub., Plenum Press, ed. I.J. Gruverman).
176. Herzenberg, C.L. and Riley, D.L. (1970) *Möss. Effect Method*, **6**, 177.
177. Herzenberg, C.L. and Riley, D.L. (1970) *Science*, **167**, 683.
178. Herzenberg, C.L. and Riley, D.L. (1970) *Proc. Apollo II Lunar Sci. Conf. Geochim. Cosmochim. Acta.* Suppl. 1, **3**, p. 2221.
179. Housley, R.M., Blander, M., Abdel–Gawad, *et al.* (1970) *Proc. Apollo II Lunar Sci. Conf., Geochim. Cosmochim. Acta*, Suppl. 1, **3**, 2251.
180. Appleman, D.E., Nissen, H.U., Stewart, D.B. *et al.* (1971) *Proc. 2nd Lunar Sci. Conf., Geochim. Cosmochim. Acta* Suppl., 2, **1**, 117.
181. Hafner,. S.S., Virgo, D. and Warburton, D. (1971) *Earth Planet Sci. Lett.*, **12**, 159.
182. Huffman, G.P, Dunmyre, G.R., Fisher, R. *et al.* (1970) *Möss. Effect Method.*, **6**, 209 (annual Pub. Plenum Press, ed. I.J. Gruverman).
183. Muir, A.H., Housley, R.M., Grant,. R.W. *et al.* (1970) *Science*, **167**, 688.
184. Muir, A.H., Housley, R.M., Grant, R.W. *et al.* (1970) *Möss. Effect Method.*, **6**, 163 (annual Pub., Plenum Press, ed. I.J. Gruverman).
185. Gibb, T.C., Greenwood, N.N. and Battey, M.H. (1972) *Proc. 3rd Lunar Sci. Conf., Geochim. Cosmochim. Acta*, Suppl. 3, **3**, 2479.
186. Malysheva, T.V. and Kurash, V.V. (1972) *Space Res.*, **12**, 137.
187. Schurmann, K. and Hafner, S.S. (1972) *Proc. 3rd Lunar Sci Conf., Geochim. Cosmochim. Acta*, Suppl. 3, **1**, 493.
188. Brecher, A., Vaughan, D.J., Burns, R.G. and Morash, K.R. (1973) *Proc. 4th Lunar Sci. Conf., Geochim. Cosmochim. Acta*, Suppl. 4, **3**, 2991.
189. Forester, D.W. (1973) *Proc. 4th Lunar Sci. Conf., Geochim. Cosmochim. Acta*, Suppl. 4, **3**, 2697.
190. Gibb, T.C., Greatrex, R. and Greenwood, N.N. (1974) *J. chem. Soc. Dalton*, **1974**, 1148.
191. Huffman, G.P. and Schwerer, F.C. (1974) *Proc. 5th Lunar Sci. Conf., Geochim. Cosmochim. Acta*, Suppl. 5, **3**, 2779.
192. Yajima, T. and Hafner, S.S. (1974) *Proc. 5th Lunar Sci. Conf., Geochim. Cosmochim. Acta*, Suppl. 5, **1**, 769.
193. Zemcik, T. and Raclavsky, K. (1974) *J. Physique*, Colloq. C6, **35**, 549.
194. Zemcik, T. and Raclavsky, K. (1974) *Geokhimiya*, **1974**, 1084.
195. Brecher, A., Menke, W.H., Adams, J.B. and Gaffey, M.J. *Proc. 6th Lunar Sci. Conf., Geochim. Cosmochim. Acta*, Suppl. 6, **3**, 3091.
196. Zemick, T. and Raclavsky, K. (1976) *J. Physique*, Colloq. C6, **37**, 833.

197. Gibb, T.C., Greatrex, R. and Greenwood, N.N. (1977) *Phil. Trans*, **284**, 157.
198. Gibb, T.C., Greatrex, R. and Greenwood, N.N. (1977) *Phil. Trans*, **285**, 235.
199. Malysheva, T.V., Polyakova, N.P. and Mishin, N.E. (1978) *Geokhimiya*, **1978**, 835.
200. Neibuhr, H.H., Zeira, S. and Hafner, S.S. *Proc. 4th Lunar Sci. Conf., Geochim. Cosmochim. Acta.*, Suppl. 4, **1**, 971.
201. Sprenkel-Segal, E.L. and Hannna, S.S. (1966) *Möss Effect Method.*, **2**, 113 (annual Pub., Plenum Press, ed. J.J. Gruverman).
202. Marzolf, J.G., Dehn, J.T. and Salmon, J.F. (1967) *Advances in Chem. No. 68* (ed. R.H. Herber), American Chemical Society, Washington.
203. Herr, W. and Skerra, B. *Meteorite Res. Proc. Symp. 1968* (ed. P.M. Milman), Reidel Publications, Dordrecht, p. 492.
204. La Fleur, L.D., Goodman, C.D. and King, E.A. (1968) *Science*, **162**, 1268.
205. Levskii, L.K., Ostanevich, Y.M. and Tomilov, S.B. (1968) *Geochim. Int.*, **5**, 867.
206. Sprenkel-Segal, E.L. (1968) *Proc. Symp. Meteorite Res.*, (ed. P.M. Milman), Reidel Publications, Dordrecht, p. 93.
207. Sprenkel-Segal, E.L. and Perlow, G.J. (1968) *Icarus*, **8**, 66.
208. Sprenkel-Segal, E.L. (1970) *J. Geophys. Res.*, **75**, 6618.
209. Malysheva, T.V., Lavrukhina, A.K., Stakeeva, S.A. and Satarova, L.M. (1974) *Meteoritika*, **33**, 34.
210. Vdovykin, G.P., Grachev, V.I., Malysheva, T.V. and Satarova, L.M. (1975) *Geokhimiya*, **1975**, 1872.
211. Malysheva, T.V., Satarova, L.M. and Polyakova, N.P. (1977) *Geokhimiya*, **1977**, 1136.
212. Albertsen, J.F., Aydin, M. and Knudsen, J.M. (1978) *Phys. Scripta*, **17**, 467.
213. Albertsen, J.F., Jensen, G.B. and Knudsen, J.M. (1978) *Nature, Lond.*, **273**, 453.
214. Florenskii, P.W., Dikow, Y.P. and Gendler, T.S. (1978) *Chem. Erde Bd*, **37**, 109.
215. Minai, Y., Wakita, H. and Tominaga, T. (1978) *Radiochem Radioanal. Lett.*, **36**, 193.
216. Oliver, F.W. (1978) *Planet. Space Sci.*, **26**, 289.
217. Ouseph, P.J., Groskreutz, H.E. and Johnson, A.A. (1978) *Meteorites*, **13**, 189.
218. Danon, J., Scorzelli, R., Souza-Azevedo, I. *et al.* (1979) *Nature, Lond.*, **277**, 283.
219. Dannon, J., Scorzelli, R., Souza-Azevedo, I. and Christophe-Michel-Levy M. (1979) *Nature, Lond.*, **281**, 469.
220. Danon, J., Scorzelli, R., Souza-Azevedo, I. *et al.* (1979) *Radiochem. Radioanal. Lett.*, **38**, 339.
221. Evans, B.J. and Leung, L.K. (1979) *J. Physique*, Colloq. C2, **40**, 489.

222. Ouseph, P.J., Groskreutz, H.E. and Johnson, A.A. (1979) *Meteorites*, **14**, 97.
223. Clarke, R.S. and Scott, E.R.D. (1980) *Am. Mineral.*, **65**, 624.
224. Danon, J., Scrozelli, R. and Souza-Acevedo, I. (1980) *J. Physique*, Colloq. C2, **41**, 363.
225. Malysheva, T.V. (1980) *J. Physique*, Colloq. C2, **41**, 405.
226. Gosselin, J.P., Simony, U., Grozins, L. and Cooper, A.R. (1967) *Phys. Chem. Glasses*, **8**, 56.
227. Lewis, G.K. and Drickhamer, H.G. (1968) *J. Chem. Phys.*, **49**, 3785.
228. Buckley, R.R., Kenealy, P.F., Beard, G.B. and Hooper, H.O. (1969) *J. appl. Phys.*, **40**, 4289.
229. Barsukov, V.L., Durhsova, N.A., Malysheva, T.V. and Bobrsergeev, A.A. (1970) *Geokhimiya*, **1970**, 758.
230. Kurkjian, C.R. (1970) *J. Non-cryst. Sol.*, **3**, 157.
231. Taragin, M.F. and Eisenstein, J.C. (1970) *J. Non-Cryst. Sol.*, **3**, 311.
232. Longworth, G. and Warren, S.E. (1979) *J. Archeol.*, **6**, 179.
233. Takeda, Masuo, Sato, Kazuo, Sato, Jun and Tominaga, Takeshi (1979) *Rev. Chim. Min.*, **16**, 400.
234. Chavez–Rivas, F., Regnard, J.R. and Chappert, J. (1980) *J. Physique*, Colloq. C1, **41**, 275.
235. Coey, J.M.D. (1974) (Review) *J. Physique*, Colloq. C6, **35**, 89.
236. Bandyapadhyay, A.K., Zarzycki, J., Auric, P. and Chappert, J. (1980) *J. Non-cryst. Solids*, **40**, 353.
237. Hofmann, U., Fluck, E. and Kuhn, P. (1967) (Glauconite) *Angew. Chem.*, **79**, 581.
238. Malden, P.J. and Meads, R.E. (Muscovite, Kaolinite and Gibbsite) (1967) *Nature, Lond.*, **215**, 844.
239. Pecuil, T.E., Wampler, J.M. and Weaver, C. (1967) *Clays, Clay Min., Proc. 15th Nat. Conf.*, Pergamon Press, p. 143 (data for 16 Clay Minerals).
240. Weaver, C.E., Wampler, J.M. and Pecuil, T.E. (Data for 15 Clay Minerals) (1967) *Science*, **156**, 504.
241. Herzenberg, C.L., Riley, D.L. and Lamoreaux, R. (Zinnwaldite and Biotite) (1968) *Nature, Lond.*, **219**, 364.
242. Bowen, L.H., Weed, S.B. and Stevens, J.G. (K depleted Micas) (1969) *Am. Mineral.*, **54**, 72.
243. Haggstrom, L., Wäppling, R. and Annersten, H. (Biotites) (1969) *Chem. Phys. Lett.*, **4**, 107.
244. Hogarth, D.D., Brown, F.F. and Pritchard, A.M. (Phlogopites) (1970) *Can. Mineral.*, **10**, 710.
245. Annersten, H., Devanarayanan, S., Haggstrom, L. and Wäppling, R. (Ferriphlogopites) (1971) *Phys. Stat. Solid.*, **48B**, K 137.
246. Pol'shin, E.V., Matyash, I.V., Tepikin, V.E. and Ivanitskii, V.P. (Biotites) (1972) *Kristallografiya*, **17**, 328.

247. Brunot, B. (Nontronites) (1973) *Neues Jahrb. Mineral. Monatsh.*, **1973**, 425.
248. Gangas, N.H., Simopoulos, A., Kostikas, A. *et al.* (Superpamag Component) (1973) *Clay Clay Min.*, **21**, 151.
249. Annersten, H. (Synthetic Phlogopite) (1975) *Fortschr. Mineral.*, **52**, 583.
250. Annersten, H. (Biotites) (1974) *Am. Mineral.*, **59**, 143.
251. Annersten, H. (Glauconites) (1975) *Neues Jahrb. Mineral. Monatsh.*, **1975**, 378.
252. Chandra, R., Tripathi, R.P. and Lokanathan, S. (Biotites) (1975) *Proc. nucl. Phys. Solid State Phys. Symp.*, Department of Atomic Energy, India, **18** C, p. 528.
253. Ershova, Z.P., Nikitina, A.P., Perfil'ev, Y.D. and Babeshkin, A. (Chamosite) (1976) *Proc. Int. Clay Conf., 1975*, Applied Publications, Wilmette, Illinois, p. 211.
254. Helsen, J.A., van Deyck, M., Langouche, G. *et al.* (Vermiculite) (1975) *Clay Clay Min.*, **23**, 332.
255. Jefferson, D.A., Tricker, M.J. and Winterbottom, A.P. (Kaolinite) (1975) *Clay Clay Min.*, **23**, 355.
256. Sharma, K.C., Chandra, R. and Lokanathan, S. (Biotite) (1976) *Proc. nucl. Phys. Solid State Phys. Symp., 1975*, Department of Atomic Energy, India, p. 531.
257. Cimbalnikova, A., Raclavsky, K., Hejl, V. and Sitek, J. (Biotites), (1976) *Proc. Conf. Clay Clay Mineral Petrol.*, Charles University, Prague, **7**, 77.
258. Eirish, M.V. and Dvorechenskaya, A.A. (Montmorillonite, Nontronite, Glauconite, Vermiculite) (1976) *Geokhimiya*, **1976**, 748,.
259. Goodman, B.A. (Calc. of efg) (1976) *J. Physique*, Colloq. C6, **37**, C6 819.
260. Goodman, B.A. (Muscovite and Biotite) (1976) *Min. Mag.*, **40**, 513.
261. Goodman, B.A., Russell, J.D., Fraser, A.R. and Woodhams, F.W.D. (Smectites, Nontronite) (1976) *Clay Clay Min.*, **24**, 53.
262. Malysheva, T.V., Kazakov, G.A. and Satarova, L.M. (Glauconite, Celadonite) (1976) *Geokhimiya*, **1976**, 1291.
263. Mkrtchyan, G.R., Saakyan, A.A., Pulatov, M. *et al.* (Nontronite) (1976) *Izv. Akad. Nauk Arm. S.S.R. Fiz.*, **11**, 390.
264. Angel, B.R., Cuttler, A.H., Richards, K.S. and Vincent, W.E.J. (Kaolinite) (1977) *Clay Clay Min.*, **25**, 381.
265. Chandra, R. and Lokanathan, S. (Biotite) (1977) *Phys. Stat. Solidi*, **B83**, 273.
266. Drago, V., Baggio-Savitch, E. and Danon, J. (Muscovite, Biotite, Phlogopite, Talc) (1977) *J. inorg. nucl. Chem.*, **39**, 973.
267. Ouseph, P.J. and Groskreutz, H.E. (Micas) (1977) *Phys. Stat. Solidi*, **A44**, K181.
268. Rolf, R.M., Kimball, C.W. and Odom, I.E. (Glauconites) (1977) *Clay. Clay. Min.*, **25**, 131.

269. Rozenson, I. and Heller–Kallai, L. (15 Smectites) (1977) *Clay. Clay Min.*, **25**, 94.
270. Tkacheva, T.V., Simakova, L.G. and Pauker, V.I. (Chamosites) (1977) *Tsvetn Met.*, **1977**, 27.
271. Bookin, A.S., Dainyak, L.G. and Drits, V.A. (Micas) (1978) *Phys. Chem. Mineral.*, **3**, 58.
272. Chandra, R., Tripathi, R.P. and Lokanathan, S. (Biotites) (1978) *Phys. Stat. Solidi*, **B88**, 633.
273. Goodman, B.A. (Nontronites) (1978) *Clay. Clay Min.*, **26**, 176.
274. Grechishkin, V.S., Zangalis, K.P., Murin, I.V. *et al.* (Glauconites, Montmorillonite) (1978) *Izv. Akad. Nauk S.S.S.R. Ser. Fiz.*, **42**, 2654.
275. Kupriyanova, I.I., Getmanskaya, T.L., Zhukhlistov, A.P. *et al.* (1978) (Micas) *Tr. Mineral. Muz. Akad. Nauk S.S.S.R.*, **26**, 77.
276. Mineeva, R.M. (efg Biotites) (1978) *Phys. Chem. Mineral.*, **2**, 267.
277. Pavlishin, V.I., Platonov, A.N., Pol'shin, A.V. *et al.* (1978) (Phlogopite, Biotites) (1978) *Zap. Vses. Mineral O-va,* **107**, 165.
278. Rozenson, I. and Heller–Kallai, L. (Glauconite, Illite, Nontronite) (1978) *Clay. Clay Min.*, **26**, 173.
279. Sanz, J., Meyers, J., Vielvoye, L. and Stone, W.E. (Biotites, Phlogopites) (1978) *Clay Min. Bull.*, **13**, 45.
280. Tricker, M.J., Jefferson, D.A., Thomas, J.M. *et al.* (1978) (Choritoid) *J. chem. Soc. Faraday II*, **74**, 174.
281. Ballet, O., Coey, J.M.D. and Massent, O. (Chlorite, Greenalite, Biotite, Vermiculite, Berthierine) (1979) *J. Physique*, Colloq. C2, **40**, C2 283.
282. Goodman, B.A. and Bain, D.C. (Chlorite) (1979) *Devel. Sedimentol.*, **1979**(1), 65.
283. Grechishkin, V.S., Zangalis, K.P., Murin, I.V. *et al.* (Glaconites, Montmorillonites) (1979) *Okeanologiya (Moscow)*, **19**, 438.
284. Gupta, D.C. and Nath, N. (Vermiculites) (1979) *Japan. J. appl. Phys.*, **18**, 2093.
285. McConchie, D.M., Ward, J.B., McCann, V.H. and Lewis, D.W. (Glauconites, Phyllosilicates) (1979) *Clay Clay Min.*, **27**, 339.
286. Rozenson, I., Bauminger, E.R. and Heller-Kallai, L. (Chamosite, Kaolinite, Dickite, Lizardite, Chrysotile Antigorite) (1979) *Am. Mineral.*, **64**, 893.
287. Blaauw, C., Stroink, G. and Leiper, W. (Talc, Chlorite) (1980) *J. Physique*, Colloq. C1, **41**, 411.
288. Coey, J.M.D. (Review) (1980) *Atomic Energy Rev.*, **18**, 73.
289. Gessa, C., Bart, J.C. and Burriesci, N. (Bentonite) (1980) *Clay Clay Min.*, **28**, 233.
290. Krishnamurthy, A., Srivastava, B. and Lokanathan, S. (Vermiculite) (1980) *Nat. Acad. Sci. Lett.*, **3**, 53.
291. Rozenson, I., Zak, I. and Spiro, B. (Kaolinite, Akaganite) (1980) *Chem. Geol.*, **31**, 83.

292. Silver, J., Sweeney, M. and Morrison, I.E.G. (Superparamag. Component in Clay) (1980) *Thermochem. Acta*, **35**, 153.
293. Bonnin, D. and Muller, S. (Muscovite) (1981) *Phys. Stat. Solidi*, **B105**, 649.
294. Komusinski, J., Stoch, L. and Dubiel, S.M. (Nacrite, Dickite, Kaolinite, Halloysite and Allophane) (1981) *Clay Clay Min.*, **29**, 23.
295. Kotlicki, A., Szcyrba, J. and Wiewiora, A. (Pelletal Glauconites) (1981) *Clay Mineral.*, **16**, 221.
296. Krishnamurthy, A., Srivastava, B.K. and Lokanathan, S. (Phlogopite) (1981) *Pramaña* **1**, 39.
297. Stucki, J.W. and Banwart, W.L. (Review) (1980) *Adv. Chem. Meth. Soil and Clay Min. Res.*, NATO Adv. Studies Series.
298. Coey, J.M.D. (1980) *Atomic Energy Rev.*, **18**, 73 (Review with almost all necessary references).
299. Van der Giessen, A.A., Rensen, T.G. and Van Wieringen, J.S. (1968) *J. inorg. nucl. chem.*, **30**, 1739.
300. Janot, C., Chabanel, M. and Herzog, E. (1968) *Comp. Rend.*, **266C**, 103.
301. Janot, C., Chabanel, M. and Herzog, E. (1968) *Bull. Soc. Fr. Mineral Crist.*, **91**, 166.
302. Krupyanskii, Y.F. and Suzdalev, I.P. (1973) *Zh. Eksp., Teor. Fiz.*, **65**, 1715.
303. Perlow, G.J., Potzel, W. and Edgington, D. (1974) *J. Physique*, Colloq. C6, **35**, C6, 547.
304. Gendler, T.S., Kurz'min, R.N. and Urazaeva, T.K. *Kristallografiya*, **21**, 774.
305. Goodman, B.A. and Berrow, M.L. (1976) *J. Physique,* Colloq. C6, **37**, C6, 849.
306. Readman, P.W., Coey. J.M.D., Mosser, C. and Weber, F. (1976) *J. Physique*, Colloq. C6, **37**, C6, 845.
307. Singh, A.K., Jain, B.K. and Chandra, K. (1977) *Phys. Stat. Solidi*, **A44**, 443.
308. Dumance, E.M., Meads, R.E., Ballard, R.R.B. and Walsh, J.N. (1978) *Geol. Soc. Amer. Bull.*, **89**, 1231.
309. Manning, P.G. and Ash, L.A. (1978) *Can. Mineral.*, **16**, 577.
310. Ujihara, Y., Ohyabu, M., Murakami, T. and Horie, T. (1978) *Bunseki Kagaku*, **27**, 631.
311. Barb, D., Diamandescu, L., Morariu, H. and Georgescu, I.I. (1979) *J. Physique*, Colloq. C2, **40**, C2, 445.
312. Carlson, I. and Schwertmann, U. (1980) *Clay Min.*, **28**, 272.
313. Henmi, T., Wells, N., Childs, C.W. and Parfitt, R.L. (1980) *Geochim. Cosmochim. Acta*, **44**, 365.
314. Murad, E. and Schwertmann, U. (1980) *Am. Mineral.*, **65**, 1044.
315. Coey, J.M.D. and Readman, P.W. (1973) *Earth Planet Sci. Lett.*, **21**, 45.
316. Petersen, L. and Rasmussen, K. (1980) *Clay Min.*, **15**, 135.

317. Coey, J.M.D. (Review) (1975) *Proc. 5th Int. Conf. Möss. Spectr.* Institute of Nuclear Physics, Cracow, **2**, 333.
318. Brady, G.W., Kurkjian, C.R., Lyden, E.F.X. *et al.* (1968) *Biochemistry*, **7**, 2185.
319. Coey, J.M.D. (1975) *Geochim. Cosmochim. Acta*, **39**, 401.
320. Blume, M. and Tjön, J.A. (1968) *Phys. Rev.*, **165**, 446.
321. Manning, P.G., Williams. J.O.H., Charlton, M.N. *et al.* (1979) *Nature, Lond*, **280**, 134.
322. Manning, P.G. and Ash, L.A. (1979) *Can. Mineral*, **17**, 111.
323. Coey, J.M.D., Schindler, D.W. and Weber, F. (1974) *Can. J. Earth Sci.*, **11**, 1489.
324. Gager, H.M. (1968) *Nature, Lond.*, **220**, 1021.
325. Herzenberg, C.L. and Riley, D.L. (1969) *Nature, Lond.*, **224**, 259.
326. Johnson, C.E. and Glasby, G.P. (1969) *Nature, Lond.*, **222** 376.
327. Georgescu, I.I. and Nistor, C. (1970) *Rev. Roum. Phys.*, **15**, 819.
328. Hrynkiewicz, A.Z., Sawicka, B. and Sawicki, J. (1970) *Phys. Stat. Solidi*, **3A**, 1039.
329. Hrynkiewicz, A.Z., Pustowka, A.J., Sawicki, B.D. and Sawicki, J. (1972) *Phys. Stat. Solidi*, **10A**, 281.
330. Eissa, N.A., Sallam, H.A., El-Kerdani, H.A. and Taiel, F.M. (1976) *J. Physique*, Colloq. C6, **37**, C6, 857.
331. Johnston, J.H. and Glasby, G.P. (1978) *Geochem. J.*, **12**, 153.
332. Sozanski, A.G. and Cronan, D.S. (1979) *Can. J. Earth Sci.*, **16**, 126.
333. Bancroft, G.M., Maddock, A.G. and Strens, R.G.J. (1966) *Nature, Lond.* **212**, 913.
334. Bancroft, G.M. and Burns, R.G. (1968) *Proc. 5th Gen. Meeting Int. Mineral Assoc.*, 1966, Mineralogical Society, London, p. 36.
335. Bancroft, G.M. and Burns R.G. (1967) *Earth Planet. Sci. Lett.*, **3**, 125.
336. Bancroft, G.M. (1967) *Phys, Lett.*, **26A**, 17.
337. Bancroft, G.M., Burns, R.G. and Howie, R.A. (1967) *Nature, Lond.*, **213**, 1221.
338. Bancroft, G.M., Burns, R.G. and Maddock, A.G. (1967) *Am. Mineral.*, **52**, 1009.
339. Dundon, R.W. and Walter, L.S. (1967) *Earth Planet. Sci. Lett.*, **2**, 372, 648.
340. Evans, B.J., Ghose, S. and Hafner, S. (1967) *J. Geol.*, **75**, 306.
341. Ghose, S. and Hafner, S. (1967) *Z. Krist.*, **125**, 109, 157.
342. Marzolf, J.G., Dehn, J.T. and Gould, R.F. (1967) *Mössbauer Effects and Applications in Chemistry* (ed. R.F. Gould), American Chemical Society, Washington, p. 61.
343. Dundon, R. and Lindsley, D.H. (1968) *Carnegie Inst. Year Book*, **66**, 366.
344. Bancroft, G.M. and Burns, R.G. (1969) *Mineral Soc. Amer., Spec. Paper*, **2**, 137.
345. Bancroft, G.M., Williams P.G.L. and Essene, E.J. (1969) *Mineral Soc. Amer., Spec. Paper*, **2**, 59.

346. Häggström, L., Wäppling, R. and Annersten, H. (1969) *Phys. Stat. Solidi*, **33**, 741.
347. Herzenberg, C.L. and Riley, D.L. (1969) *Acta Cryst.*, **25A**, 389.
348. Malysheva, T.V., Kurash, V.V. and Ermakov, A.N. Ermakov, (1969) *Geokhimiya*, **1969**, 1405.
349. Malysheva, T.V., Ermakov, A.N., Aleksandrov, S.M. and Kurash V.V. (1971) *Proc. Conf. Appls. Möss. Spec., 1969*, Hungarian Academy of Science, Hungary, p. 745.
350. Mokeeva, V.I. and Aleksandrov, S.M. (1969) *Geokhimiya*, **1969**, 428.
351. Müller, R.F. (1969) *Mineral. Soc. Amer., Spec. Paper*, **2**, 83.
352. Thompson, J.B. (1969) *Am. Mineral.*, **54**, 341.
353. Virgo, D. and Hafner, S.S. (1969) *Mineral Soc. Amer., Spec. Paper*, **2**, 67.
354. Bancroft, G.M. (1970) *Chem. Geol.*, **5**, 255.
355. Bush, W.R., Hafner, S.S. and Virgo, D. (1970) *Nature, Lond.*, **227**, 1339.
356. Ernst, E.G. and Chein Moo Wai (1970) *Am. Mineral.*, **55**, 1226.
357. Perchuk, L.L. and Surikov, V.V. (1970) *Izv. Akad Nauk S.S.S.R., Ser. Geol.*, **1970**, 3.
358. Valter, A.A., Gorogtskaya, L.I., Zverev, N.D. and Romanov, V.P. (1970) *Dokl. Akad. Nauk S.S.S.R.*, **192**, 629.
359. Virgo, D. and Hafner, S.S. (1970) *Am. Mineral.*, **55**, 201.
360. Matsui, Yoshito Maeda, Yonezo and Syono, Yasuhiko (1970) *Geochem. J.*, **4**, 15.
361. Greaves, C., Burns, R.G. and Bancroft, G.M. (1971) *Nature, Lond.*, **229**, 60.
362. Saxena S.K. and Ghose, S. (1971) *Am. Mineral.*, **56**, 532.
363. Walters, D.S. and Wirtz, G.P. (1971) *J. Amer. Ceramic Soc.*, **54**, 563.
364. Williams, P.G.L., Bancroft, G.M., Bown, M.G. and Turnock, A.C. (1971) *Nature, Lond.*, **230**, 149.
365. Barabano, A.V. and Tomilov, S.B. (1973) *Geokhimiya*, **1973**, 1669.
366. Litvin, A.L., Michnik, T.L., Ostapenko, S.S. and Pol'shin, E.V. (1973) *Geol. Zh.*, **33**, 49.
367. Shinno, I., Hayashi, M. and Kuroda, Y. (1974) *Mineral. J.*, **7**, 344.
368. Nikitina, L.P., Ekimov, S.P., Krizhanskii, L.M. and Khristoforov, K.K. (1976) *Mineral Sb. (Lvov)*, **30**, 18.
369. Seifert, F. (1977) *Phys. Chem. Min.*, **1**, 43.
370. Aldridge, L.P., Bancroft, G.M., Fleet, N.F. and Herzberg, C.T. (1978) *Am. Mineral.*, **63**, 1107.
371. Brovkin, A.A., Pol'shin, E.V. and Brovkina, V.S., (1978) *Kristallografiya*, **23**, 107.
372. Johnston, J.H. and Knedler, K.E. (1979) *Min. Mag.*, **43**, 279.
373. Yamanaka, T. and Sadanaga, R. (1979) *J. Physique*, Colloq. C2, **40**, C2, 475.
374. Ivanitskii, V.P., Kalichenko, A.M., Matyash, I.V. *et al.* (1980) *Mineral Zh.*, **2**, 34.

375. Seifert, F. and Virgo, D. (1975) *Science*, **188**, 1107.
376. Chambaere, D., Vochten, R., Desseyn, H. and De Grave, E. (1980) *J. Physique*, Colloq. C1, **41**, C1-407.
377. McCammon, C.A. and Burns, R.G. (1980) *Am. Mineral.*, **65**, 361.
378. Whitfield, H.J. and Freeman, A.G. (1967) *J. inorg. nucl. chem.*, **29**, 903.
379. Ershova, P.Z., Babeshkin, A.M. and Perfil'ev, Y.D. (1970) *Geokhimiya*, **1970**, 252.
380. Ershova, P.Z., Perfil'ev, Y.D. and Babeshkin, A.M. (1978) *Term. Anal. Mineral*, **1978**, 105.
381. Vochten, R., De Grave, E. and Stoops, G. (1979) *Neues Jahrb. Mineral Abh.*, **137**, 208.
382. Khromov, V.I., Fedorov, V.N., Plachinda, A.S. *et al.* (1977) *Radiochem. Radioanal. Lett.*, **29**, 301.
383. Khromov, V.I., Fedorov, V.N., Plachinda, A.S., *et al.* (1978) *Zh. Fiz. Khim.*, **52**, 245.
384. Marusak, L.A. and Mulay, L.N. (1979) *J. appl. Phys.*, **50**, 7807.
385. Gupta, V.P., Singh, A.K., Chandra, K., and Nair, N.G.K. (1981) *Thermochem. Acta*, **48**, 175.
386. Le Corre, C. (1975) *Circ. Inf. Tech. Cent. Doc. Sider*, **32**, 559.
387. De Sitter, J., Govaert, A., De Grave, E. and Chambaere, D. (1977) *Bull. Soc. Chim. Belg.*, **86**, 841.
388. Cole, R.D., Ho Liu, J., Smith, G.N. *et al.* (1978) *U.S.A. Fuel*, **57**, 514.
389. Jacobs I.S., Levinson, L.M. and Hart H.R. (1978) *J. appl. Phys.*, **49**, 1775.
390. Russel, P.E. and Montano, P.A. (1978) *J. appl. Phys.*, **49**, 1573.
391. Capes, C.E., Sproule, G.I. and Taylor, J.B. (1979) *Fuel Process Tech.*, **2**, 323.
392. Male, S.E. (1980) *J. Phys. D*, **13**, 267.
393. Amirkhanov, K.I. and Anokhina, L.K. (1974) *Dokl. Akad. Nauk S.S.S.R.*, **218**, 919.
394. Amirkhanov, K.I. and Anokhina, L.K. (1975) *Dokl. Akad. Nauk S.S.S.R.*, **225**, 659.
395. Ivanitskii, V.P., Matyash, I.V. and Kakovich, F.I. (1975) *Geokhimiya*, **1975**, 850.
396. Kuzlovskii, A.A., Syromyatnikov, N.G. and Ivanov, A.I. (1975) *Izv. Akad. Nauk Kaz. S.S.S.R., Ser Geol.*, **32**, 78.
397. Anokhina, K., Chalabov, R.I. and Batyrmurzaev, A.S. (1980) *Dokl. Akad. Nauk, S.S.S.R.*, **225**, 440.
398. Armikhanov, K.I., Anokhina, L.K. and Chalabov, R.I. (1980) *J. Physique*, Colloq. C1 **41**, C1-413.
399. Eissa, W.A., Sallam, H.A. and Ashi, B.A. (1979) *J. Physique*, Colloq. C2, **40**, C2-449.
400. Boitsuva, E.P., Sagruzina, I.A. and Komaruva, N.I. (1980) *Sov. Geol.*, **2**, 75.

401. Brown, D.B. (ed.) (1980) *Proc. NATO, Inst. Adv. Studies. Conf.*, Reidel Publications, London.
402. Burns, R.G., Nolet, D.A., Parkin, K.M. *et al.* (1980) *Proc. NATO Inst. Adv. Studies Conf.*, (ed. D.B. Brown), Reidel Publications, London, p. 295.
403. Gérard, A. and Grandjean, F. (1971) *Solid State Comm.*, **9**, 1845.
404. Burns, R.G. (1972) *Can. J. Spectr.*, **17**, 51.
405. Warner, B.N., Shire, P.N. and Allen, J.L. (1972) *J. Geomag. Geoelec*, 24, 253.
406. Pollak, H. and Bruyneel, W. (1974) *J. Physique*, Colloq. C6, **35**, C6-571.
407. Gérard, A. and Grandjean, F. (1975) *Solid State Comm.*, **16**, 553.
408. Dambly, M., Pollak, H., Quartier, R. and Bruyneel, W. (1976) *J. Physique*, Colloq. C6, **37**, C6-807.
409. Fatseas, G.A., Dorman, J.L. and Blanchard, H. (1976) *J. Physique*, Colloq. C6, **37**, C6-787.
410. Lotgering, F.K. and Van Diepen, A.M. (1976) *J. Phys. Chem. Solids*, **38**, 565.
411. Scorzelli, R.B., Baggio-Saitovitch, E. and Danon, J. (1976) *J. Physique*, Colloq. C6 **37**, C6-801.
412. Amthauer, G., Annersten, H. and Hafner, S.S. (1977) *Phys. Chem. Min.*, **1**, 299.
413. Heilmann, I.V., Olsen, N.B. and Staun, J. (1977) *Phys. Scripta*, **15**, 285.
414. Huggins, F.E., Virgo, D. and Huckenholz, H.G. (1977) *Am. Mineral.*, **62**, 475, 646.
415. Parkin, K.M., Loeffler, B.M. and Burns, R.G. (1977) *Phys. Chem. Min.*, **1**, 301.
416. Abu-Eid, R.M., Langer, K. and Seifert, E. (1978) *Phys. Chem. Min.*, **3**, 271.
417. Amthauer, G. and Evans, B.J. (1978) *Phys. Chem. Min.*, **3**, 55.
418. Goldman, D.S. and Rossman, G.R. (1978) *Am. Mineral.*, **63**, 490.
419. Goldman, D.S., Rossman, G.R. and Parkin K.M. (1978) *Phys. Chem. Min.*, **3**, 225.
420. Nolet, D.A. (1978) *Solid State Comm.*, **28**, 719.
421. Govaert, A., De Grave, E., Quartier, H. *et al.* (1979) *J. Physique*, Colloq. C2, **40**, C2-442.
422. Nolet, D.A. and Burns, R.G. (1979) *Phys. Chem. Min.*, **4**, 221.
423. Pollak, H., Quartier, R., Bruyneel, W. and Walter, P. (1979) *J. Physique*, Colloq. C2, **40**, C2-455.
424. Tricker, M.J., Ash, L.A. and Jones, W. (1979) *J. inorg. nucl. Chem.*, **41**, 891.
425. Tricker, M.J. and Manning, P.G. (1979) *J. Physique*, Colloq. C2, **40**, C2-477.
426. Amthauer, G., Langer, K. and Schleistedt, M. (1980) *Phys. Chem. Min.*, **6**, 19.
427. Amthauer, G. (1980) *Am. Mineral.*, **65**, 157.

428. Evans, B.J. and Amthauer, G. (1980) *J. Phys. Chem. Solids*, **41**, 985.
429. Long, G.J., Longworth, G., Day, P. and Beveridge, D. (1980) *Inorg. Chem.*, **19**, 821.
430. Schwartz, K.B., Nolet, D.A. and Burns, R.G. (1980) *Am. Mineral.*, **65**, 142.
431. Varishnava, P.P., Tricker, M.J. and Manning, P.G. (1980) *Phys. Stat. Solidi.*, **63A**, K89.
432. Grandjean, F. and Gérard, A. (1981) *J. appl. Phys.*, **52**, 2164.
433. Pollak, H., Quartier, R. and Bruyneel, W. (1981) *Phys. Chem. Min.*, **7**, 10.
434. Tricker, M.J., Varishnava, P.P. and Manning, P.G. (1981) *J. inorg. nucl. Chem.* **43**, 1169.
435. Dziobkowski, C.T., Wrobleski, J.T. and Brown, D.B. (1981) *Inorg. Chem.*, **20**, 679.

6

Electron Spin Resonance and Nuclear Magnetic Resonance Applied to Minerals

William R. McWhinnie

It does not require great powers of advocacy to successfully argue the case that the application of the resonance spectroscopies, particularly electron spin resonance (ESR) and nuclear magnetic resonance (NMR), has been of major significance to the advance in the depth of understanding of chemical science which has occurred over the past two decades. Thus not only have the techniques been of value in a purely analytical context, but they have provided much structural information; they have given insight to the field of molecular dynamics, they have provided invaluable data for the determination of reaction kinetics and mechanism, and they produce data which can provide ever more stringent tests of theoretical models.

It may be argued that a possible disadvantage of NMR spectroscopy for the study of minerals is that the bulk of data in the chemical literature relate to measurement on solutions. Attempts to determine NMR spectra of solids have, until recently, resulted in the observation of broad line spectra. However, as will be shown later in the chapter, much data of interest may be extracted from such studies even if they are somewhat limited in scope. Two major developments of the recent past have been of particular significance. First, the increasingly general availability of Fourier Transform (FT) multinuclear spectrometers which have made possible routine determination of NMR spectra for all suitable nuclei within the periodic table. Secondly, the development of 'Magic Angle' Spinning (MAS) which has made possible the measurement of narrow line spectra of solids (see below). The fact that well resolved spectra of ^{29}Si for solid minerals are now accessible will surely confirm the advent of FTMAS NMR spectroscopy as a major landmark in the spectroscopic examination of minerals.

By contrast, well resolved ESR spectra of solids have been available for many years, hence within the literature of mineral chemistry the data base for ESR is currently much greater than for NMR. Electron spin resonance spectroscopy is applicable only to the study of paramagnetic species. Thus the mineral must contain an appropriate paramagnetic ion for study. Even the presence of a paramagnetic ion will not guarantee that data will be readily obtained, for example the presence of nickel(II) may not be readily detected whereas the presence of copper(II) is easily confirmed. Given that most minerals will contain even small quantities of paramagnetic ions, or that by way of, say, ion exchange, may be 'doped' with such ions, the technique obviously has wide applications.

In this chapter, both ESR and NMR spectroscopy will be considered in the context of their application to mineral chemistry. In each case the basis of the technique will be presented in an essentially descriptive fashion. This will be followed by illustrations of the type of problem in mineral chemistry that might be solved by application of these spectroscopies. No effort will be made to offer a comprehensive review of the literature since this would both duplicate much of the existing review literature and would also blur the purpose of this chapter. An exception will be made for ^{29}Si FTMAS NMR spectroscopy where, given the recent development of the technique and the potential importance, a more comprehensive coverage would be appropriate.

6.1 ELECTRON SPIN RESONANCE SPECTROSCOPY

A number of excellent texts give full theoretical details of ESR. A good general text is that by Atherton[1] whereas that by Abragam and Bleaney[2] refers more specifically to transition metal ions. Recently a most welcome introductory text by Symons[3] became available which, although it does not cover mineral applications as such, provides a good basic introduction for the newcomer to the technique.

The fundamental principle of ESR spectroscopy is that an electron has a spin together with an associated magnetic moment. If the electron is present in a molecule with an S ground state (i.e. $L = 0$), an applied magnetic field (B) will lift the degeneracy ($\pm\frac{1}{2}$) by causing the magnetic moment to align either parallel or antiparallel to the applied field (Fig. 6.1). If the system is irradiated with electromagnetic energy of the correct frequency, transitions occur between the two spin orientations. Since the lower state is slightly more populated than the upper state at thermal equilibrium, a net absorption of energy occurs (loss of power from incident radiation) because absorptive transitions ($-\frac{1}{2} \rightarrow +\frac{1}{2}$) outnumber radiative transitions ($+\frac{1}{2} \rightarrow -\frac{1}{2}$). Equation (6.1) relates the observed frequency to the applied field:

$$h\nu = g\beta B \tag{6.1}$$

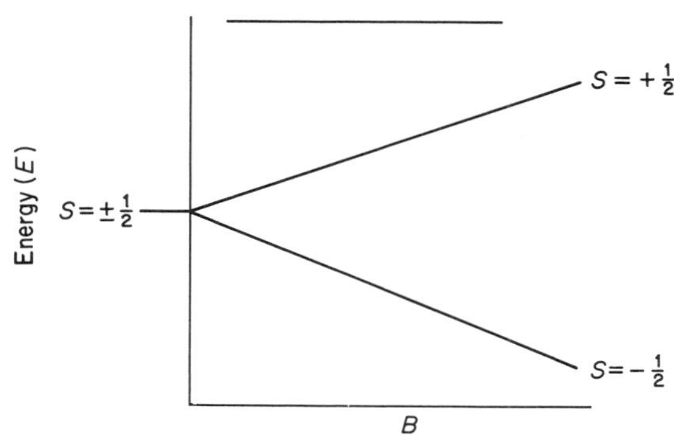

Fig. 6.1 Resolution of spin degeneracy in applied magnetic field (the spin and magnetic vectors are opposed for the electron, hence the state for which $S = -\frac{1}{2}$ is lower).

where v is the frequency, β the electronic Bohr Magneton ($eh/4\pi m_e = 9.27410(20) \times 10^{-24}$ Am^2(JT^{-1})), B is the applied field and g is the *g-value*. ESR spectra are commonly measured at the so-called X-band frequency (approximately 9 GHz) which would correspond to B of around 0.35 Tesla. Some measurements are made at Q-band frequency (approximately 36 GHz) in which case B will be around 1.32 Tesla.

6.1.1 The g-value

In a free atom, i.e. one not subjected to any external field arising from a molecular or crystalline environment, or to any electromagnetic field, the spin and orbital angular momenta of the electron couple to give a resultant (J) which is a constant of the motion (the familiar Russell–Saunders coupling) (Equation (6.2))

$$J = L + S \tag{6.2}$$

(S relates to the spin angular momentum, L to the orbital angular momentum).

If the atom is now perturbed by an external magnetic field each state characterized by a given J-value will split into $(2J + 1)$ equally spaced levels according to Equation (6.3)

$$\Delta E_J = g_J \beta B \tag{6.3}$$

where

$$g_J = 1 + \frac{J(J+1) + S(S+1) - L(L+1)}{2J(J+1)} \tag{6.4}$$

g_J is the Landé g factor. The resonant field (B_{res}) for magnetic dipole transitions between adjacent levels is then (Equation (6.5))

$$B_{\text{res}} = \frac{h\nu}{g_J\beta} \tag{6.5}$$

Now, for a paramagnetic *molecule* with a non-degenerate ground state, the orbital angular momentum of the unpaired electron is not free to respond to the aligning force of the Zeeman interaction with the external field since it will normally be strongly aligned along a specific molecular axis by the intra-molecular electric field. This results in a so-called 'quenching' of the orbital contribution. From (6.2) and (6.4) it is seen that if the quenching were total, $g = 2$ – the free electron value (actually 2.0023 after relativity effects are considered). In practice the quenching may not be total since spin-orbit coupling can produce a slight breakdown of the coupling between the orbital angular momentum and the electric field. Formally, the effect is described by an admixture of various excited states which can be coupled to the ground state by components of the orbital angular momentum operator along various molecular axes.

Since some degree of orbital angular momentum quenching will generally occur, (6.3) may be rewritten:

$$\Delta E = h\nu = g\beta B$$

which is of course Equation (6.1). The g-value is now a proportionality constant derivable from a spectrum. It may be rather precisely determined (± 0.001), hence small difference arising from differing molecular environments are readily detectable.

Given that deviations (Δg) from the free electron value of g arise from an admixture of excited states coupled to the ground state by components of the orbital angular momentum along the various molecular axes, it follows that, if the symmetry is less than cubic, Δg may assume different values for each axis. That is, the g-value may be anisotropic. The g-value may be represented as a second rank tensor, hence it is always possible to define a set of axes such that off diagonal components are zero. These are the principal axes of the g-tensor which may, or may not, coincide with the molecular axes.

In axial symmetry they will be two g-values denoted by parallel ($g_{\parallel}$) and perpendicular ($g_{\perp}$). For cases of lower symmetry three values may be determined which are designated g_1, g_2, g_3 or g_{xx}, g_{yy}, g_{zz}.

The g-values can be determined directly from powder spectra (see later) or by means of measurements on single crystals. An appropriate single crystal is rotated about each of a set of orthogonal axes and the angular (θ) dependence of g is determined, and the maximum and minimum values noted. For crystals g(observed) will be given by:

$$g^2 = \alpha + \beta \cos 2\theta + \gamma \sin 2\theta \tag{6.6}$$

where α, β and γ have to be determined. Actually, for each axis of rotation:

$$2\alpha = g_{max}^2 + g_{min}^2$$
$$2\beta = (g_{max}^2 - g_{max}^2)\cos 2\theta_{max}$$
$$2\gamma = (g_{max}^2 - g_{min}^2)\sin 2\theta_{max} \tag{6.7}$$

Contributions to Δg may arise from many atoms in the molecule. However, for transition metal ions, particularly in relatively weak ligand fields, the major contribution to the g shift is from the metal. This is the most likely situation to be encountered in the study of minerals, hence some brief remarks concerning the calculation of Δg are in order.

The Hamiltonian for the problem may be written as

$$H = H_0 + H' \tag{6.8}$$

where the Zeeman interaction between the electron spin (S) and the magnetic field (B) is:

$$H_0 = g_e \beta B S \tag{6.9}$$

(where $g_e = 2.0023$, the free electron value).

The perturbation, H', is the sum of orbital Zeeman interactions and the spin orbit interaction:

$$H' = \Sigma_\mu(\beta B l_\mu + \lambda_\mu l_\mu S_\mu) \tag{6.10}$$

where l_μ is the orbital angular momentum of the μth electron which has spin S_μ. λ_μ is the spatial part of the spin–orbit interaction. A first order perturbation treatment shows that for any non-degenerate electronic wavefunction, e.g. the ground state, $|0\rangle$, all matrix elements of the form $\langle 0|l_\mu|0\rangle$ vanish. This is a mathematical statement of the quenching of the orbital contribution. Thus, calculation of the first contribution to Δg requires second order perturbation theory. Terms linear in B and S are selected from the complete expression for the second order energy and this part of the energy is written as $\Delta g \beta B M_s$ where Δg is the g shift for a particular orientation of B, and M_s is the projection of the total spin on B. If one particular excited state gives the major contribution to Δg, the result of the calculation assumes the form:

$$\Delta g = \left(\frac{2}{M_s}\right)\frac{\langle 0|\lambda l_B S_B|1\rangle\langle 1|l_B|0\rangle}{E_0 - E_1} \tag{6.11}$$

where $|1\rangle$ denotes the wavefunction of the low lying excited state of energy E, and L_B, S_B are the components of l_μ and S_μ along B.

In general, for d^1 systems for example, the expression for Δg may be written

$$\Delta g = \pm\frac{n\lambda}{\Delta E} \tag{6.12}$$

where the negative sign is used if the coupling is to an empty level, and the positive sign if it is to a filled level. Taking the example of an axial copper(II) complex, which in 'hole' terminology we could write as $d^1_{x^2-y^2}$:

$$\Delta g_{\parallel} = \frac{8\lambda}{\Delta E(x^2 - y^2 \leftrightarrow xy)}$$

$$\Delta g_{\perp} = \frac{2\lambda}{\Delta E(x^2 - y^2 \leftrightarrow xz, yz)} \tag{6.13}$$

If the metal–ligand bond were considerably covalent, then λ may depart from the free atom (ion) value. Thus, if the electronic transitions may be confidently assigned, measurement of g gives a 'reduced' λ from which valuable information on the covalency of the metal–ligand bond is obtained.

If a molecule is in solution, it will tumble rapidly. Since the trace of a tensor is invariant under a rotation of axes, an average value of g is determined (properly the *g-factor*) from an isotropic spectrum. If the rate of tumbling is less than the linewidth in cycles per second (typically 10^7–10^8 cps), the anisotropy may be only partially quenched.

6.1.2 Zero field splitting

For many transition metal ions of interest in mineral chemistry, several spin states are accessible, i.e. when the number of unpaired electrons exceeds unity. The field generated by the other electrons adds a constant energy to all Zeeman terms in cubic symmetry, but when the symmetry of the environment is low, polarization of the secondary field may occur which causes splitting in the absence of a magnetic field. No electrostatic field will completely resolve the spin degeneracy, e.g. for iron(III) where $S = 5/2$, separation of the $\pm 1/2$, $\pm 3/2$, $\pm 5/2$ states may be caused by zero-field splitting. These states are known as *Kramers doublets.* It is an experimental fact that, for transition metal ions – particularly at the X-band frequency, ESR spectra are only readily observed if the ground state is a Kramers doublet, e.g. Cu(II), Mo(V), V(IV), Cr(III), Fe(III), Mn(II).

In the case where the crystal field effects are greater in magnitude than the magnetic effects, the spin Hamiltonian is:

$$H = g\beta BS_z + DS_z^2 + E(S_x^2 - S_y^2) \tag{6.14}$$

D effectively describes an axial distortion, E a rhombohedral distortion. For the case of iron(III) in an axial field, $E = 0$ and the energy levels, W, are given by:

$$W = Dm^2 \tag{6.15}$$

where $m = \pm 5/2, \pm 3/2, \pm 1/2$ (Fig. 6.2). Small distortions result in a five-line spectrum. For large values of D a limiting case is reached where a '$g_{\perp}$' feature is seen close to 6, whereas '$g_{\parallel}$' remains at 2. Sometimes iron(III) gives an isotropic feature at $g \simeq 4.3$ – this occurs when $E \neq 0$ and the ratio of E to D is $\simeq 1/3$.

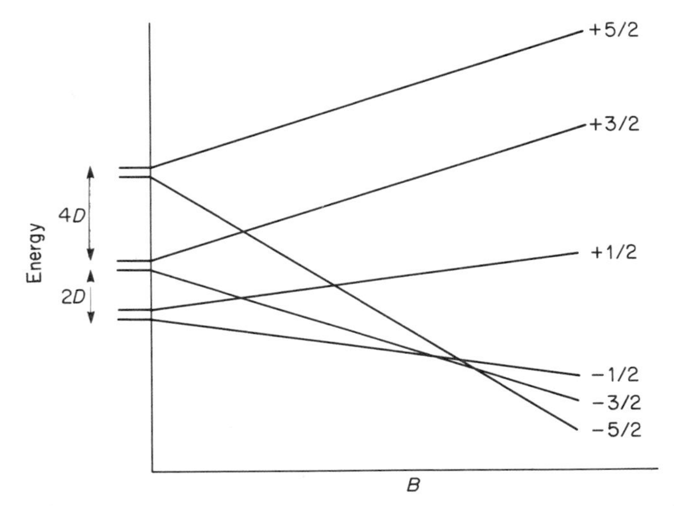

Fig. 6.2 Zero field splitting, and divergence of levels in an applied magnetic field (B) for an $S=\frac{5}{2}$ complex, axially distorted.

6.1.3 Hyperfine interactions

Nuclei with spin (I) > 0 have magnetic moments, hence an interaction between the electron and nuclear magnetic moments is to be expected (the electronic orbital magnetic moment may be ignored since quenching by the molecular field makes the term very small). The hyperfine interaction may be divided into an isotropic and an anisotropic component. The isotropic part arises from unpaired electron density at the nucleus which can only be non-zero for spherically symmetrical s-type orbitals.

The anisotropic term corresponds to the classical part of the magnetic dipole interaction. An appropriate Hamiltonian is:

$$H = IAS \tag{6.16}$$

where

$$A_{ij} = A_o\delta_{ij} + B_{ij} \tag{6.17}$$

where A_o is the isotropic component and B the anisotropic contribution. Often it is valid to modify Equation (6.1):

$$h\nu = g\beta(B + M_I A) \tag{6.18}$$

This defines A as it is usually measured – as the separation of peaks in an experimental spectrum. The selection rules governing hyperfine interactions are $\Delta M_S = \pm 1, \Delta M_I = 0$, thus a magnetic nucleus of spin I will split the resonance into $(2I + 1)$ components. When g is anisotropic, all components

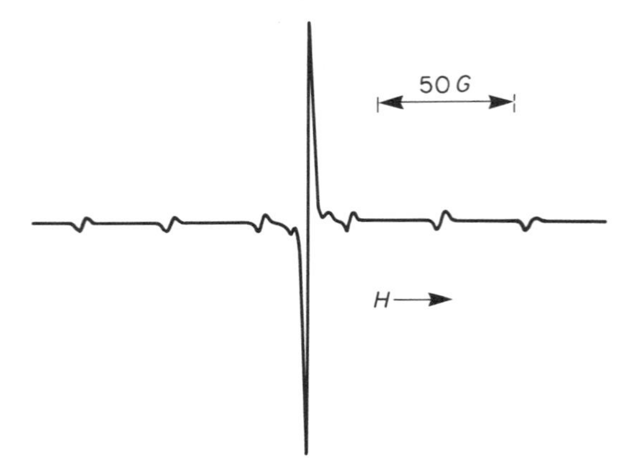

Fig. 6.3 First derivative of absorption spectrum of $K_3\{Mo(CN)_8\}$ in aqueous solution (after McGarvey, B.R. (1966) *Trans. Metal Chem.*, **3**, Edward Arnold (publishers) Ltd.).

will be split, thus, for the axial case, both $g_\perp$ and $g_\parallel$ will be split into $(2I + 1)$ components separated by $A_\perp$ and $A_\parallel$ respectively. The difference in $A_\perp$ and $A_\parallel$ will arise from the anisotropy of B in Equation (6.17). In some cases both isotopes with magnetic nuclei and others with non-magnetic nuclei may be present in significant abundance. For example the odd isotopes ^{95}Mo and ^{97}Mo both have spin 5/2, have very similar nuclear magnetic moments, and show a combined natural abundance of 25.18%. The even isotopes (^{92}Mo, ^{94}Mo, ^{96}Mo, ^{98}Mo) have $I = 0$. A solution spectrum of $K_3[Mo(CN)_8]$ is shown in Fig. 6.3. The abundance ratio of odd to even isotopes is 1:3, hence each hyperfine line has intensity (1/3) (1/6), i.e. 1/18 of the central line.

6.2 PRACTICAL ASPECTS OF ESR

Much of the above may be summarized in an equation for the Spin Hamiltonian of a molecule for which $S = 1/2$, the applied field is B and magnetic hyperfine interaction is with one nucleus of spin I:

$$H_{\text{spin}} = \beta B g S + IAS - \gamma\hbar \text{BI} \tag{6.19}$$

The third term is the nuclear Zeeman interaction which is included for completeness (γ is the nuclear gyromagnetic ratio). Since in many ESR transitions only the electron spin changes orientation, this term may usually be ignored.

The objective of an ESR experiment is to determine the principal components of g and A and then to relate those to features of molecular structure. It is now necessary to turn to some practical aspects before illustrating some typical applications of the technique.

6.2.1 The spectrum

ESR spectra are recorded at fixed microwave frequency by scanning the magnetic field. Instrument design features are such that it is usual to record the first derivative of the absorption versus field function. Measurement of the g-value requires accurate knowledge of the frequency and of the resonant field, B_{res}, since from Equation (6.1):

$$h\nu = g\beta B_{res} \tag{6.20}$$

A proton resonance gaussmeter could be used to measure B_{res}. Normal practice involves the use of 'g-markers'. These are compounds of accurately known g, for example diphenylpicrylhydrazyl (DPPH) which has $g = 2.0036 \pm 0.0003$, thus for fixed ν (Equation (6.20)):

$$B_1 g_1 = B_2 g_2 \tag{6.21}$$

where B_1, g_1 may refer to the unknown and B_2, g_2, to the marker. Manganese(II) in a matrix of magnesium oxide is useful since a six-line spectrum is seen (^{55}Mn, $I = 5/2$ natural abundance 100%). Thus one line may serve as a g-marker and the separation of lines serves as a convenient field calibration.

Hyperfine constants (see Equation (6.18)) are determined from the separation of lines in the spectrum – as illustrated by the manganese(II) marker above.

6.2.2 Powder spectra

Many minerals are not available for study as suitable single crystals, nor are solution measurements feasible. Hence investigation of microcrystalline powders is necessary. Actually, provided that the g- and A-tensors share the same principal axes (which is not always the case), data may be extracted more rapidly from a powder study than from single crystal work.

In a glass, a frozen solution, or in a fine powder, paramagnetic species will be statistically distributed over all orientations to the applied magnetic field. However, no resonance can be possible beyond the field values characterizing the outermost features of the spectrum (usually the parallel features since generally $A_{\parallel} > A_{\perp}$), thus there is a sharp onset of absorption of microwave energy. It is reasonable to expect spectral changes at other turning points in the spectrum and these will be accentuated by the normal first derivative presentation. Background theory of powder line shapes may be found in the classic papers of Sands[4] and Kneubühl.[5] A review by Adrian[6] is worth consulting on this topic and others covered here. In Fig. 6.4 are illustrated a number of typical line shapes that may be encountered from studies of microcrystalline powders.

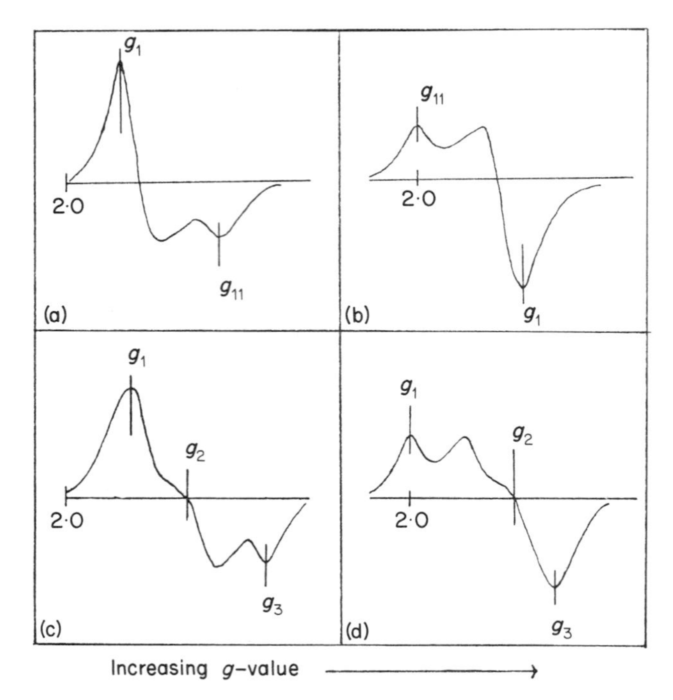

Fig. 6.4 Typical powder ESR spectra for spin $= \frac{1}{2}$ system such as copper (II). (a) Axial distortion ($g_{11} > g_1$; (b) axial distortion ($g_1 > g_{11}$); (c) rhombic distortion; (d) rhombic distortion, one g-value close to 2.0.

6.2.3 Linewidths

A number of factors may contribute to experimental linewidths. One quite common cause of a broad line is unresolved hyperfine splitting which confers a Gaussian outline. Quite often the perpendicular component of g in the spectrum of an axially distorted molecule has unresolved hyperfine splitting. Two other factors contributing to linewidths are spin–lattice relaxation and spin–spin relaxation.

For a spectrometer operating in the X-band at normal temperature, the difference in population of the spin ground and excited states is no more than 0.0014%. If transitions are stimulated between the two states, it can be shown that the population difference will diminish with time, hence the magnitude of the electromagnetic energy absorbed will decrease ultimately to zero, when the spin system is said to be 'saturated'. Normally the intensity of an ESR spectrum is constant with time, thus there must be some alternative to radiative transitions by which the system 'relaxes' to the ground state. This will be by transfer of energy to other degrees of freedom within the lattice, i.e. spin–lattice relaxation. The spin–lattice relaxation time, T_1, is related to the width of the absorption. If T_1 is large, low powers may be used to avoid saturation. If, however, T_1 is small, as may be deduced from a consideration of the Heisenberg Uncertainty Principle, this will lead to broad lines.

Spin–spin interaction arises from interaction between spin dipoles. The field experienced by an individual spin is the sum of the applied field and those fields due to neighbouring unpaired spins, hence small variations in the value of B_{res} may occur across the spin array. This phenomenon is of more significance for crystals than for dilute solutions.

6.3 SOME APPLICATIONS OF ESR IN MINERAL CHEMISTRY

Selection of examples will be made such that the relevance of the various points covered in Sections 6.1 and 6.2 becomes apparent. For convenience an attempt is made to list applications under a range of headings. To some extent this is an arbitrary classification since many areas of overlap occur. Also it is sometimes questionable whether a particular experiment is designed primarily to study the mineral or the properties of some sorbed species. In this chapter, properties of sorbed species are only considered if they provide information on the mineral, thus a great deal of literature dealing with catalytic applications of minerals is consciously omitted.

6.3.1 ESR from bulk mineral structures

Minerals may be paramagnetic by virtue of exchanged paramagnetic ions, radiation damage or through the absorption of paramagnetic molecules. Here we are concerned with some examples of paramagnetism associated with the bulk (non-exchangeable) structure. Samples of natural calcite can be relatively pure with respect to calcium carbonate and can be used in ceramic formulations. An ESR spectrum may however be obtained (Fig. 6.5(a)) which arises from small quantities of isomorphically substituted manganese(II). The spectrum is complex showing both some g-anisotropy and also some spin forbidden transitions as weak features. The latter arise when the nuclear quantum number changes ($\Delta M_I = \pm 1$) simultaneously with the resonance absorption (see Equation (6.17)). When rhombohedral calcite decomposes at 800° C to cubic calcium oxide, a simple isotropic six line pattern is seen – Fig. (6.5(b)). The manganese(II) spectrum can be useful to monitor the fate of calcite in high temperature reactions with other ceramic materials.[7]

All natural specimens of the clay mineral kaolinite show ESR spectra which, although similar, reflect in their detail the place of origin of the clay. Features at $g = 2$ and $g = 4$ are common to all specimens. The $g = 4$ feature is due to lattice iron(III). Meads and Malden[8] carefully removed iron contaminants from kaolinite and showed by Mössbauer spectroscopy that the iron(III) was in the octahedral layer (the possibility that the $g = 4$ signal arose from iron(III)

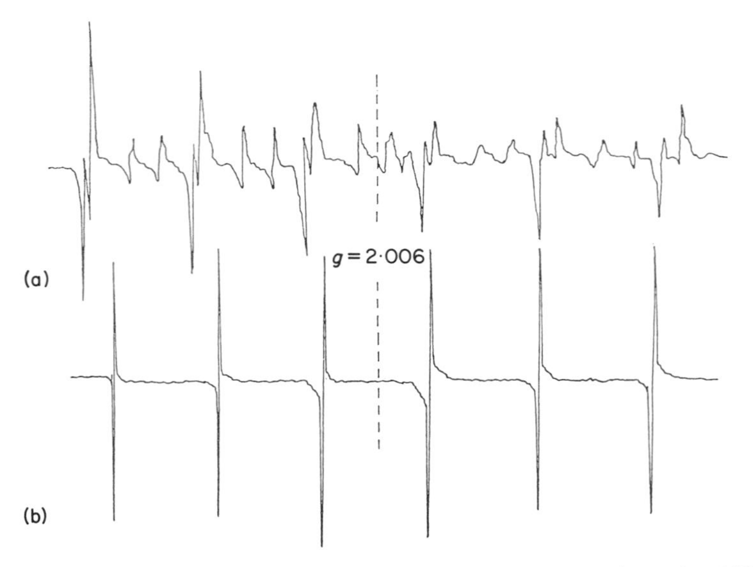

Fig. 6.5 (a) Calcite with manganese(II) impurity. (b) Same specimen heated to 800° C.

substituting silicon(IV) in distorted tetrahedral sites had previously been suggested[9]). The origin of iron(III) resonances at $g = 4$ has been covered in Section 6.1.2.

More detailed investigation of the $g = 4$ resonance led to the conclusion that two overlapping signals were present.[10] One signal was isotropic with $g = 4.2$, the other anisotropic with three g-values of 4.9, 3.7 and 3.5. The respective signals can be related to iron(III) in sites of complete ($E/D \simeq 1/3$) and 'partial' orthorhombic symmetry. The complex spectrum in the $g = 4$ region has also been related to degrees of crystallinity[8,10] in that iron(III) may be present both in layers with stacking disorder and in sites of high crystallinity with ordered stacking. Treatment of the clay with a reducing agent such as hydrazine modifies the $g = 4$ resonance, possibly implying that the iron(III) in one site is more readily reduced than is the other.[11]

When kaolinite is heated beyond 500° C (to the metakaolinite transition), the initially complex $g = 4$ resonance becomes more intense and isotropic ($g = 4.2$).[12] Possibly the signal may now be related to iron(III) in distorted tetrahedral sites after dehydroxylation of the octahedral layer.[13]

The major feature of the kaolinite spectrum at $g = 2$ arises from an $S = 1/2$ system showing both $g_{\perp}$ (~ 2.0) and $g_{\parallel}$(~ 2.05) features. The unique axis is parallel to the mineral c axis.[12] It is now generally accepted that the origin of the resonance is a defect. Synthetic kaolinites doped with iron(III) fail to give the signal[15] but doping with magnesium(II)[14] or iron(II)[15] followed by X-irradiation reproduces the spectrum of the natural material. The magnesium(II) or iron(II) are 'pre-centres' required to stabilize the 'hole' defect.

A range of other iron containing minerals show resonances at $g = 4$, for example smectites,[16] micas and vermiculites,[17–19] which can be interpreted in terms of two non-equivalent octahedrally disposed iron(III) ions. One possible distinction in these triple layer minerals is the distinction between *cis*- and *trans*-$FeO_4(OH)_2$ arrangements, but other workers noted a sensitivity of the higher field components to the nature of the exchangeable ions on smectites. Thus sodium and potassium ions at zero humidity cause significant decreases in the intensity of the anisotropic high field resonance, whereas the strongly hydrated lithium and calcium ions leave the line shape unaffected.[20–22] It was concluded that the high field anisotropic resonance was associated with negative charge centres in the silicate framework, hence the differing environments for the iron(III) may result from one set being adjacent to magnesium(II) sites and the other set to aluminium(III) sites.

6.3.2 Electron transfer reactions

The addition of a foreign substance to the mineral may include a redox (electron-transfer) reaction. In several such cases, analysis of associated ESR data may provide information about the mineral as such.

Nitrogen(II) oxide (nitric oxide, NO) is a useful reagent. If a Y-type zeolite is treated with the gas, broad ESR signals are obtained. However, if the specimen is left for several days at room temperature under nitric oxide and subsequently quickly pumped free of the gas, no ESR signal is seen. Reintroduction of a small quantity of NO immediately gives a sharp, well resolved spectrum.

In dehydrated zeolites many of the cations are located close to the inner surface of the large channels and cavities. They are thus unevenly shielded as is, in its turn, the negatively charged aluminosilicate network. The material resembles an expanded ionic crystal having irregular arrangements of ions. The Mådlung energy of such a system should be reduced by filling the intracrystalline cavities with either appropriately polarized molecules or with suitable ions. It has therefore been proposed that the above observations with NO may be accounted for in terms of a disproportionation reaction:

$$3NO \rightarrow N_2O + NO_2$$
$$NO + NO_2 \rightarrow N_2O_3 \rightarrow NO^+ + NO_2^-$$

The NO^+ and NO_2^- ions position themselves in the cavities such as to lower the Mådlung energy and thus account for the 'NO pre-treatment' phenomenon.[13]

A further example arises when NaY zeolite is exposed to sodium vapour at 300–500°C. A bright red material is produced giving an ESR spectrum showing 13 hyperfine components ($A = 32.3$ Hz, $g = 1.999$).[24] The spectrum arises from Na_4^{3+} clusters. The spin, I, of ^{23}Na is 3/2 and the isotope is 100% abundant. The appropriate expression for the hyperfine splitting for a cluster of four sodium units is $[(2\sum_{i=1}^{4} I_i) + 1]$, i.e. 13. Such clusters may also be produced by γ-irradiation of NaY.[25] Spin density measurements reveal the presence of one electron per α–cage. Again the electrolytic properties of the zeolite may be invoked, as for N_2O_3 above, since the following ionization will be encouraged:

$$Na \rightarrow Na^+ + e$$

An optical absorption band is seen at 500 nm which is assigned to the transition to the lowest lying excited state. Since the electron is effectively confined to a cubic box defined by a sodium species at alternate corners, particle in a box theory may be used to calculate the side dimension to be 0.67 nm which agrees well enough with expectation for Na_4^{3+} in an α-cage.

The octahedrally coordinated nickel(II) ion has a $^3A_{2g}$ electronic ground state, and is therefore a non-Kramers ion which will generally be dead in an ESR experiment. Treatment of NiY zeolite with NO at 100° C over 3 h gave an ESR spectrum typical of a $3d^9$ ion ($g_{\parallel} = 2.430$, $g_{\perp} = 2.171$) which may unambiguously be assigned to Ni(I).[26] Ions on site II of the Y-zeolite should be most readily reduced and it was shown that spin density measurements of the concentration of nickel ions on site II were in exact agreement with an X-ray study of NiY.[27]

Treatment of kaolinite with nitric acid should either remove, or at least oxidize, all surface iron to iron(III). However treatment of such a specimen with 2, 2′-bipyridyl (bipy) gives the familiar colour of the diamagnetic complex ion $[Fe(bipy)_3]^{2+}$.[11] The complex is present in amounts too small for detection by Mössbauer spectroscopy which does however show the presence of lattice iron(II). A new ESR signal at $g = 2.115$ is seen for the bipyridyl treated clay which may conceivably arise from iron(III) generated by electron transfer from iron(II), across an oxide layer, to the surface.

6.3.3 Paramagnetic probes

Although NO is an odd electron molecule, it shows no paramagnetism in the electronic ground state $^2\pi_{\frac{1}{2}}$ since the degeneracy of the pi-orbitals leads to exact cancelling of the spin and orbital moments. Quenching the orbital moment, e.g. by removal of the degeneracy of the pi-orbitals should lead to an ESR signal which is sensitive to the separation of the levels. Accordingly, NO should be a sensitive probe with which to study electric fields associated with cations in minerals. An example from zeolite chemistry illustrates the point.

Treatment of 'NO-pretreated' (see Section 6.3.2 above) NaY of BaY zeolites with NO gives sharp spectra. It is anticipated that for the perturbed NO molecule: $g_z < g_y \simeq g_x = 2.0023 = g_e$. Also $\Delta g_z (= g_z - g_e)$ is inversely proportional to the separation induced within the pi-antibonding set.[28] Thus since Δg_z is less for BaY than for NaY, it follows that the electric field must be greater at Ba^{2+} than at Na^{+}.[23]

Nitroxides are stable organic radicals which may also be used as probes. Dilute solutions of nitroxides such as protonated 4-amino-2,2,6,6-tetramethylpiperidine-*N*-oxide in solvents of low viscosity should give a three line ESR spectrum since rapid tumbling will average the anisotropy of the g and A tensors (Section 6.1.1). As the correlation time (τ_c) for tumbling increases from values associated with fast tumbling ($\sim 10^{-10}$ s) to those for moderately fast tumbling ($\sim 5 \times 10^{-9}$ s), the line shape alters and τ_c can be evaluated from the line broadening. Since τ_c is also related to the viscosity of the medium via Stokes' law, it appeared that nitroxide spin probes might give information on the viscosity in clay interlayers.

In the event, the spin probes are partially orientated in the interlayers even when fully hydrated. On K. hectorite for example, the nitroxide molecules are orientated such that the nitrogen p-orbital makes an angle of 45° to the silicate sheets.[29]

6.3.4 Transition metal ions

Transition metal ions in minerals may be studied by a variety of techniques many of which are covered in other chapters. If the ion is paramagnetic with a Kramers doublet as ground state, ESR provides a powerful method of investigating both the stereochemical and electronic environment of the species as well as, in some cases, the dynamic behaviour. Examples in this section will be taken mainly from clay chemistry.

Clays are not readily investigated by single crystal X-ray studies, they also have significant to large ion exchange capacities and may thus contain a spectrum of transition metal constituents. Against this background well designed ESR experiments may provide virtually unique information on at least the micro-environment, of selected species. Given that the purpose of the chapter is to illustrate applications of ESR, it is suggested that powerful example may be culled from this area of mineral chemistry.[30]

(a) *Copper(II)*

Copper(II) ($3d^9, S = \frac{1}{2}$) is a useful probe. An elegant study by McBride, Pinnavaia and Mortland[31] provides a good illustration of information to be obtained on ions located in the interlamellar regions of expandable lays such as those of the smectite group. The structure of the hydrated interlayer of the mineral depends to some extent on the exchangeable cation. Thus for hectorite, complete copper(II) exchange gives a species which can show a well-defined monolayer of water in the interlamellar region ($d_{001} = 1.24$ nm), but increase in relative humidity gives a continuous transition to a multilayer

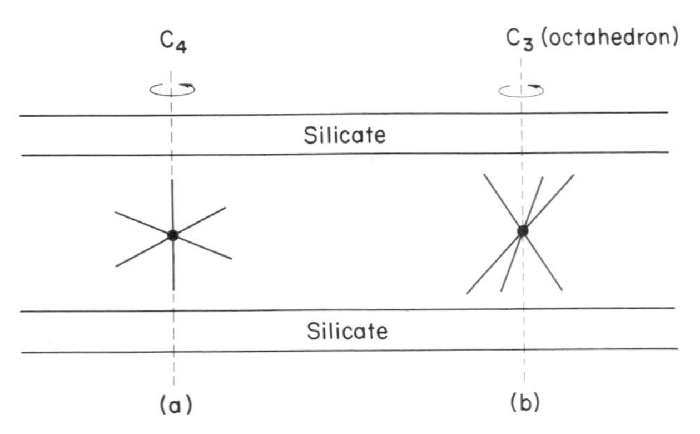

Fig. 6.6 Tetragonal copper(II) in interlamellar region of a sheet silicate. (a) Tetragonal (C_4) axis perpendicular to silicate sheet. (b) Tetragonal axis inclined to plan of silicate sheet (for a regular octahedron, the C_3 axis would be perpendicular to the silicate sheet as indicated).

phase ($d_{001} = 2.0$ nm). If magnesium(II) is the exchangeable cation, a definite intermediate hydration phase is defined ($d_{001} = 1.5$ nm) which is retained on doping with copper(II) to a level of 5%. The copper(II) ions give a good ESR spectrum which shows strong dependence on the orientation[32] of the silicate sheets to the magnetic field. Thus when the silicate sheets were orientated perpendicular to the field only a $g_{\parallel}$ component was seen ($g_{\parallel} = 2.335$) which was clearly split into four components by the spin ($I = 3/2$) of ^{63}Cu and ^{65}Cu.

Since $g_{\parallel} > g_{\perp}$, it is probable that the ion has a $d_{x^2-y^2}$ ground state. This fact coupled with the orientation dependence of the spectra strongly implies the elongated tetragonal axis to be perpendicular to the silicate sheets (see Fig. 6.6) and, by implication, the magnesium(II) ions have a similar orientation. By contrast, ESR spectra of copper(II) doped into the $d_{001} = 1.43$ nm hydration phase of vermiculite show no orientation dependence, hence the major symmetry axis must be orientated at 45° to the silicate layers. This difference in behaviour doubtless reflects the different ways in which the excess negative charge arises in the minerals. For hectorite, the major cause is M(I) or M(II) substitution for Al(III) in the octahedral layer; for vermiculite Al(III) substitution for Si(IV) in the tetrahedral layer. The structure of the interlayer water in vermiculite maximizes hydrogen bonding interactions with the silicate sheets.

Full hydration of the Cu(II) (5%)/Mg(II) hectorite results in a broad signal analogous to that observed for copper(II) in methanol/water. Rapid tumbling of the ion may be the cause. Interestingly the fully copper(II) exchanged, fully hydrated, hectorite gives an isotropic line which is somewhat narrower than that at the 5% doping level. This may be an example of spin exchange narrowing.

Hectorite is an attractive mineral for ESR studies of exchanged ions since it is low in structural iron. Montmorillonite, by contrast, may contain significant quantities of structural iron. If montmorillonite is exchanged with copper(II) and heat treated, significant loss of signal intensity is observed which may however be regenerated on reswelling in 95% ethanol.[31] There is infra-red evidence that the water layer collapses at 156° C and that dehydrated copper(II) ions move into hexagonal holes in the lattice. This may lead to enhancement of interaction with lattice iron(III) giving very short spin–lattice relaxation times and broadening the signal.[33]

Not surprisingly, exchange of copper(II) with calcium montmorillonite may be eliminated in the presence of large (5 M) concentrations of sodium chloride. However, a copper(II) species which gives $g_{\parallel} = 2.00$($A_{\parallel} = 80$ gauss) and $g_{\perp} = 2.20$, is adsorbed onto the clay surface in small amounts.[34] This is an example of a d_z2 ground state ($g_{\perp} > g_{\parallel} \simeq 2.0$).[35] The stereochemistry of the copper(II) ion must therefore be consistent with the unpaired spin in the $3d_z2$ orbital. It has been suggested that a trigonal bipyramidal species such as $CuCl_5^{3-}$ may be responsible.[34]

Clays and micas are not the only species with layered structures offering interlamellar locations for exchangeable ions. The synthetic material α-zirconium phosphate, $Zr(HPO_4)_2 \cdot H_2O$ provides another example. Copper(II) ions exchanged onto this species give a further instance of $g_{\perp} > g_{\parallel}$ and a d_z2 ground state.[36] The explanation may be that the ion is in a compressed tetragonal environment, or that the $[Cu(H_2O)_6]^{2+}$ species has a *cis*-distortion. In either case strong hydrogen bonding of two coordinated water molecules to phosphate is likely.

(b) *Manganese(II) and iron(III)*

In many respects copper(II) is an excellent paramagnetic probe. However, when the ion gives an isotropic spectrum the interpretation is ambiguous. Whereas the spectrum may imply a freely tumbling aquo-ion, there is always the possibility that a dynamic Jahn–Teller effect is involved. Hexaaquomanganese(II) ($3d^5$) will, in most environments, have $g_1 = g_2 = g_3$ hence it is not useful for orientation studies, but it can be a useful probe of ion mobility. Generally, studies of linewidth are involved.

The linewidth can be considered to be made up of two contributions, namely the intrinsic linewidth (related to T_1, see Section (6.2.3) and a further component due to dipolar interactions between neighbouring manganese(II) ions.[37] The latter contribution is concentration dependent because of an r^{-3} dependence, where r is the average distance between manganese(II) centres. A manganese(II) ESR spectrum consists of six hyperfine lines, the fourth highest field line being the $M_I = -\frac{1}{2}$ transition. The correlation time for the collision of manganese(II) with bulk water is directly proportional to the width of this line. ESR studies of manganese(II) in dilute solution, and doped into fully hydrated magnesium hectorite show that the line width is only 30% greater in the mineral, hence the interlayers offer a very solution-like environment.[38]

Manganese(II) has no crystal field stabilization energy and should not show strong site preference when exchanged in small quantities onto a zeolite previously treated with a diamagnetic ion. In this way site preferences for ions on Linde-X[39] and – Y[40] (synthetic faujasites) have been examined, thus for Linde-X type I sites were shown to be especially favourable for electropositive ions with radius 0.095–0.135 nm, whereas small ions are strongly bound on type II sites.[39]

Iron(III) which is isoelectronic with manganese(II) is less attractive as a probe due to the usual presence of structural iron in many silicate minerals. However, a recent study[41] of iron(III) exchanged HNaY zeolites claims to recognize hydrated iron(III) as giving a sharp signal at $g = 2.0$ which vanishes on dehydration. A broad band, also centred on $g = 2.0$, is attributed to Fe–O clusters and the usual $g = 4.0$ resonance is assigned to structural iron.

(c) *Vanadyl (VO^{2+})*

The vanadyl ion combines some advantages of copper(II) and manganese(II). The presence of the V = O bond ensures a strongly tetragonal species with large g-anisotropy which may be valuable in orientation studies. At the same time an isotropic spectrum in a solution environment indicates rapid tumbling. The isotope ^{51}V is 99.76% abundant with spin 7/2. An anisotropic powder spectrum can appear quite complex; however $A_{\parallel}$ is usually greater than $A_{\perp}$, hence the outermost parallel features are easily recognized and the extraction of components of the spin Hamiltonian is generally straightforward.

The ion lends itself to line broadening studies of ion mobility[42] similar to those described above for manganese(II) with, in this instance, the width of the $M_I = 7/2$ transition being proportional to the correlation time for collision of VO^{2+} with bulk water.

A recent study of vanadyl ions exchanged onto kaolinite showed overlapping isotropic signals with $g = 1.940$, $\langle A \rangle = 118$ gauss and $g = 1.996$, $\langle A \rangle = 95$ gauss. The first set of parameters imply a site on which VO^{2+} is freely tumbling in an aqueous solution-type environment. The other site may be associated with the usual hydrated iron oxide contamination associated with many kaolinites.

6.3.5 Summary

The selection of examples of applications of ESR spectroscopy to the study of minerals reflects to a large extent the particular interests of the writer; however from consideration of the particular more general thoughts may often flow. In the final analysis, it is the type of application which is to be stressed rather more than the substrate to which the application is made.

6.4 NUCLEAR MAGNETIC RESONANCE SPECTROSCOPY

Some nuclei possess the property of 'spin' and hence have magnetic moments which may give rise to a set of enrgy levels in an applied magnetic field – as for the electron. Nuclei with odd mass numbers possess spin, I, of $n/2$ where n is an integer, e.g. $^{1}H(I = 1/2)$, $^{19}F(I = 1/2)$ $^{23}Na(I = 3/2)$, $^{27}Al(I = 5/2)$, $^{29}Si(I = 1/2)$, $^{31}P(I = 1/2)$, all of which are of obvious interest in the study of minerals. Nuclei with even mass numbers but odd charge (atomic) numbers also possess spin but the values of I are integral, e.g. $^{2}D(I = 1)$, $^{14}N(I = 1)$. If both the mass and charge numbers are even, then $I = 0$ and the nucleus is non-magnetic. Among nuclei in this class is ^{16}O.

6.4.1 A description of the experiment

NMR spectroscopists find it convenient to use both quantum mechanical and classical descriptions of the experiment. It will be convenient to consider both approaches since the quantum description will highlight similarities with ESR and the classical description will be more readily related to some newer experimental techniques.

The nuclear sub-particles have both orbital and spin motions which combine to give a resultant spin which may or may not be zero (see above). The spin angular momentum vector is measured in units of $h/2\pi(= \hbar)$ and the conventional symbol is $I\hbar$. The quantity is quantized, thus an operator $\bar{I}$ is required which acts on an eigenfunction (nuclear spin wavefunction – ψN) to generate the necessary set of eigenvalues I according to:

$$I\psi N = [I(I + 1)]^{\frac{1}{2}}\psi N \tag{6.22}$$

Where I is the nuclear spin quantum number. If $I \neq 0$, a nuclear magnetic moment, μ, will arise such that:

$$\mu = \gamma_{N} I\hbar \tag{6.23}$$

Where γ_N is the gyromagnetic ratio which may be expressed as:

$$\gamma_{N} = g_{N}\mu_{N} \tag{6.24}$$

Where g_N (dimensionless) is the nuclear g-factor and μ_N the nuclear magneton ($= e\hbar/2M$ where e and M are the charge and mass of the proton). Numerically, $g_N = 5.05095(15) \times 10^{-27}$ $Am^2(JT^{-1})$. Therefore

$$g_{N}\mu_{N}\hbar I = \gamma_{N} I \tag{6.25}$$

The argument may conveniently be narrowed to nuclei for which $I = 1/2$ placed in an external magnetic flux B_O. The energy of interaction between μ and B_O may be expressed as the eigenvalues of an operator $\bar{H}$:

$$\bar{H} = -\mu B_{O} \tag{6.26}$$

If, by convention, the direction of B is taken to be the z-axis, from (6.22):

$$\bar{I}_z\psi_N = m_z\psi_N \tag{6.27}$$

where m_z is limited in value to $I, (I-1), \ldots, -I$, i.e. to $+\frac{1}{2}$ for $I = \frac{1}{2}$. Therefore

$$\bar{H} = -\hbar\gamma_N B_O \bar{I}_z$$

The appropriate Schrödinger equation is:

$$\begin{aligned} -\hbar\gamma_N B_O I_z\psi_N &= E_N\psi_N \\ &= -\hbar\gamma_N B_O m_z\psi_N \\ &= -\hbar\gamma_N B_O(+\tfrac{1}{2})\psi_N \end{aligned} \tag{6.28}$$

Thus the resonance condition is

$$h\nu = \hbar\gamma_N B_O = g_N\mu_N B_O \tag{6.29}$$

That is, for $I = 1/2$, there are two levels separated by $g_N\mu_N B_O$. Classically the lower level corresponds to μ aligned with the field. The similar form of this condition to that for ESR, Equation (6.1), will be noted. For typical values of B_O (say 2.35 Tesla), ν will be a radiofrequency, e.g. at 2.35 Tesla, 100.04 MHz for ^1H or 40.48 MHz for ^{31}P. The gyromagnetic ratio, γ_N, is sufficiently characteristic of a particular nucleus for that nucleus to be readily selected for study.

The same problems of saturation are encountered in NMR as in ESR (see Section 6.2.3). Similar reasoning to that already given is applicable and, clearly, if the spin system were rapidly saturated, the experiment would be of little chemical interest. Relaxation effects are responsible for restoring the Boltzmann distribution and both have been met previously – spin–lattice relaxation of time constant, T_1, and spin–spin relaxation of time constant T_2.

In classical language, if a spinning charged particle is placed in a magnetic flux B_O with the magnetic vector μ inclined at angle θ to the direction of $B_O(z)$ (see Fig. 6.7), it will experience a torque L which tends to align it with the field. It can be shown that

$$L = \mu B_O \tag{6.30}$$

The torque causes μ to precess about B_O at angular frequency ω_O where

$$\omega_O = \gamma_N B_O \tag{6.31}$$

This is the Larmor frequency which is, in frequency units, ν_L

$$\nu_L = \gamma_N B_O/2\pi \tag{6.32}$$

If a small secondary field, B_1, is applied in the xy plane, then at a given point in the precessional path the magnetic dipole experiences a combination of B_1 and B_O which changes θ by $+\delta\theta$. However, π radians further on the change in θ is $-\delta\theta$, thus no net change in θ is possible. However if B_1 were to rotate at ν_L

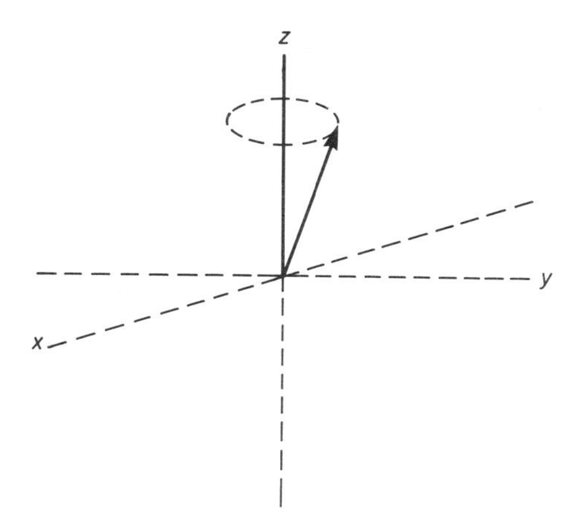

Fig. 6.7 Precession of vector about z (direction of B_O).

within the xy plane, a change of orientation (spin flip, the resonance condition) is possible. Circularly polarized light provides just such a rotating magnetic component; however in practice linearly polarized light is used since it contains two circularly polarized components of which only the one of correct 'handedness' will be active.

6.4.2 Pulse methods and Fourier transform

Modern NMR spectrometers, particularly those used to study nuclei such as ^{29}Si subject the spin system to a series of equally spaced pulses of radio frequency energy. The objective is to examine the decay in magnetization of the specimen as a function of time betwen pulses, rather than to use continuous wave excitation. The pulse method has advantages where the magnetic isotope is of relatively low abundance, e.g. ^{13}C (1.11%), ^{29}Si(4.70%).

If an external field B_O is applied along the z axis of a cartesian coordinate set that rotates about z at the Larmor frequency, $\gamma_N B_O$, the net magnetization vector M is invariant with time. B_1 rotates at the Larmor frequency in the laboratory framework and is applied in the xy plane. At resonance, B_O is cancelled along z by a component of magnetic induction, ω/γ_N, leaving only B_1 to interact with M. Since B_1 and the rotating cartesian set rotate at the same frequency, B_1 is considered directed along x, hence M will begin to precess about x. In t_p s it will precess through θ radians where

$$\theta = \gamma_N B_1 t_p \qquad (6.33)$$

A pulse of radio frequency (rf) energy is applied for t_p s. The instrument is arranged to detect only the component of M along y, hence maximum intensity is obtained if $\theta = \pi/2$ (the 90° pulse). After t_p, B_1 returns to zero and the system

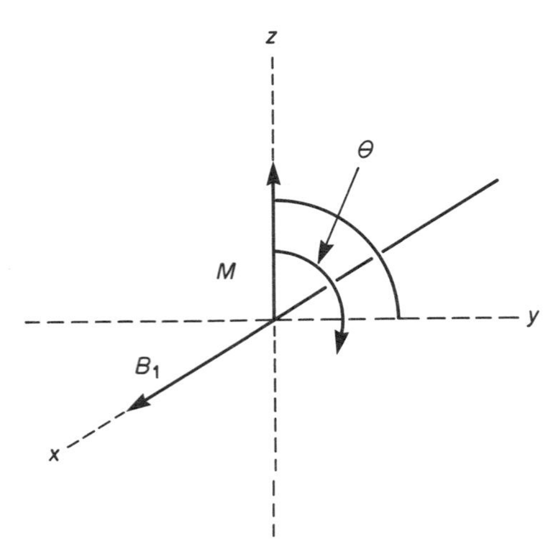

Fig. 6.8 Precession of M about yx. (After Phillips, L. (1976) *Spectroscopy*, Vol. 1, (eds Straughan, B.P. and Walker, S.), Chapman and Hall, London).

relaxes to equilibrium with M along the z axis. The component of magnetization along y decays to zero with a time constant T_2, the spin–spin relaxation time. The restoration of the z component of M to the equilibrium value occurs with time constant T_1, the spin–relaxation time.

It has been implicity assumed that all magnetic nuclei under investigation have identical chemical and magnetic environments and that one Larmor frequency is applicable. In systems of interest there will be several chemical environments for the nuclei which, through changes in shielding, will lead to a spectrum of Larmor frequencies. Also spin–spin coupling may lead to further multiplicity of bands. Thus if the rf pulses were to contain all the Larmor frequencies of the sample, it can excite all transitions simultaneously. The decay sequence following the 90° pulse is not now a simple exponential decay but rather an interferogram containing all resonance frequencies.

The problem is that the spectroscopic information is not contained in a conventional form. The information is available in a so-called 'time domain', not the familiar 'frequency domain'. Fortunately the two may be mathematically linked and the frequency information may be extracted by a Fourier transformation. This would be achieved via an interfaced computer which, in turn, leads to a convential presentation of the spectrum.

Since each pulse leads to data which contains information about the entire spectrum, the value for nuclei of low natural abundance is obvious. As many scans as are required to achieve an acceptable signal to noise ratio may be obtained. Further technical details are beyond the scope of the chapter but will be covered in the many specialist texts available.

6.5 NMR OF SOLIDS

The study of NMR spectra in solution has been well established for many years. Sharp, well resolved lines are obtained and the spectrum is interpreted in terms of the chemical shift (δ) of a particular resonant nucleus from a standard compound containing the same nucleus. Tetramethylsilane (TMS) is a convenient standard for ^{1}H, ^{13}C and ^{29}Si NMR. Further information on the environment may be obtained from hyperfine splitting of the resonance line due to coupling of the spin to those of other magnetic nuclei (spin–spin coupling, J). One reason that sharp lines are obtained for liquids is that dipolar and quadrupolar interactions are averaged to zero by molecular tumbling. NMR spectroscopists spin their sample tubes and this also contributes to the narrowing of the resonance lines.

A large number of factors contribute to the chemical shift, the most obvious of which would be differing chemical environments for two resonant nuclei showing differing values δ, e.g. in ^{29}Si NMR differences in δ would be expected for ^{29}Si–O–Si and ^{29}Si–O–Al. Spin–Spin coupling may provide further information on the detailed environment of a particular group. For example in ^{1}H NMR a methyl group adjacent to oxygen will give a singlet, but when adjacent to a methylene group a triplet is seen. For ^{29}Si NMR which will be of particular concern at the end of the chapter, spin–spin coupling is less important since the nearest neighbours in many mineral environments are non-magnetic atoms such as oxygen. The power of the technique is well illustrated by a recent study of potassium silicate in aqueous solution.[43] ^{29}Si enrichment was used in conjuction with ^{29}Si, ^{29}Si homonuclear spin decoupling. No fewer than eleven distinct silicate species were identified providing possibly for the first time true insight into the nature of aqueous silicate solutions (Fig. 6.9).

Following Griffen[44] we may usefully consider a typical nuclear spin Hamiltonian, $\bar{H}$:

$$\bar{H} = \bar{H}_{cs} + \bar{H}_{J} + \bar{H}_{D} + \bar{H}_{Q} \tag{6.34}$$

where $\bar{H}_{cs}$ relates to the chemical shift, $\bar{H}_{J}$ to spin–spin coupling, $\bar{H}_{D}$ to dipolar terms and H_{Q} to quadrupolar terms. Now consider the magnitude (in Hz) of these terms for both solids and liquids:

	$\bar{H}_{cs}$	$\bar{H}_{J}$	$\bar{H}_{D}$	$\bar{H}_{Q}$
Liquid	10^{3}	10	0	0
Solid	10^{3}	10	5×10^{4}	10^{5}–10^{6}

An inevitable consequence of the sequence in solids is that very broad lines will be observed in a 'normal' NMR experiment thus masking the valuable chemical shift data. Two approaches suggest themselves. First to attempt to understand factors affecting line shapes in solids and consequently to extract

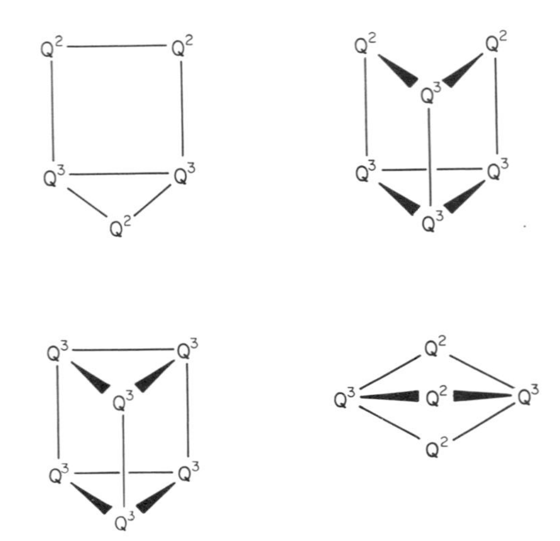

Fig. 6.9 Some of the silicate species identified by ^{229}Si NMR of potassium silicate solutions (after ref. 43). Q^2, SiO_4 unit with two siloxane linkages; Q^3, SiO_4 unit with three siloxane linkages.

information. Secondly, to design modifications to the experiment to achieve high resolution NMR in solids. Both approaches are now briefly considered before applications are illustrated.

6.5.1 Absorption lines

Within a solid a spin I_i is likely to be close to and in fixed spatial relation with a spin I_j. Dipole coupling between such spins is the major cause of line broadening in solids (term H_D in (6.34)). We may write:

$$H_D = \sum_{i>j} g_i \mu_i g_j \mu_j (1 - 3\cos^2\theta_{ij})/r^3{}_{ij} \times F(I_i I_j) \qquad (6.35)$$

$$\leftarrow \text{Spatial part} \rightarrow \qquad \leftarrow \text{Spin part} \rightarrow$$

where θ_{ij} is the angle between the laboratory field and internuclear vector, and r_{ij} the distance between I_i and I_j. Rapid tumbling in solution causes $\langle 1 - 3\cos^2\theta \rangle = 0$, since the time averaged value of $\cos^2\theta$ becomes 1/3. Each spin neighbour will contribute differently to the linewidth and the spectrum will be an unresolved sum of many lines and, hence, structureless. In principle, however, it is still possible to extract information by means of a second moment analysis. The second moment is the mean square width (ΔB^2) of the normalized line measured from the centre. Following Van Vleck,[45] when n identical nuclei interact:

$$\Delta B^2 = 3/4 \ g^2\mu^2 I(I+1)^{1/n} \sum_{i>j} [(1 - 3\cos^2\theta_{ij})/r_{ij}^3] \qquad (6.36)$$

Thus very accurate values of r_{ij} are attainable.

In some cases definite structure is seen in the spectrum. For example if two spins, $I = 1/2$, interact only one term of Equation (6.35) is conserved and the spectrum consists of a doublet, the separation of which is $1.5\,g\mu r^{-3}(1 - \cos^2\theta)$. The separation is θ-dependent and each component of the doublet will be broadened by interaction with more distant spins. Clearly the condition will only be fulfilled when the interaction between the given spins is significantly greater than that of each with other spins in the system. Depending upon the experiment either θ or r may be determined.

Several nuclei of NMR interest have spin > 1/2, e.g. $^2D(I = 1)$, $^3Li(I = 3/2)$. Such nuclei have quadrupole moments (Q) which can interact with local electrostatic field gradients (eq). If the magnetic field, B_O, is large then the Zeeman energy will be greater than the quadrupole interaction, in which case the first $2I$ levels arising in the magnetic field are given by:

$$E = -g\mu B_O M + 1/8e^2qQ(3\cos^2\theta - 1)\frac{3M^2 - I(I+1)}{I(2I-1)} \tag{6.37}$$

where M is a quantum number arising from the projection of I along B_O, θ is the angle between B_O and eq. The selection rule is $M = \pm 1$. For the deuteron, a two-line spectrum is seen. If the quadrupole coupling constant, e^2qQ, is known, structural information may be forthcoming.

If paramagnetic centres are present in the solid note must be taken of the magnetic properties of the unpaired electrons. In fact the magnetic moment of the electron is some 10^3 times greater than a nuclear magnetic moment; however the electron spin relaxes quickly at room temperature and the nucleus senses a time averaged value of the elecron magnetic moment, μ_e

$$\bar{\mu}_e \text{ (time average)} = (\mu_e)^2\frac{B_o}{3k(T - \theta)} \tag{6.38}$$

where k is the Boltzmann constant and θ is a constant. The nuclear resonance line may be shifted some tens of Hertz.

6.5.2 'High resolution' solid state NMR spectra

(a) *Magic angle spinning*

Inspection of Equation (6.35) reveals that if $\theta = 54.74°$ the term $(1 - 3\cos^2\theta)$ is zero and H_D should vanish. Rapid spinning of the sample tube about an axis inclined at 54.74° to B_o should then narrow the resonance line.[46,47] This is the so-called 'magic angle'. The technique is not so good at removing homonuclear coupling since, generally, adequate sample spinning speeds are not attainable without disintegration of the rotor (spinning rates of 3 kHz are often used with rotors manufactured for Delrin, an acetal resin). MAS alone is valuable when small interactions are present. Recently Fyfe *et al.*[48] have

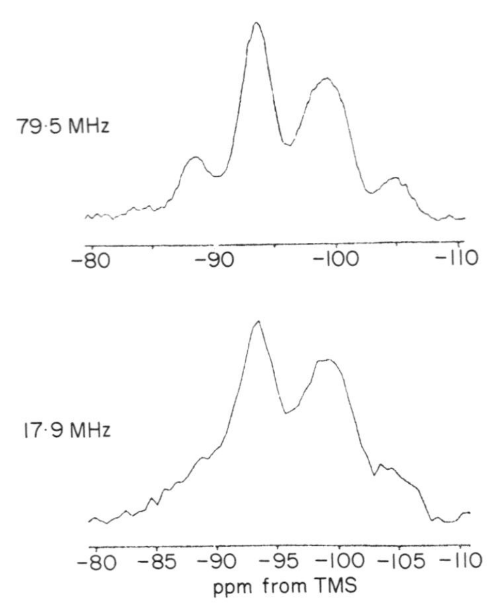

Fig. 6.10 ^{29}Si MAS NMR of zeolite Y (5858 scans at 79.5 MHz, 3762 scans at 17.9 MHz) (after ref. 48).

shown that better resolution of spectra can be obtained using MAS with a high field instrument. Figure 6.10 shows the MAS ^{29}Si NMR spectrum for zeolite *Y* at 79.5 MHz (Bruker WH400) and 17.9 MHz (Bruker CXP100).

(b) *Motional averaging in spin space*

H_D in Equation (6.35) may also be reduced to zero if the spin part of the expression were to vanish. One way in which this has been achieved is to use properly chosen rf pulse sequence so that the spins are stirred in the rotating frame.[49,50] This technique will not be of significant interest in this chapter.

(c) *Magnetic dilution and cross polarization*

Some nuclei have low abundance e.g. ^{13}C (1.1%), ^{29}Si (4.7%) thus the separation of the magnetic nuclei should, on average, be large. Since H_D has a r^{-3} dependence, dipolar broadening may be reduced in studies of these systems.[51] In many cases, however, an abundant spin species, often hydrogen, co-exists with the dilute species. The abundant spins (I_1 say) are polarized in a high field and cooled to a low spin temperature in the rotating frame (disturbing the Boltzmann distribution is akin to altering the temperature). Contact between abundant spins (I) and rare spins (S) is established for time τ

(this is cross polarization) and finally the free induction decay of S is recorded whilst decoupling I.[51] Trick pulse sequences are used to achieve the cross polarization of I and S spins.[51–53]

In some experiments a combination of the above techniques can be used.[54] For many minerals ^{29}Si MAS NMR is sufficient to achieve spectra which clearly show resolved chemical shifts between chemically and magnetically distinct ^{29}Si atoms. If however a significant number of immobilized protons were present in the mineral structure, in addition to MAS the use of cross polarization may be considered. One practical consequence is that any silicon bearing an OH group will give a significantly enhanced signal,[55] which means that resonances from silicon atoms bearing OH groups can be identified.

6.6 APPLICATIONS

These will be dealt with under two broad headings. In the first section applications based on line shape analysis are considered; in the second section primarily ^{29}Si MAS NMR applications will be described.

6.6.1 Strongly hydrated systems

A recent review by Stone[56] covers this and other topics relating to broad line NMR studies of clay minerals, hence the subject will not be treated in great detail here.

Anisotropic interactions between water molecules in bulk water average to zero on the NMR time scale. If however a surface is present, this may give rise to an anisotropic potential which remains constant over several molecular motions and will therefore be apparent within the NMR experiment. This is of especial importance in the study of clay minerals since not only are clay–water systems of interest in the mineralogical sense but they also happen to be of technological importance and of interest to soil chemists.

If a water molecule were preferentially oriented with respect to a surface the dipolar coupling between the protons is non-zero and two lines should be seen in the NMR spectrum (Section 6.5.1). If proton exchange between water molecules is fast, the doublet may be damped out since the spin of the incoming proton may differ from that of the leaving proton. Some advantages accrue from using heavy water, D_2O, in these studies since the doublet spectrum arises from the quadrupole interaction which will be unaffected by exchange and which will not be averaged to zero for the orientated molecules.

Studies of the line shape of broad line NMR spectra of a range of hydrated clays,[57] particularly with Na hectorite, have shown that the preferential ordering does not extend much beyond one layer. Also, because of rapid diffusion, all molecules in this layer experience the same average orientation. A somewhat surprising result of the investigation was that the behaviour of the water molecules was similar for a range of clay types, both expandable (e.g.

hectorite, montmorillonite, vermiculite) and non-expandable (e.g. kaolinite). Thus the presence of a static interface is the key factor in determining the preferred orientation of the first layer of water molecules. The electrical and chemical nature of the surface are secondary.

Another interesting phenomenon which can be revealed by NMR line shape studies is that some water associated with clays either does not freeze, or only does so at low temperatures. Thus either ice-like spectra are seen at temperatures well below 0° C or definite evidence for two phase systems is seen.[58]

6.6.2 Interlamellar water of expandable clays

It will be interesting to complement the discussion of ESR studies of paramagnetic ions in the interlamellar region of clays with some brief remarks on results obtained from NMR studies.

A relevant study[59] has been made of the somewhat atypical Na–Llano vermiculite which shows two well-defined hydration steps corresponding to a single and to a double ($d_{001} = 1.48$ nm) water layer in the interlamellar region. ^1H and ^2D NMR studies show doublet spectra indicating that the water molecules are strongly orientated. The rather large individual flakes of the clay available enable orientation experiments to be carried out and the expected relationship between positions of the two components of the doublet and the angle with the magnetic field was observed. Analysis of the line shape was consistent with ocatahedrally coordinated sodium ions with the C_3 axis of the octahedron perpendicular to the silicate sheet. It is believed that the aquated ion rotates about the axis and that individual coordinated water molecules rotate about a C_2 axis defined by the dative bond. The agreement between these conclusions and those drawn from ESR (Section 6.3.4(a)) using paramagnetic probes on vermiculite will be noted. The ^1H spectra show a central line in addition to the doublet; this however is absent in the ^1D spectra. Probably some water molecules exchange protons rapidly thus collapsing the doublet. Exchange will not influence the deuteron quadrupolar doublet.[59]

6.6.3 Micas – paramagnetic shifts

NMR spectra can yield structural information. For example the trioctahedral mica biotite may show iron(II) replacement for magnesium(II) to up to 20%. Part of the coordination shell comprises OH groups, each OH group being associated with one M_1 site (*trans* di-hydroxo-) and with two M_2 sites (*cis*-dihydroxo-). The ^1H NMR spectra of biotites typically show a central line associated with satellites the intensity and number of which vary with iron(II) content. Further the resonant positions of the satellites showed an orientation dependence in accord with the theory of paramagnetic shifts. These extra bands arise from OH groups which may be first, second or third neighbours of iron(II) ions on M_1 and/or M_2 sites. Analysis revealed that iron(II) was randomly distributed between the sites and further that the parmagnetic ions have a tendency to cluster around an OH site.[60]

6.6.4 Second moment analysis

Applications above have referred to minerals which are available in crystalline form. When a powder is available information may still be extracted from the structureless spectrum by second moment analysis (Section 6.5.1). As a brief example an early study of kaolinite may be quoted where the observed second moment of the proton resonance was in reasonable agreement with a value calculated for H–H and H–Al interactions using lattice parameters obtained from X-ray studies.[61]

6.6.5 Other nuclei

It would be unfortunate if the impression were given that studies were restricted to protons and deuterons. Broad line studies of other nuclei are clearly possible. For example, the 92.58% abundant $^7Li(I = 3/2)$ is quite commonly used. The line shape can be affected by quadrupolar coupling of the ^7Li nuclei with the electric field gradient of the crystal. Studies of a lithium zeolite A[62] have been used to demonstrate isotropic diffusion of lithium ions through the lattice at temperatures above 230° C.

6.7 HIGH RESOLUTION NMR STUDIES OF MINERALS

In Section 6.5.2 a brief outline was given of possible experimental approaches for the measurement of narrow line NMR spectra of solids. Thus fast spinning of the sample is helpful in its own right;[47,63] further narrowing of the resonance line is achieved if the sample is spun at the 'magic angle'. Indeed MAS may be sufficient to produce adequate resolution in some cases, particularly if used in conjunction with a high field.[48] When a significant concentration of protons is present in the mineral specimen MAS may be combined with cross polarization.[64,65] Some care is required to ensure that the mineral is pure, particularly paramagnetic contaminants should be absent. For example samples of sanidine ($[K, Na]AlSi_3O_8$) contaminated with a ferromagnetic impurity (demonstrated by ESR) gave significantly sharper spectra when the contaminant was magnetically separated.[66]

6.7.1 ^{23}Na NMR

In some cases another phenomenon, motional narrowing, may contribute to the observation of relatively narrow lines from solids. A sodium exchanged Y-zeolite (Si/Al = 2.36) has seven sodium ions per cage. Four ions occupy the S2 positions near the centre of six-membered rings, and in the hydrated zeolite the signals from these ions are motionally narrowed reflecting a pseudo-solution-like environment. The signal from the remaining three ions remains too broad to observe. The four ions in the S2 positions may be exchanged with calcium(II); no ^{23}Na NMR spectrum was discernible from the Ca_2Na_3–Y zeolite.[67].

6.7.2 ^{1}H NMR

A range of calcium and barium silicates have been studied:[68] $CaNaHSiO_4$; $(CaOH)Ca(HSiO_4)$; $(CaOH)_2Ca_4Si_6O_{17}$; BaH_2SiO_4; $[(CaOH)_2SiO_3]_x$. The solid state proton NMR spectra showed distinct signals for the acidic and basic protons, for example $(CaOH)Ca(HSiO_4)$ shows two signals in 1:1 intensity ratio. The acidic protons are strongly hydrogen bonded and show greater chemical shifts from hexamethyldisiloxane (used as reference) than do the basic groups.

6.7.3 ^{29}Si NMR

The bulk of the examples in Section 6.7 are, understandably, taken from experiments on ^{29}Si. A significant amount of work has been done on solutions of silicon compounds,[43] and a useful survey and compilation of chemical shift data is available.[69] The range of observed ^{29}Si chemical shifts is some 300 ppm; however the range spanned by silicon resonances in minerals is much narrower, perhaps some 30–40 ppm. The usual reference compound is tetramethylsilane (TMS) and the silicate and aluminosilicate ^{29}Si resonances occur at approximately 100 ppm to low field of this standard.

(a) *Silicates*

Orthosilicic acid, $Si(OH)_4$, is conveniently regarded as the basic unit from which silicates are built. Salts of this acid comprise the orthosilicates. Two molecules of the acid may condense with the elimination of water to give a pyrosilicate, $O_3SiOSiO_3^{6-}$, and continuation of this condensation in a linear fashion will give pyroxene chains (or rings), $(SiO_3)_x^{2x-}$. Cross-linking will produce amphiboles, then clays and micas (ladder and sheet like polymers), and finally three dimensional cross-linking will lead to feldspars, zeolites and ultramarines. Thus a given SiO_4 tetrahedron may share vertices with either zero-Si(0Si), one-Si(1Si), two-Si(2Si), three-Si(3Si) or four-Si(4Si) other SiO_4 tetrahedra. If some degree of isomorphous substitution of silicon(IV) by aluminium(III) should occur, then depending on the degree of cross-linking, given SiO_4 tetrahedra may share vertices with a number of AlO_4 tetrahedra, i.e. with zero-Si(0Al), one-Si(1Al), two-Si(2Al), three-Si(3Al), or four-Si(4Al). The ability of silicon and aluminium atoms to scatter X-rays is similar, hence

Table 6.1 ^{29}Si chemical shifts for silicates and silica after ref. 71 (Spectra were measured at 39.74 MHz with MAS and polarization transfer)

Mineral	*^{29}Si chemical shift (ppm) w.r.t. TMS*				
	Si(OSi)	*Si(1Si)*	*Si(2Si)*	*Si(3Si)*	*Si(4Si)*
NaH_3SiO_4	−66.4				
$Na_2H_2SiO_4 \cdot 8.5H_2O$	−67.8				
$CaNaHSiO_4$	−73.5				
$(CaOH)CaHSiO_4$	−72.5				
$Zn_4(OH)_2 \cdot [Si_2O_7] \cdot H_2O$(hemimorphite)		−77.9			
$Ca_6(OH)_6[Si_2O_7]$		−82.6			
$K_4H_4[Si_4O_{12}]$			−87.5		
$Ca_2(OH)_2[SiO_3]$(hillerbrandite)			−86.3		
$Ca_4(OH)_2[Si_3O_9]$(foshagite)			−86.5		
$Ca_3[Si_3O_9]$(wollastonite)			−88.0		
$Ca_2NaH[Si_3O_9]$(pectolite)			−86.3		
$Ca_6(OH)_2[Si_6O_{17}]$(xonotlite)			−86.8	−97.8	
$Mg_3(OH)_2[Si_4O_{10}]$(talc)				−98.1	
$(Me_4N)_8[Si_8O_{20}]$				−99.3	
$(Et_4N)_6[Si_6O_{15}]$				−90.4	
SiO_2 (low quartz)					−107.4
SiO_2 (low cristobalite)					−109.9

even when single crystal X-ray studies have been carried out, the ordering of the silicon and aluminium atoms has to be inferred, often from bond lengths. Thus within a series of feldspars it appears that the Si–O and Al–O bond lengths are approximately 0.16 nm and 0.174 nm respectively.[70]

Clearly a technique that could identify a particular environment for a SiO_4 tetrahedron would be invaluable. A paper[71] published by Lippmaa *et al.* in 1980 indicated that ^{29}Si NMR was such a technique. Table 6.1 gives some chemical shift data for a range of silicate minerals (the observed chemical shifts are within a few ppm of the values in solution for the soluble species). The interpretation is very straightforward, thus the number of vertices shared by one SiO_4 tetrahedron with others is clearly differentiated. In the mineral xonotlite, the presence of both Si(2Si) and Si(3Si) is evident from the spectrum. Even more subtle environmental differences for SiO_4 tetrahedra in quartz and cristobalite give rise to a small chemical shift difference.

Table 6.2 gives data for some aluminosilicates. Pyrophyllite is related to talc in that aluminium has replaced magnesium in the octahedral layer; however, given the charge difference only 2/3 of the octahedral holes are occupied. Here, as with muscovite, the ^{29}Si resonance is sensitive to the occupancy of the

Table 6.2 ^{29}Si NMR chemical shift data for some aluminosilicates (from ref. 71)

		Si(3Si)...Al(oct)[a]		*Tetrahedral Layer*[b]				
Mineral	*Al:Si*	*Si(3Si)Al*	*Si(3Si)0Al*	*Si(4Al)*	*Si(3Al)*	*Si(2Al)*	*Si(1Al)*	*Si(0Al)*
$Al_2(OH)_2[Si_4O_{10}]$	0		− 91.5					
(pyrophyllite)			− 95.0					
$KAl_2(OH)_2[AlSi_3O_{10}]$	1 : 3	− 84.6						
(muscovite)		− 86.7						
SiO_2	0							− 107.4
(low quartz)								
$Na[AlSi_3O_8]$	1 : 3					− 92.5	− 96.7	
(albite)							− 104.2	
$K[AlSi_3O_8]$	1 : 3					− 95	− 98	
(orthoclase)							− 101	
$K[AlSi_3O_8]$	1 : 3					− 95	− 98.2	
(adular)							− 100.5	
$K[AlSi_3O_8]$	1 : 3					− 95.7	− 96.8	
(sanidine)							− 100.9	
$Na_2[Al_2Si_3O_{10}]$	2 : 3				− 87.7	− 95.4		
(natrolite)								
$KNa_3[AlSiO_4]_4$	1 : 1			− 84.8	− 88.4			
(nepheline)								
$Ca[Al_2Si_2O_8]$	1 : 1			− 83.1				
(anorthite)								

[a] Data relate to aluminium in octahedral layer (six coordinate Al)
[b] Data relate to the silicate (for coordinate Al).

octahedral layer. The spectra also satisfactorily distinguish the SiO_4 tetrahedra sharing vertices with AlO_4 tetrahedra. Although the data of Table 6.2 are limited in scope they do warn that some overlap of the chemical shift ranges may occur. Thus Si(1Al) can give a resonance at − 104.2 ppm (albite) which is very close to the Si(0Al) range as illustrated by quartz (− 107.4 ppm). It is likely, as more data become available, that overlap of other ranges may occur.

(b) *Zeolites*

Zeolites are cage-like aluminosilicates of exceptional importance, particularly as ion exchangers, in their ability to separate molecules of differing size and/or geometry (molecular sieves) and, possibly most important, as catalysts. Many occur naturally but a vast range of synthetic zeolites has been prepared. Quite often, within an iso-structural series, differing only in Si/Al ratio, marked differences in catalytic properties are seen. The question of the distribution of silicon and aluminium within the framework structure is then one of importance and ^{29}Si MAS NMR is proving a valuable aid to the study of this problem.

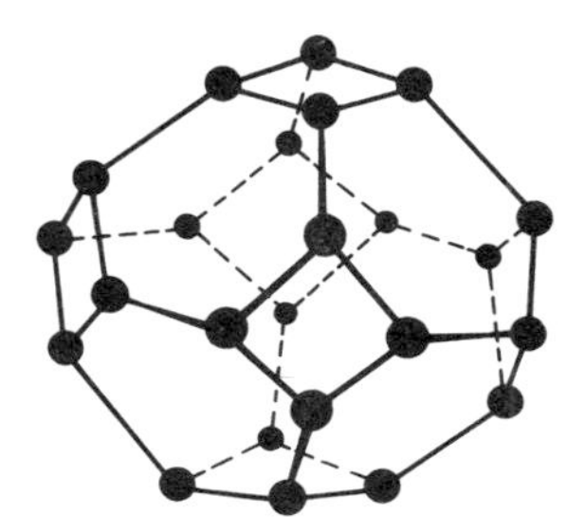

Fig. 6.11 Sodalite cage.

The majority of published work to date relates to the synthetic zeolites A, X, and Y. The basic structural unit is a cubo-octahedron comprising 24SiO_4/AlO_4 tetrahedra. This is often termed a sodalite cage. In the mineral sodalite these units are close packed such that each four- and each six-membered ring is shared between two cages. In zeolite A the sodalite units are linked octahedrally via double four rings. In zeolites X and Y the units are linked tetrahedrally via double six rings giving a diamond like structure. X and Y zeolites differ in their Si/Al ratios (Si/Al 1.18 → 1.5, X zeolite; Si/Al 1.5 → ~ 3, Y zeolite).

Much work on the structural chemistry of aluminosilicates has been influenced by Lowenstein's rules.[72] The first rule states that no two AlO_4 tetrahedra should share a vertex, i.e. Al–O–Al bridges should not occur. The second rule relaxes this constraint in the event that the coordination numbers of aluminium atoms may differ, i.e. Al(4-coordinate)-O-Al(5 or 6-coordinate) is possible. The rules were justified by reference to Pauling[73] and, if true, imply that isomorphous substitution of silicon by aluminium will not be totally random. The rules have come to be so well accepted that many workers dealing with structural problems have eliminated from consideration any proposal which is at variance with Lowenstein. ^{29}Si NMR spectra have recently provided a quite stringent test of the rules.

(c) *Natural zeolites*

Table 6.3 gathers some ^{29}Si chemical shift data for natural zeolites and zeolitic systems. The data for gmelinite and chabazite are of interest. The framework structure is one of double six-membered rings linked by tilted four-membered rings (NB this terminology is imprecise; 'member' is a MO_4 unit, not a single atom). Given the Si/Al ratio of 2:1, aluminium atoms may be in meta or para positions within the six-membered rings. The former arrangement is in accord with Lowenstein's rule, the latter is in better agreement with the NMR spectrum. Meta substituted rings predict a Si(3Al):Si(2Al):Si(0Al) = 2:1:1,

Table 6.3 ^{29}Si chemical shift data for zeolite materials after ref. 55

		^{29}Si chemical shift from TMS				
Mineral	*Si/Al ratio*	*Si(4Al)*	*Si(3Al)*	*Si(2Al)*	*Si(1Al)*	*Si(0Al)*
$Na_6[(AlO_2)_6(SiO_2)_6]\cdot 7.5H_2O$ (hydrated sodalite)	1:1	−83.5				
$Na_8[(AlO_2)_6(SiO_2)_6]Cl_2\cdot xH_2O$ (sodalite)	1:1	−84.8				
$Na_4Ca_8[(AlO_2)_{20}(SiO_2)_{20}]\cdot 24H_2O$ (thomsonite)	1:1	−83.5				
$Na_{16}[(AlO_2)_{16}(SiO_2)_{24}]\cdot 16H_2O$ (natrolite)	1.5:1		−87.7	−95.4		
$Na_8[(AlO_2)_8(SiO_2)_{16}]\cdot 24H_2O$ (gmelinite)	2:1		−92.0†	−97.2	−102.5	
$Ca_2[(AlO_2)_4(SiO_2)_8]\cdot 13H_2O$ (chabazite)	2:1		−94.0†	−99.4	−104.8	−110†
$Na_{16}[(AlO_2)_{16}(SiO_2)_{32}]\cdot 16H_2O$ (analcime)	2:1		−92†	−96.3	−101.3	−108†
$Na_2Ca_4[(AlO_2)_{10}(SiO_2)_{26}]\cdot 28H_2O$ (stilbite)	2.6:1			−98	−101.5 −103.6‡	−108
$Ba_2[(AlO_2)_4(SiO_2)_{12}]\cdot 12H_2O$ (harmotome)	3:1			−95	−98.6 −102.6	−108†
$Ca_4[(AlO_2)_8(SiO_2)_{28}]\cdot 24H_2O$ (heulandite)	3.5:1			−95	−99.0 −105.3‡	−108
$Na_3K_3[(AlO_2)_6(SiO_2)_{30}]\cdot 24H_2O$ (clinoptilolite)	5:1				−100.6	−106.9‡ −112.8

†Weak resonance.
‡Tentative assignment.

para substitution should give Si(2Al): Si(1Al) = 3: 1. The experimental spectrum shows the Si(2Al) resonance to be the strongest. Thus either the Lowenstein rule is broken or the − 104.8 resonance must be reassigned to Si(0Al) and doubt be cast on the validity of intensity measurements.

The accepted structure for analcime has silicon containing four rings linked via AlO_4 tetrahedra, i.e. only Si(2Al) should be present. The NMR spectrum clearly shows that the specimen studied has a less ordered structure. The specimen of harmotome gives a spectrum similar to the feldspar albite (Table 6.2), but the lines are significantly broader implying less ordering of the silicon and aluminium tetrahedra. Henlandite and stilbite also show this line broadening implying that the Si/Al distribution is not very regular (although the major element of the chemical shift will arise from nearest neighbour tetrahedra, the ordering of more remote units will have a second order effect. If long term order is low, a small range of chemical shifts will be seen which will not be resolved and which results in a broadish line).

Whilst the data for the natural minerals are broadly in accord with expectation based on current views of structure, there is the interesting possibility that Lowenstein's rule may sometimes be violated. This should not,

Table 6.4 ^{29}Si NMR chemical shifts for X and Y synthetic zeolites (data taken from refs. 75 and 77)

Si/Al (analysis)	*Si/Al (NMR)*	^{29}*Si chemical shift from TMS*				
		Si(4Al)	*Si(3Al)*	*Si(2Al)*	*Si(1Al)*	*Si(0Al)*
1.19 ↑	1.14	−83.9	−88.5	−93.4	−98.1	−102.4
1.27 X	1.26	−86.5	−90.5	−95.2	−100.0	−103.0
1.33	1.32	−85.4	−89.8	−94.6	−99.2	−103.0
1.35	1.39	−84.1	−88.4	−93.3	−98.5	−102.6
1.50 ↓	1.50	−86.0	−90.3	−93.9	−98.9	
1.53 ↑	1.40	−84.9	−89.3	−93.4	−98.5	
1.59	1.57	−83.8	−88.1	−93.1	−97.7	−101.9
1.67	1.71	−83.9	−88.3	−93.2	−98.0	−102.1
1.82	1.75	−85.2	−89.6	−94.4	−99.3	
1.87	1.85	−83.8	−88.2	−93.3	−98.1	−102.8
2.00	1.98	−84.0	−88.6	−93.7	−98.6	−103.6
2.16	2.31		−89.5	−94.7	−100.0	
2.24 Y	2.22		−90.5	−95.1	−100.3	−106.2
2.35	2.46		−89.0	−94.1	−99.6	−105.3
2.39	2.52		−89.0	−94.1	−99.3	−104.8
2.49	2.50		−89.4	−94.6	−99.7	−105.5
2.52	2.67		−89.9	−94.2	−99.3	−104.7
2.53	2.64		−90.5	−95.2	−100.5	
2.56	2.69		−88.6	−94.6	−99.1	−104.6
2.61	2.56		−88.6	−94.8	−99.3	−105.1
2.62	2.66		−90.1	−95.2	−100.8	−106.4
2.75 ↓	2.69		−88.7	−94.0	−100.4	−106.0
$\Delta\delta = (\delta_{MAX} - \delta_{MIN})$		2.7	2.4	2.1	3.1	4.5

as yet, be taken as proven as a recently resolved controversy over zeolite A has shown (see below).

(d) *Zeolites X and Y*

Three major research groups have recently published extensively on the NMR spectra of the X and Y zeolites. Fortunately, the conclusions of Fyfe, Thomas *et al.*,[74–76] Melchior *et al.*[77] and Lippmaa *et al.*[55,78] are in accord and given that three very extensive papers are available,[75,77,78] only a brief summary will be given here. Table 6.4 records chemical shift data for the X and Y synthetic zeolites (synthetic faujasites).

The assignments of Table 6.4 are quite definite since an evolution of spectra with varying Si/Al ratio in an isostructural series is obtained. There is also clear evidence that Lowenstein's rule is obeyed. This conclusion may be reached in two ways. First the use of areas under the resonance peaks to calculate the Si/Al ratio, working on the assumption that Lowenstein's rule is obeyed. The excellent agreement obtained justifies the initial assumption.[75] Alternatively, it is possible to consider a range of model structures and calculate the NMR spectra and demonstrate agreement with experiment only if Lowenstein's rule is obeyed.[77]

The model of ordering in the X and Y zeolites requires that the sodalite cages are linked tetrahedrally through six-membered rings. There is the additional requirement that the Lowenstein rule is obeyed, i.e. that Al–O–Al linkages are absent. The distribution of Si and Al should correspond to the lowest, or nearly lowest, electrostatic energy which will be achieved if the number of next-nearest-neighbour Al, Al pairs across the 4-rings and 6-rings is minimized for a given Si/Al ratio.

The range of chemical shifts for a particular silicon environment (expressed as $\Delta\delta$ in Table 6.4) arises from a next-nearest-neighbour effect as explained above (Section 6.7.3(c)). The fact that data from two laboratories has been combined in Table 6.4 may have enhanced the ranges. For example, for Si(4Al) the Lowenstein rule will ensure that the next nearest neighbour is *not* Al, hence a small $\Delta\delta$ would be anticipated. The value -86.5 ppm (Si/Al $= 1.27$) is mainly responsible for enhancing the range observed.

The availability of high resolution ^{29}Si NMR spectra has revolutionized the understanding of X and Y zeolite structural chemistry and the reader who requires more specific detail is urged to consult the excellent papers of Klinowski *et al.*[75] and Melchior.[77] Previously, it was considered that the structure of zeolite A was significantly better understood. Ironically the initial ^{29}Si MAS NMR studies of that zeolite shed more darkness than light; however the situation is now reasonably clear.

(e) *Zeolite A*

Lippmaa *et al.*[55] pointed out that sodium zeolite A ($Na_{12}[(AlO_2)_{12}(SiO_2)_{12}]\cdot 27H_2O$), in which the Si/Al ratio is 1:1, gave one major resonance at $\delta = -89.6$ ppm with respect to TMS, together with two extremely weak features at -85.0 ppm and -94.5 ppm. If, therefore, direct comparison is made with other data in Tables 6.2, 6.3 and 6.4 it seems inescapable that the major silicon environment is Si(3Al) in which case the Lowenstein rule must be violated.

Bursill *et al.*[79] independently were led to doubt the accepted cubic Fm3c space group implying strictly alternating Si and Al in accordance with Lowenstein via re-examination of X-ray data and electron diffraction measurements. Also a neutron diffraction study revealed a rhombohedral distortion in polycrystalline samples of dehydrated sodium zeolite A. A new model space group $R\bar{3}$ was advanced to account both for the rhombohedral distortion and the Si(3Al) arrangement.[79] Klinowski *et al.*[80] confirmed the Lippmaa *et al.* results for zeolite A ($\delta = -88.9$ ppm) and further found that Losod ($Na_{11.5}Al_{12}Si_{12}O_{48} \cdot 17.8H_2O$) gave a similarly placed single resonance ($\delta = -88.9$ ppm). So well do these figures agree with expectation for Si(3Al) that it seemed further examples of the breakdown of the Lowenstein rule had been found.

Melchior *et al.*[81] investigated another range of materials (ZK4) of Si/Al ratios 1.18 and 1.40. These materials obey the Lowenstein rule and give the resonance for Si(4Al) at -89.1 ppm. The importance of an evolutionary approach to assignment was stressed and the single resonance for zeolite A reassigned as Si(4Al), i.e. in accord with Lowenstein. In the meantime Adams and Haselden found no evidence for rhombohedral distortions of Na zeolite A with Si/Al ratios of 1.03, 1.09 and 1.12 from a neutron diffraction study.[82] Also Bursill *et al.*, having exchanged the sodium zeolite A previously used[79] with thallium(I) found the material to be cubic.[83] There is now general agreement that the zeolite is cubic, and that the Lowenstein rule is obeyed with strict Si–O–Al–O–Si alternation.

The story of zeolite A advises caution with assignments as new classes of material are examined. It may be that other samples in which the initial interpretation of ^{29}Si MAS NMR suggest the Lowenstein rule to be broken will, on re-examination, cease to be renegade.

6.7.4 ^{27}Al NMR

The nucleus ^{27}Al($I = 5/2$) is also suitable for NMR studies. Although the nucleus is quadrupolar, generally only the $1/2 \leftrightarrow -1/2$ transition is seen and this, to the first order at least, is not affected by quadrupolar interactions.

The catalytic activity of the hydrogenic form of zeolites increases with increasing Si/Al ratio. Hydrothermal treatment of ammonium exchanged zeolite Y causes aluminium to leave framework positions and migrate to the interstitial positions, the framework vacancies being filled by silicon. More recently it has been shown that treatment of sodium Y with gaseous silicon(IV) chloride gives a virtually aluminium-free faujasite. The parent sodium Y zeolite gave four ^{29}Si resonances ($\delta = -88.5, -93.7, -99.2, -105.0$ ppm for TMS), the dealuminated material gave a single sharp band at -107.0 ppm (cf. quartz, Table 6.1).[84] The 2 ppm shift and sharpening of the line may be attributed to removal of aluminium next nearest neighbours.

^{27}Al NMR spectra were measured at 104.22 Mh.[84] The parent NaY gave a single line at 61.3 ppm (relative to $Al(H_2O)_6^{3+}$) attributed to aluminium in a tetrahedral environment. The dealuminated (framework) material gave two resonances at 54.8 ppm (broad, tetrahedral aluminium), and 0.2 ppm (octahedral). No explanation is currently available to account for the dealuminated species.

There is little doubt that ^{27}Al studies will expand rapidly. Here, as for ^{29}Si, advantages accrue from using high fields.[48,84] In practical terms, it is clear that the combination of ^{29}Si and ^{27}Al NMR spectroscopies should provide an extremely powerful probe in areas such as cement chemistry.

6.7.5 ^{17}O NMR

The ^{17}O isotope has an abundance of only 0.037% and has spin 5/2. Very recently the first study of high resolution ^{17}O NMR spectra of a solid was reported.[85] The spectra were obtained at 67.8 MHz for static samples and also for samples spinning at the magic angle (5 KHz). The specimens required enrichment with ^{17}O.

Cubic magnesium(II) oxide gave a sharp line at + 46.1 ppm versus $H_2{}^{17}O$. Corundum (α-Al_2O_3) gave a relatively narrow line in the static spectrum at + 66 ppm, whereas silica (quartz) gave a well resolved quadrupolar pattern. Diopside, $CaMgSi_2O_6$, a pyroxene gave three signals arising from the two non-bridging and one bridging oxygen all of which are crystallographically distinct.

Clearly this development is exciting since many mineral species of interest may be synthesized and hence the problem of ^{17}O enrichment does not become insurmountable.

6.8 CONCLUSION

Although examples have been selected mainly for clay and zeolite chemistry, the typical ranges of application of ESR and NMR spectroscopies have been demonstrated. As is usual with spectroscopic techniques they become more powerful when combined with other experimental methods such as X-ray crystallography or, indeed, with other spectroscopies. The potential for solid state NMR is clearly considerable and the brief survey above represents no more than the beginning of a story.

ACKNOWLEDGEMENT

The writer wishes to thank Dr Parisa Monsef-Mirzai for her help in assembling material for this chapter.

REFERENCES

1. Atherton, N.M. (1973) *Electron Spin Resonance*, Halsted Press, London.
2. Abragam, A. and Bleaney, B. (1970) *Electron Paramagnetic Resonance of Transition Metal Ions*, Oxford University Press, London.
3. Symons, M. (1978) *Chemical and Biochemical Aspects of Electron Spin Resonance Spectroscopy*, Van Nostrand Reinhold, Wokingham.
4. Sands, R.H. (1955) *Phys. Rev.*, **99**, 1222.

5. Kneubühl, F.K. (1968) *J. Chem. Phys.* **33**, 1074.
6. Adrian, F.J. (1968) *J. Colloid Interface Sci.*, **26**, 317.
7. Gillespie, P.A. (1979) Ph.D. Thesis, University of Aston in Birmingham.
8. Meads, R.E. and Malden, D.J. (1975) *Clay Miner.*, **10**, 313.
9. Boesman, E. and Schoemaker, D. (1961) *Compt. Rend.*, **252**, 1931.
10. Jones, J.P.E., Angel, B.R. and Hall, P.L. (1974) *Clay Miner.*, **10**, 257.
11. Monsef-Mirzai, P. and McWhinnie, W.R. (1982) *Inorg. Chim. Acta*, **58**, 143.
12. Angel, B.R. and Hall, P.L. (1973) in *Proc. Int. Clay. Conf.*, (1972) (ed. J. Serratosa). CSIO, Madrid, p. 71.
13. Brindley, G.W. and Nakahira, M. (1959) *J. Am. Ceram. Soc.*, **42**, 311.
14. Angel, B.R., Jones, J.P.E. and Hall, P.L. (1974) *Clay Miner.*, **10**, 247.
15. Angel, B.R., Cuttler, A.H., Richards, K.S. and Vincent, W.E.J. (1977) *Clays Clay Miner.*, **25**, 381.
16. Olivier, J., Vedrine, J.C. and Pezerat, H. (1975) *Bull. Group Franc. Argiles*, **27**, 153.
17. Olivier, D., Vedrine, J.C. and Pezerat, H. (1975) *Proc. Int. Clay Conf., Mexico*, p. 231.
18. Olivier, D., Lauginie, P. and Fripiat, J.J. (1976) *Chem. Phys. Lett.*, **40**, 131.
19. Olivier, D., Vedrine, J.C. and Pezerat, H. (1977) *J. Solid State Chem.*, **20**, 267.
20. McBride, M.B., Pinnavaia, T.J. and Mortland, M.M. (1975) *Clays Clay Miner.*, **23**, 103.
21. McBride, M.B., Pinnavaia, T.J. and Mortland, M.M. (1975) *Clays Clay Miner.*, **23**, 162.
22. Berkheiser, V. and Mortland, M.M. (1975) *Clays Clay Miner.*, **23**, 404.
23. Kasai, P.H. and Bishop, R.J. (1972) *J. Am. chem. Soc.*, **94**, 5560.
24. Vedrine, J.C. and Naccache, C. (1973) *Chem. Phys. Letters*, **18**, 190.
25. Kasai, P.H. (1965) *J. Chem. Phys.*, **43**, 3322.
26. Kasai, P.H. and Bishop, R.J. Jr. (1973) *J. Phys. Chem.*, **77**, 2308.
27. Olsen, D.H. (1968) *J. Phys. Chem.*, **72**, 4366.
28. Kasai, P.H. and Bishop, R.J. Jr. (1976) in *Zeolite Chemistry and Catalysis* (ed. J.A. Rabo), American Chemical Society Monograph 171, p. 350.
29. McBride, M.B. (1977) *Clays Clay Miner.*, **25**, 6.
30. Pinnavaia, T.J. (1982) *Devel. Sedimentol.*, **34** (Adv. Tech. Clay Miner. Anal.), 139.
31. McBride, M.B., Pinnavaia, T.J. and Mortland, M.M. (1975) *J. Phys. Chem.*, **79**, 2430.
32. Clementz, D.M., Pinnavaia, T.J. and Mortland, M.M. (1973) *J. Phys. Chem.*, **77**, 196.
33. McBride, M.B. and Mortland, M.M. (1974) *Proc. Soil Sci. Soc. Amer.*, **38**, 408.
34. Monsef-Mirzai, P. and McWhinnie, W.R. (1984) *Inorg. Chim. Acta*, in press.

35. Hathaway, B.J. and Billing, D.E. (1970) *Co-ord. Chem. Rev.*, **5**, 143.
36. Clearfeld, A. and Quayle, L.R. (1982) *Inorg. Chem.*, **21**, 4197.
37. Rubinstein, M., Baram, A. and Luz, Z. (1971) *Molec. Phys.*, **20**, 67.
38. McBride, M.B., Pinnavaia, T.J. and Mortland, M.M. (1975) *Am. Mineral.*, **60**, 66.
39. Barry, T.I. and Lay, L.A. (1966) *J. Phys. Chem. Solids*, **27**, 1821.
40. Barry, T.I. and Lay, L.A. (1968) *J. Phys. Chem. Solids*, **29**, 1395.
41. Wichterlova, B., Kubelkova, L., Jiru, P. and Kolikova, D. (1980) *Coll. Czech. Chem. Commun.*, **45**, 2143.
42. McBride, M.B. (1979) *Clays, Clay Miner.*, **27**, 91.
43. Harris, R.K., Knight, C.T.G. and Hull, W.E. (1981) *J. Am. chem. Soc.*, **103**, 1577.
44. Griffen, R.G. (1977) *Anal. Chem.*, **49**, 951.
45. Van Vleck, J.H. (1948) *Phys. Rev.*, **74**, 1168.
46. Lowe, I.J. (1959) *Phys. Rev. Lett.*, **2**, 285.
47. Andrew, E.R., Bradbury, A. and Eades, R.G. (1959) *Nature, Lond.*, **183**, 1802.
48. Fyfe, C.A., Gobbi, G.C., Hartman, J.S. *et al.* (1982) *J. magn. Reson.*, **47**, 168.
49. Osteroff, E.D. and Waugh, J.S. (1966) *Phys. Rev. Lett.*, **16**, 1097.
50. Waugh, J.S., Huber, L.M. and Haeberlen, U. (1968) *Phys. Rev. Lett.*, **20**, 180.
51. Pines, A., Gibby, M. and Waugh, J.S. (1972) *J. Chem. Phys.*, **56**, 1776.
52. Pines, A., Gibby, M. and Waugh, J.S. (1973) *J. Chem. Phys.*, **59**, 569.
53. Hartmann, S.R. and Hahn, E.L. (1962) *Phys. Rev.*, **128**, 2042.
54. Lippmaa, E.T., Alla, M.A., Pehk, T.J. and Engelhardt, G. (1978) *J. Am. chem. Soc.*, **100**, 1929.
55. Lippmaa, E., Mägi, M., Samoson, A. *et al.* (1981) *J. Am. chem. Soc.*, **103**, 4992.
56. Stone, W.E.E. (1982) *Devel. Sedimentol.*, **34** (Adv. Tech. Clay Miner. Anal.), 77.
57. Woessner, D.E. (1975) *Mass Spectroscopy and NMR Spectroscopy in Pesticide Chemistry* (eds R. Haque and F.J. Bines), Plenum Press, New York. p. 279.
58. Ananyan, A.A. (1978) *Colloid J. U.S.S.R.*, **40**, 1165.
59. Hougardy, J., Stone, W.E.E. and Fripiat, J.J. (1976) *J. Chem. Phys.*, **64**, 3840.
60. Sanz, J. and Stone, W.E.E. (1977) *J. Chem. Phys.*, **67**, 3739.
61. Gastuche, M.C., Toussaint, F., Fripiat, J.J. *et al.* (1963) *Clay Mineral Bulletin*, **5**, 227.
62. Freude, D., Pribylov, A.A. and Schmiedel, H. (1973) *Phys. Stat. Sol.(b)*, **57**, K73.
63. Andrew, E.R. (1971) *Progr. NMR Spectr.*, **8**, 1.
64. Yannoni, C.S. (1982) *Acc. chem. Res.*, **15**, 201.
65. Lyerla, J.R., Yannoni, C.S. and Fyfe, C.A. (1982) *Acc. chem. Res.*, **15**, 208.

66. Oldfield, E., Kinsey, R.A., Smith, K.A. *et al.* (1983) *J. magn. Reson.*, **51**, 325.
67. Basler, W.D. (1980) *Z. Naturforsch. A*, **35**, 645.
68. Rosenberger, V.H. and Grummer, A. -R. (1979) *Z. Anorg. allgen. Chem.*, **448**, 11.
69. Marsmann, H. (1981) in *NMR Basic Principles and Progress*, Vol. 17 (eds P. Diehl, E. Fluck and R. Kosfeld), Springer-Verlag, Berlin, Heidelberg, New York, p. 65.
70. Smith, J.V. (1975) *Feldspar Minerals*, Springer-Verlag, Heidelberg.
71. Lippmaa, E., Mägi, M., Samoson, A. *et al.* (1980) *J. Am. chem. Soc.*, **102**, 4889.
72. Lowenstein, W. (1954) *Am. Mineral.*, **39**, 92.
73. Pauling, L. (1929) *J. Am. chem. Soc.*, **51**, 1010.
74. Ramdas, S., Thomas, J.M., Klinowski, J. *et al.* (1981) *Nature, Lond.*, **292**, 228.
75. Klinowski, J., Ramdas, S., Thomas, J.M. *et al.* (1982) *J. chem. Soc. (Faraday Trans. 2)*, **78**, 1025.
76. Klinowski, J., Thomas, J.M., Fyfe, C.A. *et al.* (1983) *Inorg. Chem.*, **22**, 63.
77. Melchior, M.T., Vaughan, D.E.W. and Jacobson, A.J. (1982) *J. Am. chem. Soc.*, **104**, 4859.
78. Engelhardt, V.G., Lohse, U., Lippmaa, E. *et al.* (1981) *Z. anorg. allgem. Chem.*, **482**, 49.
79. Bursill, L.A., Lodge, E.A., Thomas, J.M. and Cheetham, A.K. (1981) *J. Phys. Chem.*, **85**, 2409.
80. Klinowski, J., Thomas, J.M., Fyfe, C.A. and Hartman, J.S. (1981) *J. Phys. Chem.*, **85**, 2590.
81. Melchior, M.T., Vaughan, D.E.W., Jarman, R.H. and Jacobson, A.J. (1982) *Nature, Lond.*, **298**, 455.
82. Adams, J.M. and Haselden, D.A. (1982) *J. chem. Soc. Chem. Commun.*, **1982**, 822.
83. Cheetham,. A.K., Fyfe, C.A., Smith, J.V. and Thomas, J.M. (1982) *J. chem. Soc. Chem. Commun.*, **1982**, 823.
84. Klinowski, J., Thomas, J.M., Fyfe, C.A. *et al.* (1983) *Inorg. Chem.*, **22**, 63.
85. Schramm, S., Kirkpatrick, R.J. and Oldfield, E. (1983) *J. Am chem. Soc.*, **105**, 2483.

7

Spectroscopy and Chemical Bonding in the Opaque Minerals

D.J. Vaughan

7.1 INTRODUCTION

The rock-forming minerals which are abundant at the Earth's surface are chiefly translucent silicates and carbonates. Opaque minerals occur as minor components in many rocks and the *ore minerals*, raw materials for the world supplies of metals, are mostly opaque phases. The major categories of opaque minerals are: (i) native metals, semimetals and alloys; (ii) metal sulphides and other chalcogenides (compounds of metals with Se, Te, As, Sb, Bi); (iii) oxides of certain elements, notably the transition metals: (iv) certain hydroxides and oxysalts.

The opaque minerals are worthy of separate discussion for several reasons apart from their importance as ores. As well as their distinctive optical properties, many exhibit interesting electrical and magnetic properties. They range from insulators to semiconductors and metallic conductors, and include materials which show not only diamagnetic and paramagnetic behaviour but the various forms of ordered magnetism (ferromagnetism, antiferromagnetism, ferrimagnetism). Many of the experimental methods for the study of chemical bonding which are discussed in the other chapters of this text are equally applicable to the opaque minerals. However, there are additional or alternative techniques which are important and which will be outlined here. These include diffuse and specular reflectance spectroscopy and certain electrical and magnetic measurements. The development of chemical bonding models in these materials also poses particular problems which will be discussed throughout the chapter.

Emphasis, in this chapter, will be placed on the transition metal sulphides and the metal oxides. Metals and alloys will not be considered as they are of limited mineralogical importance and discussed in great detail in the metallurgical literature; the other chalcogenides, hydroxides and oxysalts are also of limited importance and will receive less attention. Discussion in this chapter will be of: (i) compositions and crystal structures of the major opaque mineral groups; (ii) approaches to chemical bonding models; (iii) experimental methods for the study of bonding in opaque minerals; (iv) bonding and the properties of some major opaque mineral groups.

7.2 COMPOSITIONS AND CRYSTAL STRUCTURES OF THE MAJOR OPAQUE MINERALS

The opaque minerals can be divided into groups on the basis of chemical composition and crystal structure as shown in Table 7.1. Following the division into the major categories of metals, sulphides and sulphosalts, oxides, hydroxides and oxysalts, various groups can be recognized which are based on

Table 7.1 Major groups of opaque minerals and their crystal structures

Major subdivision and group	*Structure-type*	*Mineral examples*
Native metals and semimetals		
Gold group	Face-centred cubic close-packed	Native Au, Ag, Cu and their alloys
Platinum metal group	Cubic close-packed	Native Pt, Pd
	Hexagonal close-packed	Native Os, Ir and their alloys
Arsenic group	Hexagonal layer structure	Native As, Sb, Bi
Sulphides and sulphosalts		
Disulphide group (all containing dianion units in the structure)	Pyrite-type (cubic)	Pyrite (FeS_2), cattierite (CoS_2), vaesite (NiS_2), etc.
	Marcasite-type (orthorhombic)	Marcasite (FeS_2), Ferroselite ($FeSe_2$), etc.
	Arsenopyrite-type (orthorhombic)	Arsenopyrite (FeAsS), gudmundite (FeSbS), safflorite ($CoAs_2$)
	Loellingite-type (orthorhombic)	Loellingite ($FeAs_2$, $FeSb_2$, etc.),
Galena group	NaCl-type (cubic)	Galena (PbS), clausthalite (PbSe), altaite (PbTe), alabandite (α-MnS)

Table 7.1 (Contd.)

Major subdivision and group	*Structure-type*	*Mineral examples*
Sphalerite group	Sphalerite (zinc-blende)-type (cubic)	Sphalerite†(β-ZnS) hawleyite†(CdS), etc.
	Derivative of sphalerite-type	Chalcopyrite ($CuFeS_2$), stannite (Cu_2FeSnS_4), talnakhite ($Cu_9Fe_8S_{16}$), etc.
Würtzite group	Würtzite-type (hexagonal)	Wurtzite†(α-ZnS), greenockite†(CdS), etc.
	Derivatives of würtzite-type	Cubanite ($CuFe_2S_3$), enargite (Cu_3AsS_4), etc.
Niccolite group	NiAs-type (hexagonal)	Niccolite (NiAs), breithauptite (NiSb), etc.
	Derivatives of NiAs-type	Troilite (FeS), monoclinic pyrrhotite (Fe_7S_8), etc.
Thiospinel group	Spinel-type (cubic)	Linnaeite (Co_3S_4), polydymite (Ni_3S_4), greigite (Fe_3S_4), violarite ($FeNi_2S_4$), etc.
Layer sulphides group (various layer structures)	Molybdenite-type (hexagonal)	Molybdenite (MoS_2), tungstenite (WS_2)
	Covellite-type (hexagonal)	Covellite (CuS), idaite ($\sim Cu_3FeS_4$)
	$Cd(OH)_2$-type (hexagonal)	Berndtite (SnS_2), melonite ($NiTe_{2-x}$).
	Tetragonal PbO-type	Mackinawite $(Fe, Co, Ni, Cu)_{1+x}S$
Metal-excess group (various unusual structures adopted by metal-rich sulphides)	Pentlandite-type (cubic)	Pentlandite ($(Ni, Fe)_9S_8$), cobalt pentlandite (Co_9S_8)
	Chalcocite-type (monoclinic)	Chalcocite (Cu_2S), acanthite (Ag_2S)
	Argentite-type (cubic)	Argentite (Ag_2S), crookesite (Cu_2Se)
	Digenite-type (cubic)	Digenite (Cu_9S_5), bornite (Cu_5FeS_4)
Chain structure group	Stibnite-type	Stibnite (Sb_2S_3),

(Contd.)

Table 7.1. (Contd.)

Major subdivision and group	*Structure-type*	*Mineral examples*
(various structures containing chains of atoms)	(orthorhombic)	bismuthinite (Bi_2S_3), etc.
	Cinnabar-type (hexagonal)	Cinnabar (HgS)
Sulphosalt group (s) (complex minerals of formula AmTnXp where A = Ag, Cu, Pb; T = As, Sb, Bi; X = S)	A range of complex structures with TS_3 pyramidal groups present (T = As, Sb or Bi)	Pyrargyrite (Ag_3SbS_3) tetrahedrite ($(Cu, Fe)_{12}Sb_4S_{13}$), boulangerite ($Pb_5Sb_4S_{11}$), seligmannite ($PbCuAsS_3$), etc.
Oxides		
Rutile group	Rutile-type (tetragonal)	Rutile (TiO_2), pyrolusite (MnO_2), cassiterite (SnO_2)
	Derivatives of rutile-type	columbite-tantalite ($(Fe, Mn(Nb, Ta)_2O_6$)
Uraninite group	Fluorite-type (cubic)	Uraninite (UO_2), thorianite (ThO_2)
Haematite group	Corundum-type (hexagonal)	Haematite (Fe_2O_3), ilmenite ($FeTiO_3$), geikielite ($MgTiO_3$),
Spinel group	Spinel-type (cubic)	Magnetite (Fe_3O_4), chromite ($FeCr_2O_4$), ulvöspinel ($TiFe_2O_4$), maghaemite (γ-Fe_2O_3), hausmannite (Mn_3O_4), etc.
Hydroxides and oxysalts		
Diaspore group	Diaspore (α-AlO·OH) type (orthorhombic)	Goethite (α-FeO·OH), manganite (MnO·OH)
Boehmite group	Boehmite (γ-AlO·OH) type	Lepidocrocite (γ-FeO·OH)
Wolframite group	Wolframite-type (monoclinic)	Huebnerite ($MnWO_4$)-ferberite ($FeWO_4$) series (wolframites)

[†]Rendered opaque by the presence of impurities (e.g. Fe).

a particular crystal structure or set of related structures. In many cases, these are the classic structures of crystalline solids which are well known to all chemists and mineralogists and described in detail elsewhere (e.g. Wells[1]). As well as the minerals which exhibit the actual structure, there are often other minerals which have structures directly based on the 'parent' structure which can be considered as *derivatives* of that structure. Examples include chalcopyrite ($CuFeS_2$), derived from ordered substitution of Cu and Fe for Zn in the sphalerite-structure form of ZnS, and the accommodation of additional metals in the chalcopyrite structure to give the *stuffed derivative*, talnakhite ($Cu_9Fe_8S_{16}$). Removal of metal atoms to leave vacancies which then adopt an ordered arrangement is the manner in which monoclinic pyrrhotite (Fe_7S_8) is derived from FeS with the niccolite structure, whereas troilite (FeS) is simply a *distorted* form of the niccolite structure parent.

Some of the groups listed in Table 7.1 have a diverse membership linked by a common factor such as a layer-type structure. The information in Table 7.1 is certainly not intended to be exhaustive but does indicate the structures and compositions of the important opaque minerals. Further structural and mineralogical data are available in such references as Wells,[1] Dana and Dana,[2] Vaughan and Craig,[3] Rumble.[4]

7.3 APPROACHES TO CHEMICAL BONDING MODELS

In the development and testing of models of chemical bonding in minerals it is possible to identify three major approaches:

(i) The *phenomenological approach* in which knowledge of the crystal structure of the material and of properties such as magnetic and electrical behaviour are used as the basis for developing models.
(ii) The *experimental approach* in which an attempt is made to determine electronic energy levels directly from spectroscopic data.
(iii) The *calculational approach* in which a first principle or more approximate method of quantum mechanical calculation is used to compute the energies and spatial distributions of electrons in the system.

Ideally, models should be developed on the basis of combining all three approaches to achieve a unified understanding of a particular opaque mineral or system. Whatever approach is used, the problem has to be tackled from the standpoint of one of the major theories of chemical bonding, and to put this in perspective it is worth recounting how this has been done in the past.

The early descriptions of chemical bonding in minerals utilized purely ionic models in which the ions are charged spheres of a particular radius. Such models give no information on the electronic structure of materials or on properties influenced by the behaviour of electrons. The developments in chemistry in the early part of this century were largely concerned with the description of bonding in molecules and with the electron pair bond and valence bond theory. Such theories of covalent bonding have limited application to minerals and offered no real challenge to the ionic model, although attempts have been made to apply valence bond theory to, for example, the sulphide minerals.[5]

A major development occurred with the application to mineralogy of crystal field theory which, unlike the simple ionic models, has a quantum mechanical basis. The interpretation of the properties of minerals containing transition elements using crystal field theory has been reviewed by Burns[6] (see also this book, Chapter 3). The modification of crystal field theory which treats the anions surrounding the transition metal not as point charges but as atoms which may overlap and mix with the metal (ligand-field theory) has not been widely applied in mineralogy, although Nickel[7,8] has used it to discuss the pyrite-type and related sulphides. Both theories are, of course, limited to the description of only the d orbitals of transition metals.

Because of their importance in solid state physics, a number of the binary metal oxides and chalcogenides have been the subject of band structure calculations undertaken over the past few decades. Although band theory has been considered by many to provide the only accurate description of the electronic structures of large lattice framework solids, it is difficult to apply to crystals with more than a few atoms per unit cell and the use of a reciprocal-space representation makes it conceptually difficult. Most band structure calculations cannot be easily carried out to self-consistency in the electronic charge densities and potential energy. For these reasons, quantitative band theory has not been much employed by mineralogists who have preferred to apply the molecular orbital (MO) theory in developing electronic structure models. Because the MO theory considers only a small cluster of atoms and has been extensively applied to molecules and complexes, certain misconceptions have arisen regarding its versatility. It is capable of describing bond polarities from the totally covalent to the totally ionic and can be applied to describe many of the properties of large lattice solids. If the size of the molecular cluster used in an MO calculation is increased, the molecular orbitals broaden into bands and the band model is approached. Qualitative MO/band models have been applied to metal oxides and sulphides by Goodenough[9,10] and to sulphides by Vaughan *et al.*[11] and Vaughan and Craig.[3] The 1970s witnessed the rapid development of quantitative molecular orbital methods in the description of chemical bonding in the silicate minerals, and the use of calculations on 'molecular' cluster units representative of the extended crystal is now well established (see Gibbs[12]). The mid-1970s also saw the first successful attempts to perform MO calculations on metal oxide and sulphide mineral systems and a substantial number of systems have since been studied (see also Tossell, Chapter 1, this volume).

The molecular orbital theory appears to be the most versatile approach to chemical bonding in mineral systems, capable of dealing with ionic, covalent or metallic bonds and of interpreting and predicting a range of spectral and energetic properties. Such versatility is essential when dealing with systems such as the metal oxides and sulphides in which the valence electrons may be *delocalized*, i.e. not associated with a particular atom but free to move throughout the lattice giving rise to metallic conductivity, or may be *localized* on the atom as in those oxides and sulphides which are semiconductors or insulators. Sometimes both localized and delocalized valence electron behaviour is shown in the same compound in these materials. In this chapter, the discussions of bonding will centre on the application of molecular orbital theory, combined where appropriate with elements of simple band models.

7.4 EXPERIMENTAL METHODS FOR THE STUDY OF BONDING

Many of the techniques described elsewhere in this book for the study of bonding in minerals are applicable to the opaque minerals. In particular, X-ray emission and absorption, X-ray photoelectron and Auger spectroscopies are powerful methods of study which have already been outlined (see Chapter 2). Mössbauer spectroscopy (see Chapter 5) is also widely applicable to the opaques, many of which contain Mössbauer active isotopes (^{57}Fe, ^{125}Te, ^{121}Sb, ^{119}Sn, ^{61}Ni).

The interaction of visible light (and of radiation in the near-IR and near-UV) with the opaque minerals obviously differs from its interaction with translucent phases (discussed in Chapter 3). Light which is incident on opaque minerals is partly absorbed and partly reflected by them. There are two kinds of reflection process: that occurring when light is reflected from a flat polished surface of the mineral (*specular reflectance*), and that occurring when the light is reflected from the mineral after it has been finely powdered (*diffuse reflectance*). The latter arises from radiation which has penetrated the crystals (rather as in an electronic absorption spectrum) and reappeared at the surface after multiple scatterings; in this case there will also be a specular component to the reflectance from light which is reflected from the surfaces of the particles. The distinction between the specular and diffuse components of reflectance is familiar to the mineralogist as the difference between the colour of an opaque mineral in hand specimen and its 'streak.'

The specular reflectance of a flat polished surface of an opaque mineral measured at normal incidence can be related to the n and k terms of the complex refractive index (N) in which:

$$N = n + ik \tag{7.1}$$

where n = refractive index, k = absorption coefficient, i = complex conjugate. This relationship is related to reflectance through the Fresnel equation:

$$R = \frac{(n - N)^2 + k^2}{(n + N)^2 + k^2} \tag{7.2}$$

where R = reflectance ($R = 1 = 100\%$ reflectance), N = refractive index of the medium (when the medium is air $N \approx 1$). This is the Fresnel equation for the special case of normal incidence; the equations for non-normal incidence are much more complex. Using specular reflectance data obtained at different angles of incidence, the values of n and k as a function of wavelength (λ) can be obtained through a Kramers–Kronig calculation (see Wendlandt and Hecht[13]). The plot of k against λ approximates more closely the information obtained from a transmission measurement on a translucent solid than does a direct plot of R against λ. The variations in n and k as a function of λ for the pyrite-type disulphide minerals (after Bither *et al.*[14]) are shown in Fig. 7.3(a) and the interpretation of these data is discussed later in this chapter. The commercial spectrometers designed primarily for undertaking transmittance measurements in the near-IR/visible/near-UV spectral range, commonly have attachments which enable specular reflectance measurements to be made (see Wendlandt and Hecht,[13] Hedelman and Mitchell[15]). Measurements of the specular reflectance at normal incidence of very small grains are made using photometer attachments to the ore microscope (see Craig and Vaughan,[16] Piller[17]). Such measurements are usually restricted to the visible range and, from measurements made in air and under an immersion oil, it is possible to solve the Fresnel equation to obtain values for n and k. The errors associated with such determinations have been discussed by Embrey and Criddle[18].

Measurements of diffuse reflectance involve finely grinding the mineral with a white powder such as MgO so as to suppress the specular component of the reflectance. Pellets or discs of materials prepared in this way can also be measured using accessories provided for standard commercial spectrometers.[13,19] In diffuse reflectance, different approaches are adopted to obtaining the equivalent of an electronic absorption spectrum and these generally require much simpler manipulation of the raw data. A commonly employed method is to use the Kubelka–Munk function $f(r)$ which, as demonstrated by Kubelka and Munk,[20] provides a measure of the ratio of the absorption and scattering coefficients:

$$f(r) = \frac{(1-r)^2}{2r} \simeq \frac{k}{s} \tag{7.3}$$

where r is the diffuse reflectance, k the absorption coefficient and s is the scattering coefficient. For the kind of samples used in diffuse reflectance spectroscopy, the scattering coefficient is nearly independent of wavelength, so that the Kubelka–Munk function approximates directly the absorption coefficient. Examples of measurements of both specular and diffuse reflectance spectra for a range of sulphide and oxide minerals are provided in the following sections (see, for example, Figs. 7.3, 7.10, 7.18, 7.20).

The electrical and magnetic properties of the opaque minerals provide important information for use in developing bonding models. Aspects of the theoretical background and of the numerous measurement techniques relevant to studies of minerals are discussed in Shuey[21] and in Vaughan and Craig[3] and are given more complete coverage in numerous texts on solid-state physics and materials science. In this chapter, techniques can only be mentioned so as to draw attention to this very important area of study.

A wide range of techniques is available for the measurement of the electrical properties of materials;[22] the main properties of interest in studies of opaque minerals are those which characterize the material as a metal or semiconductor, give the band gap for the latter, and provide information on conduction mechanisms. *Conductivity* (or its reciprocal, *resistivity*) is readily measured and studied as a function of temperature and generally shows considerable variation between different opaques as seen in Fig. 7.1. However, since conductivity is a function both of carrier (electrons or holes) concentration and carrier mobility, information on these requires knowledge of more than just the conductivity. Measurement of the *Hall effect* provides a method of determining the mobility of charge carriers. The Hall effect is

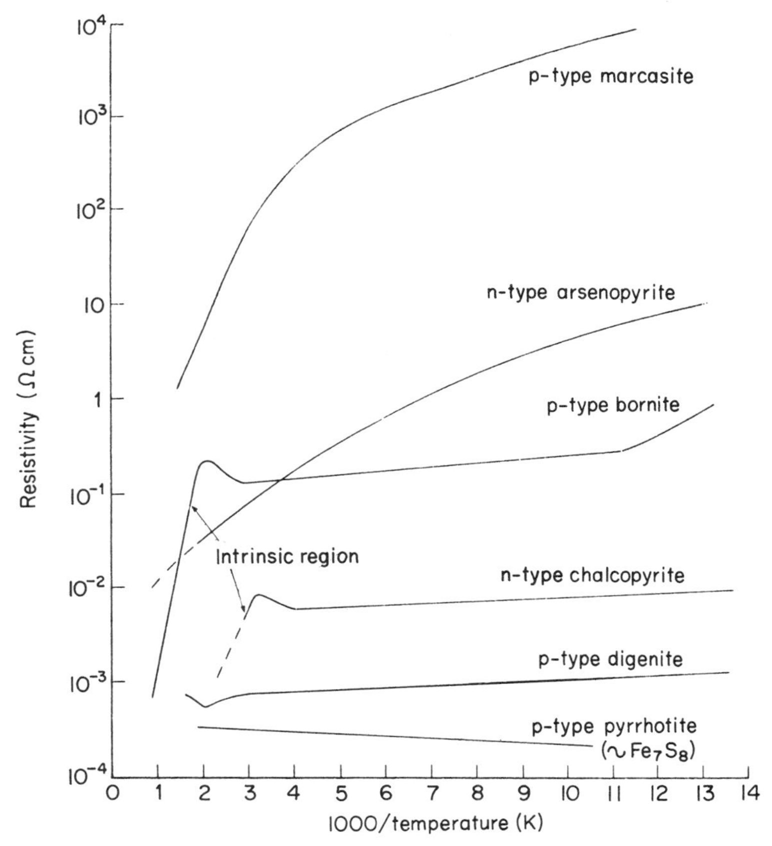

Fig. 7.1 Resistivity versus reciprocal temperature for several sulphide minerals (reproduced from Vaughan and Craig[3] with the publisher's permission).

observed when a magnetic field is applied at right angles to a conductor carrying a current. The magnetic field deflects the current carriers and a restoring force is generated in order to maintain equilibrium so that the current can continue to flow. This restoring force is experimentally determined as the *Hall voltage* and its magnitude is a measure of carrier concentration, whereas its sign indicates whether the carriers are holes (+) or electrons (−). Combined Hall effect and conductivity measurements enable the mobility of both electron or hole carriers (which may both be important in the same material) to be determined.

Thermoelectric power measurements show the tendency of mobile charge carriers to move from the hot end to the cold end of a sample which is placed in a temperature gradient. The potential difference measured across this temperature gradient is the *Seebeck voltage*; its polarity gives the sign of the majority carrier in a semiconductor and its variation with temperature gives information on the activation energy required for conduction and on the position of the *Fermi level* (and hence the energy gap).

The measurement of *magnetic susceptibility*, through determining the force exerted on a sample when it is placed in an inhomogeneous magnetic field, is normally undertaken using some form of magnetic balance. The response of the sample to the applied field enables the material to be distinguished as diamagnetic or paramagnetic, and further studies as a function of temperature enable distinction of the various forms of magnetic ordering and determination of Curie and Néel points or other magnetic transitions. In the case of ferro- or ferrimagnetic materials, studies are also undertaken as a function of the applied magnetic field.

7.5 CHEMICAL BONDING IN SOME MAJOR OPAQUE MINERAL GROUPS

The problems of understanding bonding in several of the major opaque minerals will now be considered, taking examples from some of the groups listed in Table 7.1. As noted earlier, metals and alloys will not be discussed. The groups considered first will be treated in greater detail so as to explain the techniques and the approaches which have been adopted. The application of bonding models in the interpretation and prediction of mineral properties will also be discussed.

7.5.1 The disulphides

Pyrite (FeS_2) is the most abundant natural sulphide and it has a crystal structure in which octahedrally coordinated iron atoms are at the corners and face centres of a cube unit cell and dumb-bell-shaped disulphide groups at the cube centre and midpoints of the edges. Pyrite is diamagnetic and a semiconductor with an energy gap of 0.9 eV.[3] The magnetic and electrical properties, and also the parameters derived from the Mössbauer spectrum of iron (which show a very small isomer shift and quadrupole splitting), are consistent with low-spin Fe^{2+} in the octahedral sites.

Simple qualitative MO energy level diagrams to illustrate iron–sulphur bonding in pyrite were first published by Bither *et al.*[14] and by Burns and Vaughan,[23] and a modified version is shown in Fig. 7.2(a). Here, the energies of outer atomic orbitals involved in bonding in an octahedral 'FeS_6 cluster' are shown qualitatively and strong overlap between metal and sulphur s and p orbitals is considered to result in σ(bonding) and σ^* (antibonding) molecular orbitals. The splitting of the d orbitals of Fe^{2+} by the octahedral ligand field of surrounding sulphur ions results in a t_{2g} set shown as remaining non-bonding and an e_g set which interact with the sulphurs to form a filled bonding and empty antibonding orbital set. Infilling of these molecular orbitals with electrons has the σ–bonding orbitals completely filled and the six d electrons of Fe^{2+} spin-paired and filling the t_{2g} orbitals in accordance with the properties of this material noted above. From this type of MO diagram, familiar from its widespread use in chemistry texts, to a qualitative 'one electron' energy band diagram of the type published by Goodenough,[10,24] is a simple step as seen from Fig. 7.2(b). Overlap between orbitals on adjacent clusters leads to broadening of energy levels to give bands–a filled σ valence band and corresponding empty conduction band. The e_g levels also overlap via sulphur intermediaries to form a band, but the t_{2g} levels remain localized on the cation. Conduction results from promotion of an electron from the t_{2g} levels into the e_g band and the activation energy measured for this semiconductor provides a measure of the separation between t_{2g} and e_g levels.

These models of bonding in the FeS_6 cluster have been developed entirely from the phenomenological approach and can be tested and refined through the study of spectra and through MO calculations. The valence region X-ray photoelectron spectrum of pyrite (see Fig. 7.2(c) and also Chapter 2) shows an intense peak at low binding energy arising from the six spin-paired electrons in the t_{2g} levels. Less pronounced features arise from the other valence band electrons. Examination of X-ray emission spectra aligned with the photoelectron spectrum and with each other by using more deeply buried core orbitals provides information, since the transitions involved here are governed by quantum mechanical selection rules (see Chapter 2). The FeK_β spectrum arises from orbitals which are dominantly Fe 4p in character, so maximum intensity corresponds to Fe4p contributions to the valence band with a shoulder suggesting some p orbital character in the t_{2g} levels. The SK_β spectra arise from orbitals with S3p character at the top of the valence band, and the SL spectra from orbitals with S3p character. The interpretation of the X-ray emission spectra of pyrite, especially as regards alternative interpretations of the SK_β spectrum, has been more fully discussed in Chapter 2.

The specular and the diffuse reflectance spectra of pyrite have also been studied [14,25,26] and from the specular reflectance spectra the optical constants

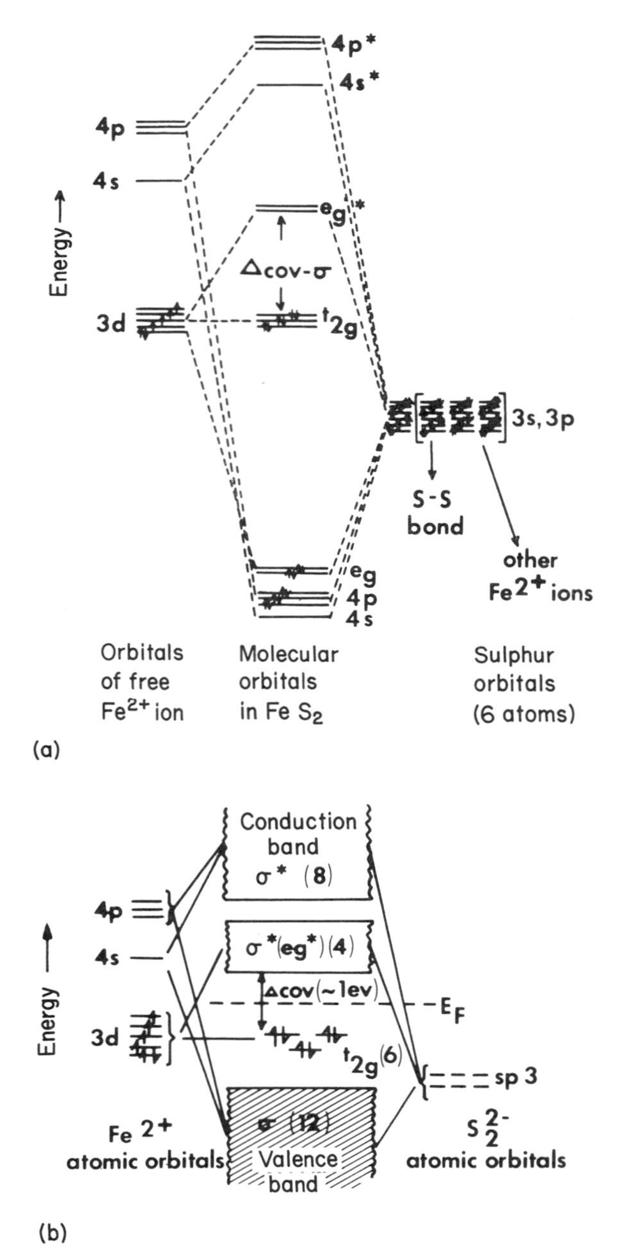

Fig. 7.2 Electronic structure models and spectroscopic data for pyrite. (a) Molecular orbital energy level diagram (modified after Burns and Vaughan[23]). (b) Schematic 'one-electron' energy band diagram. Numbers in brackets refer to the number of electron states per molecule available. E_F, Fermi level; hatched bands are filled with electrons (modified after Bither *et al.*,[14] Goodenough[24]).

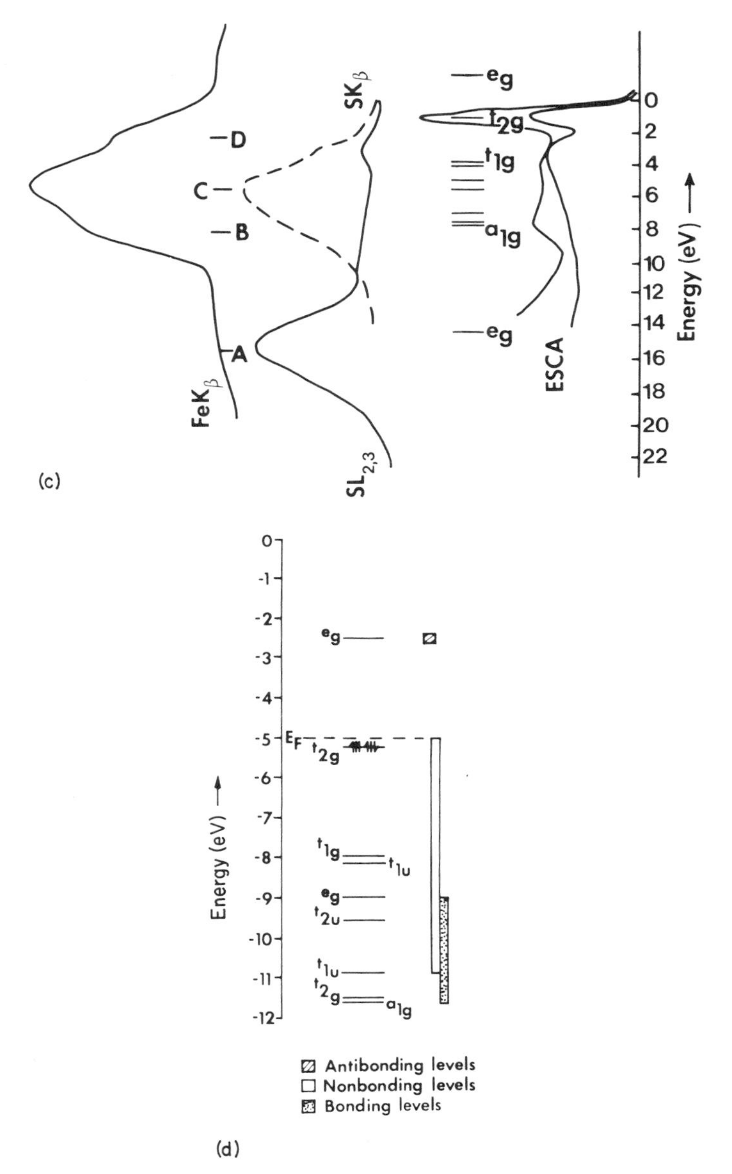

Fig. 7.2 (c) FeK_{β}, $SL_{2,3}$ and SK_{β}X-ray emission spectra for pyrite together with the uv photoelectron spectrum and energy levels from an SCF-X_{α} calculation (data from refs cited in Vaughan[30]). (d) Energy level diagram for the FeS_6^{-10} cluster from an SCF-X_{α} calculation. Energy levels have been labelled using group theory nomenclature for the O_h symmetry group (data from Li *et al*[68].).

were calculated using a Kramers–Kronig analysis (see Fig. 7.3). Feature E_1 in Fig. 7.3 (and the absorption edge in the diffuse reflectance spectrum of pyrite–see Fig. 7.3) can be attributed to the onset of $t_{2g} \rightarrow e_g^*$ transitions and the following peaks (E_2, E_3, E_4) to similar transitions or to transitions from the σ (valence band) to e_g^* levels.

The interpretation of spectra solely on the basis of qualitative models can clearly lead to errors and a better approach involves performing calculations of the electronic structure which can then be compared with spectroscopic data. Unfortunately, many of the MO methods which have yielded quite good results for hydrocarbons or simple main group oxides give poor results for transition metal sulphides. However, good results have been obtained using the SCF-X_α Scattered Wave Cluster Method (abbreviated 'X_α Method') of Slater and Johnson[27,28] which has already been discussed in Chapter 1. In Fig. 7.2(d) are shown calculated energy levels for an FeS_6^{-10} cluster (i.e. Fe^{2+} octahedrally coordinated to six S^{2-} ions and the whole cluster surrounded by a sphere of positive charge). The calculation places the filled $2t_{2g}$ levels 2.7 eV above the top of the main valence orbitals and shows them to be localized non-bonding orbitals of almost entirely Fe3d character. The empty $3e_g$ orbitals have, by contrast, appreciable S3p character as well as Fe3d character. The $1t_{1g}$ and $3t_{1u}$, $2e_g$ and $2t_{1u}$ levels are largely non-bonding S3p type, whereas the $2e_g$, $1t_{2g}$ and $2a_{1g}$ orbitals are the main bonding orbitals of the system with substantial iron and sulphur character. The set of orbitals labelled $1a_{1g}$, $1t_{1u}$ and $1e_g$ are essentially S3s non-bonding orbitals. The calculations therefore provide a much more detailed picture of the valence band structure which differs in several important respects from the qualitative model. The calculations also show good agreement with experiment as Fig. 7.2 illustrates.

Fortunately, there is a complete series of pyrite-structure disulphides extending across the periodic table from FeS_2 to ZnS_2 which is ideal for making comparisons. The properties of these disulphides are well known and a substantial body of data is available from reflectance and X-ray spectra (see Bither *et al.*,[14] Vaughan and Tossell[29]). In Fig. 7.4 are presented one-electron MO/band models based largely on the results of calculations by the X_α method on MS_6^{-10} clusters with some information also derived directly from spectroscopic studies. In addition to energy levels from the calculations, also shown are the positions of the main group of metal–sulphur bonding levels and information inferred from magnetic and electrical properties regarding the localized or delocalized behaviour of electrons at the top of the valence band. Thus, whereas in FeS_2 the completely filled t_{2g} levels are separated from the empty e_g^* band by a gap of ~ 0.9 eV, in CoS_2 the additional electron gives rise to a partly-filled e_g^* band and metallic conductivity. In CuS_2, the three-quarters filled e_g^* band similarly results in metallic conductivity, whereas in NiS_2 spin-splitting of the orbitals separates an e_g^* band into a filled band and an empty band so that semiconducting properties are observed with ordered magnetism. In all of these sulphides, the orbitals at the top of the valence band

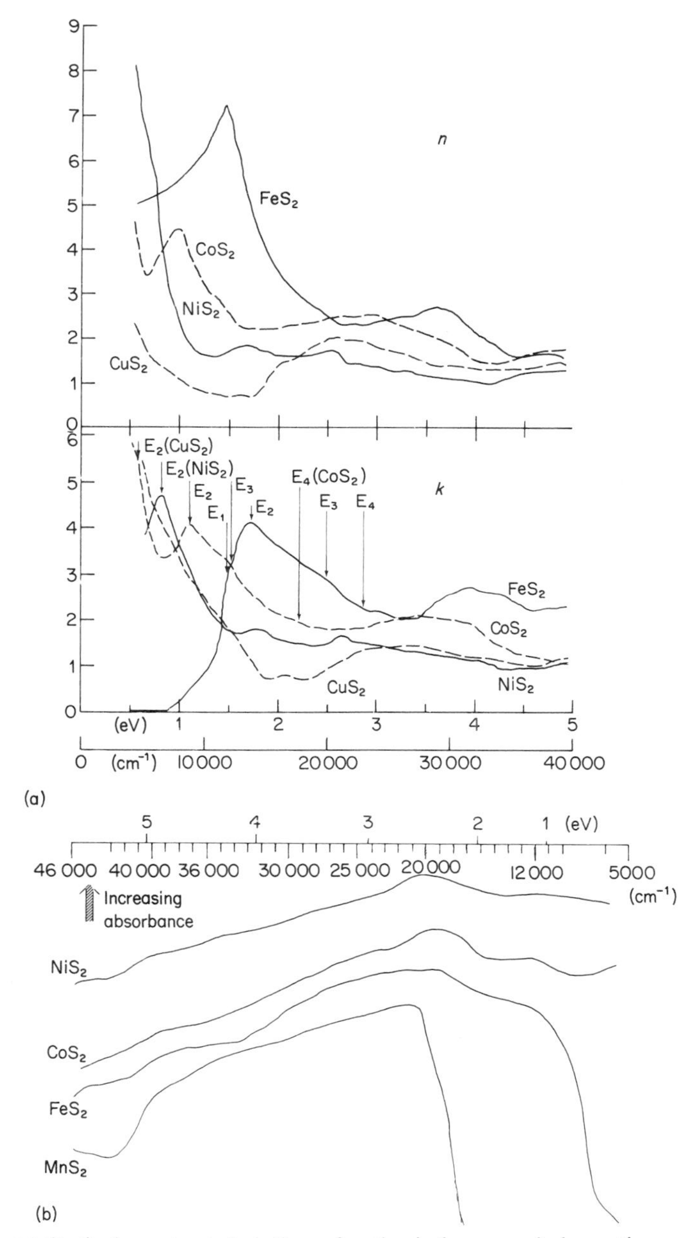

Fig. 7.3 (a) Optical constants (relative refractive index, *n*, and absorption coefficient, *k*) as a function of energy for the pyrite-type disulphides FeS_2, CoS_2, NiS_2, CuS_2 (based on Bither *et al.*[14]). (b) Diffuse reflectance spectra of pyrite-type disulphides (after Vaughan[25]) (diagrams reproduced from Vaughan and Craig[3] with the publishers' permission).

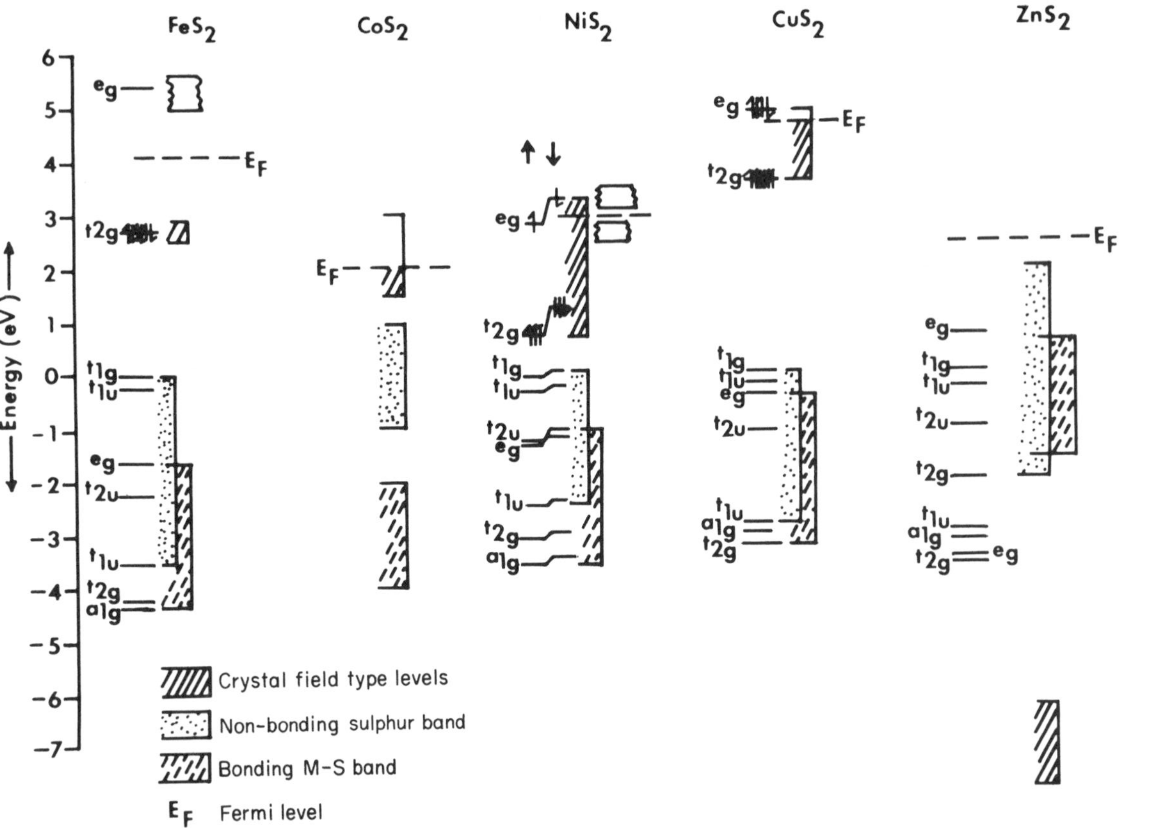

Fig. 7.4 Molecular orbital energy diagram for MS_6 clusters representing the pyrite-type disulphide minerals based on SCF-X_α calculations (the discrete energy levels) and spectroscopic data (the broad bands) (reproduced from Tossell *et al.*[31] with the publisher's permission).

are metal 3d in character and lie above a group of sulphur 3p non-bonding orbitals. In ZnS_2, however, the sulphur 3p non-bonding orbitals lie at the top of the valence band and the zinc 3d orbitals are much more deeply buried. This material is not in fact opaque, but a colourless insulator.

The disulphide series is also one which enables variations in properties to be correlated with changes in electronic structure. Properties of these phases which are of mineralogical interest are shown in Table. 7.1. Thus, in proceeding across this series a systematic increase in unit cell dimensions and, so far as data are available, a systematic decrease in Vickers hardness values are observed. This can be related to the successive addition of electrons to the e_g orbitals which are antibonding in character; greater occupancy of these orbitals, which are proximal to the ligands, can be regarded as forcing the ligands away and so increasing the cell dimensions and decreasing the hardness. However, the calculations and spectra also indicate a systematic destabilization of the metal–sulphur bonding orbitals across the series as shown in Fig. 7.4. The implied weakening of metal–sulphur bonding not only correlates with decreasing hardness and increasing unit-cell parameters but also with increasing metal–sulphur bond distances and decreasingly negative values of ΔG^{o}_{298}.

Colour and reflectance variations in the disulphide series have also been interpreted on the basis of these models (see Burns and Vaughan,[23] Vaughan[30]). The comparatively high reflectance of pyrite arises from the empty e_g orbitals into which t_{2g} electrons may be excited. The reduction in reflectance in moving across the transition series to CuS_2 is due to filling of these e_g levels with electrons, making fewer of them available for excited electrons. Pyrite has a higher reflectance at the red-end of the visible spectrum than at the blue-end accounting for the yellow colour of the mineral. This is because, as seen in Fig. 7.3, the absorption (and reflectance) maximum corresponding to the $t_{2g} \rightarrow e_g^*$ transitions is located at about 1.7 eV (730 nm). This absorption maximum is observed at progressively lower energies in the sequence FeS_2–CoS_2–NiS_2–CuS_2. In pure ZnS_2, vacant energy levels are not available just above the highest energy levels which contain electrons, so that the excitations which cause absorption in the visible region do not occur and the material is transparent.

7.5.2 Disulphides, diarsenides and sulpharsenides and their relationships

Although attention has so far been focused on the nature of metal–sulphur bonding in the disulphides, this is not meant to suggest that S–S interactions are unimportant. In fact, the S–S interactions may well introduce orbital energy splittings in addition to those arising from M–S interactions, and approaches emphasizing the S–S bonding in these systems have been employed by Tossell *et al.*[31] and Tossell.[32] In this case, the approach centres on the use of simple

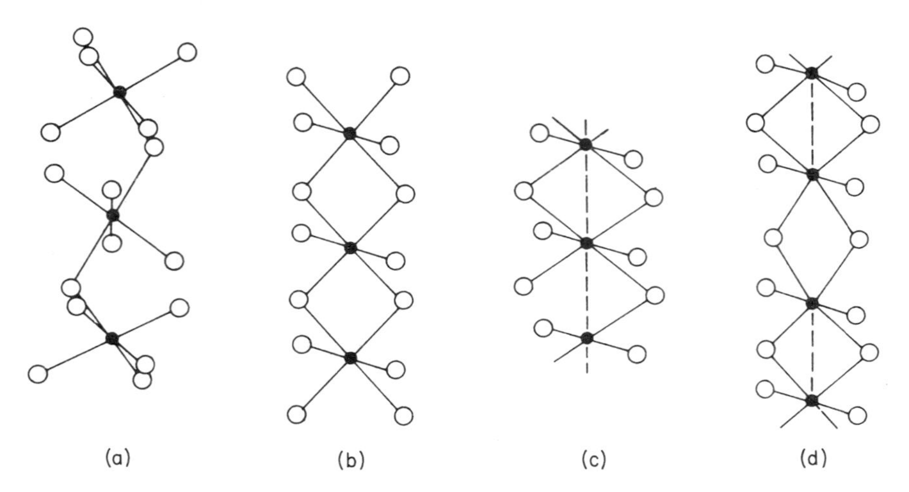

Fig. 7.5 The linkage of MX_6 octahedra in disulphides with the (a) pyrite, (b) marcasite, (c) loellingite and (d) arsenopyrite structures (reproduced from Vaughan and Craig[3] with the publisher's permission).

qualitative 'molecular' theories to describe the bonding in solids and on the use of qualitative perturbational MO arguments.[33,34]

Whereas in the pyrite structure, FeS_6 octahedra share corners, in the marcasite form of FeS_2 they share edges normal to the *c*-axis direction. Variants of the marcasite structure occur in $FeAs_2$ (loellingite) and FeAsS (arsenopyrite), with the former having closer metal–metal distances across the shared edge than in marcasite and the latter having alternate long and short distances (see Fig. 7.5). Partly because of these features, attention has been focused on metal–metal interactions in attempts to explain these structures. Hulliger and Mooser,[35] Hulliger[36] and Nickel[37] used a ligand-field theory approach in which it was proposed that the oxidation state of iron in $FeAs_2$ is Fe^{4+} and that the four d electrons not involved in bonding are spin-paired in two of the d orbitals. The orbital which remains empty, it was argued, is that parallel to the *c*-axis which explains the contraction along the *c*-axis in this mineral. In FeAsS, it was suggested that the iron atom has five d electrons not involved in metal–anion bonding, and in order to achieve spin-pairing, an unpaired electron in a t_{2g} orbital on one atom is paired with an equivalent electron on the adjacent metal atom across the shared octahedral edge. The alternate short and long distances between metals along the *c*-axis in arsenopyrite result from this metal–metal bonding. Goodenough[38] has criticized this ligand-field approach and used an MO/band model to suggest that the structural deformation in loellingite and arsenopyrite arises from metal–anion bonding, not metal–metal bonding. For example, in FeAsS, the t_{2g} orbital

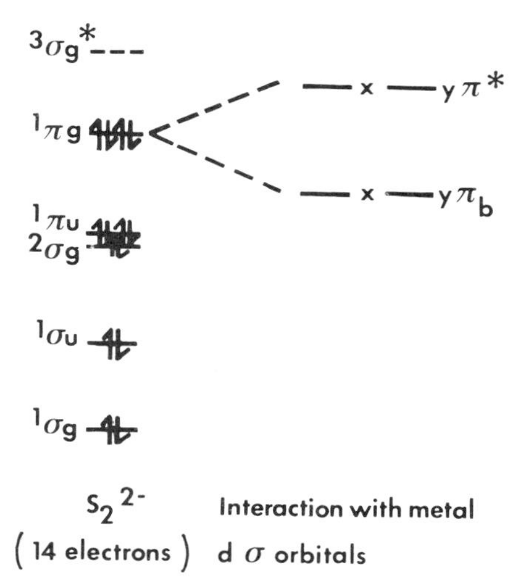

Fig. 7.6 Molecular orbital energy level diagram for the S_2^{2-} dianion and splitting of the highest-energy orbital containing electrons on interaction with metal d orbitals (reproduced from Vaughan and Tossell[29] with the publisher's permission).

parallel to the *c*-axis is shown as split by a metal–anion bonding interaction into a lower-energy filled band and a higher-energy empty band.

Tossell *et al.*[31] and Tossell[32] employed an approach in which attention is focused on the highest occupied MO of the electron donor and the lowest unoccupied MO of the electron acceptor and their energies and overlaps. A molecular orbital scheme for the S_2^{2-} ion based on X_α calculations is shown in Fig. 7.6. In S_2^{2-}, 14 valence electrons are involved in filling the orbital up to $1\,\pi_g^*$ which is antibonding in character. Mixing of this $1\,\pi_g^*$ orbital with metal d orbitals of σ symmetry with respect to M–S bonding generates two pairs of orbitals, one orientated in the *xz* plane and the other in the *yz* plane (where *z* is the internuclear axis direction). Each pair consists of an orbital stabilized compared to $1\,\pi_g^*$ and hence a metal–sulphur bonding orbital (π_b) together with a destabilized antibonding orbital (π^*). In a disulphide such as FeS_2 with a 14 electron dianion, counting the electrons added to the orbitals shows that all of the π_b orbitals will be filled and all the π^* empty. If we consider $FeAs_2$, the evidence from Mössbauer isomer shifts[3] and a number of other criteria[31] strongly suggests that the oxidation state of iron remains Fe^{2+} and hence the dianion is As_2^{2-}. The molecular orbital scheme for this dianion is the same as for S_2^{2-} (Fig. 7.6) but there are only 12 valence electrons so that only one component of the π_b orbital set would be filled (say π_{bx}). This should be reflected in a difference in the geometry of metal–anion coordination and, indeed, whereas the metals are symmetrically disposed about the anion in the pyrite structure, in loellingite the metal–anion coordination is distorted so that the metal atoms lie almost in the same (say xz) plane (see Fig. 7.5). This approach points to the electron occupancy in the dianion system as the structure-determining factor and can be extended to explain the structures adopted by other disulphides, diarsenides and sulpharsenides. In arsenopyrite, for example, the As end of the AsS group is effectively a '12 electron system' while the S end is a '14 electron system', resulting in alternately greater or less distortion of the coordination of metals around the anion. Similarly, in safflorite ($CoAs_2$), half of the As atoms (As_I) are associated with short Co–Co distances and are 12 electron in character, whereas the other half (As_{II}) are effectively like 14 electron systems and associated with longer Co–Co distances. The extra electron density at As_{II} arises from the Co d electrons so that the effective bond-types are $Co^{3+} - As_I$ and $Co^{2+} - As_{II}$.

These arguments, based on a qualitative perturbational MO theory, as well as providing insight into the principles governing the crystal structures of the disulphides and related minerals, have also been used to clarify stability relationships. Thus FeS_2, as a 14 valence electron system, should adopt the pyrite structure in which the three Fe atoms coordinated to S are symmetrically disposed about x and y. However, FeS_2 also crystallizes with the marcasite structure in which the coordination geometry is more like that in loellingite. In order to favour the marcasite structure, the S_2^{2-} anion should acquire '12 electron-like' character. The answer to this problem, and to the long debated question of the relationship between pyrite and marcasite, has been suggested to depend on the mechanism of formation of marcasite. Whereas pyrite can readily by synthesized by direct reaction between elemental iron and sulphur, marcasite can only be synthesized by precipitation from (preferably acid) solution. This is also in accordance with the occurrence of natural marcasite in low temperature assemblages. As illustrated in Fig. 7.7, interaction in solution of one of the $1\pi_g^*$ orbitals of S_2^{2-} with H^+ ions would effectively remove two electrons from the system, so that the S_2^{2-} group would then behave as a 12 electron system. The three Fe atoms coordinated to each sulphur of the S_2 group would then adopt the nearly coplanar geometry of loellingite. Subsequently, loss of H^+ in the later stages of crystallization would lead to some readjustment and adoption of the marcasite structure.

$$[S-S]^{2-} + H^+ \rightleftharpoons [S-S(H)]^-$$

$$Fe^{2+} + [S-S(H)]^- \longrightarrow FeS_2$$

MARCASITE

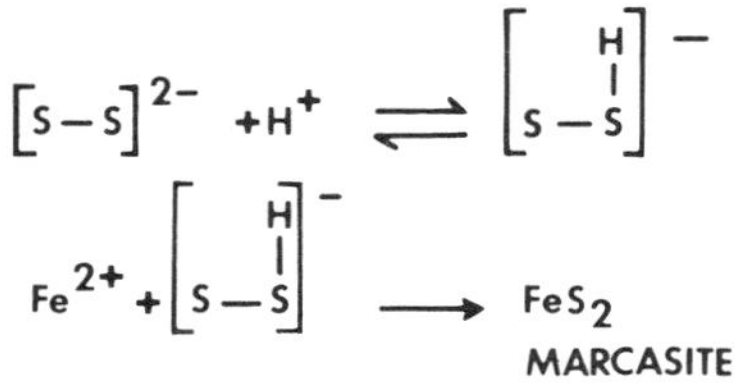

Fig. 7.7 Proposed mechanism of marcasite formation in acid solution (reproduced from Vaughan and Tossell[29] with the publisher's permission).

7.5.3 Monosulphides of the galena group

The monosulphides of the galena group have the familiar rocksalt structure with metal and anion in regular octahedral coordination. Galena (PbS) is a diamagnetic semiconductor, the electrical properties of which have been reviewed by Dalven.[39] The forbidden energy gap at 300 K is ~ 0.41 eV. For the isostructural PbSe (clausthalite) and PbTe (altaite), values of 0.27 eV and 0.31 eV respectively have been reported for the energy gaps.[40] The conduction mechanisms and carrier concentrations in all of these compounds are very sensitive to precise stoichiometry and to the presence of minor impurities. The electronic properties of galena have been extensively studied by various electrical and spectroscopic methods because of the technological applications of this compound.

The valence region X-ray photoelectron spectra of the lead chalcogenides have been reported by McFeely *et al.*,[41] and as seen in Fig. 7.8, they show a remarkable similarity between the sulphide, selenide and telluride. Both band structure calculations[42,43] and calculations using the X_α method[44] have been undertaken on these compounds which aid the interpretation of the spectra. It is also possible to compare band structure calculations, X_α calculations and a

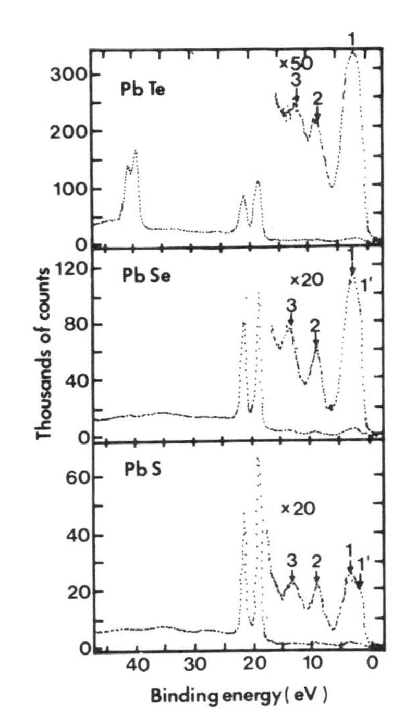

Fig. 7.8 Valence band X-ray photoelectron spectra of PbTe, PbSe, PbS (after McFeely *et al.*[41]).

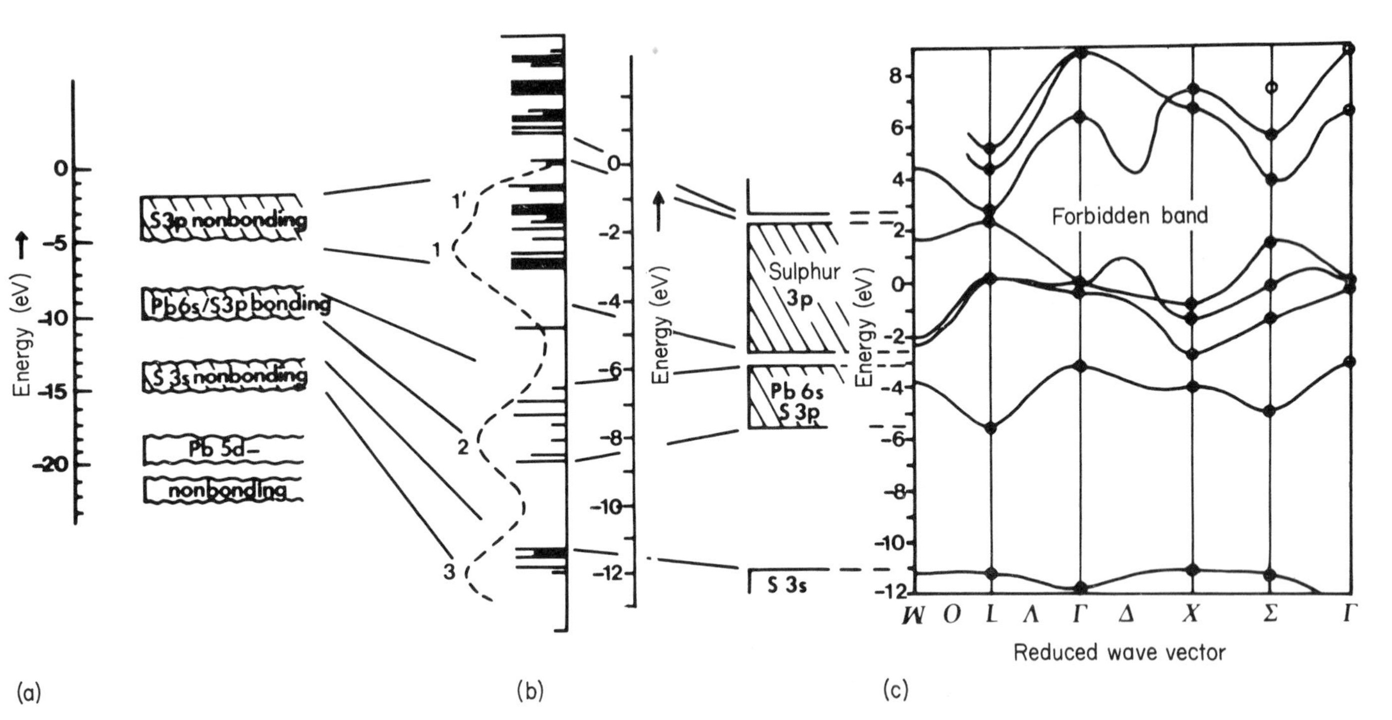

Fig. 7.9 Electronic structure models for galena (PbS). (a) Simplistic band structure representation based on spectroscopic data. (b) Molecular orbital energy levels calculated for a $SPb_6S_{12}Pb_8$ cluster using the SCF-X_α method (after Hemstreet[44]) and with features (labelled as in Fig. 7.8) from the X-ray photoelectron spectrum of PbS superimposed. (c) Band structure model for PbS calculated by the OPW method (after Tung and Cohen[42]).

simplistic MO/band model based on these and on the spectra as shown in Fig. 7.9. Hence, in examining the spectra alongside the calculations we can see that the peaks labelled 1 and 1′ at the top of the valence band are essentially sulphur (or Se or Te) p orbital in character and therefore non – bonding molecular orbitals. Below these lie the main bonding orbitals of the system which are chiefly lead 6 s/sulphur (Se, Te) p in character. Peak 3 represents sulphur 3 s orbitals which are not involved in bonding and the intense double peak below this (at ~ 20 eV binding energy) arises from the lead 5 d orbitals.

The optical properties of galena, PbSe and PbTe have been studied in considerable detail as reviewed by Schoolar and Dixon.[45] These authors performed specular reflectance measurements and transmittance measurements on a thin film of galena as shown in Fig. 7.10. Galena is translucent at long wavelengths ($< 3225\,cm^{-1}$, $\gtrsim 0.4$ eV) but is opaque beyond the fundamental absorption edge. Specular reflectance measurements on the lead chalcogenides were performed over a longer energy range by Cardona and Greenaway[46] as shown in Fig. 7.10. Again, the models presented in Fig. 7.9 can be applied to the interpretation of these data so that peaks E_1, E_2, E_3 arise from transitions into the conduction band by sulphur (Se, Te) p-type non-bonding molecular orbitals at the top of the valence band, and E_4, E_5, E_6 from lead 6 s/sulphur (Se, Te) p bonding and sulphur (Se, Te) s non-bonding orbitals. Schoolar and Dixon attribute the rise in reflectance at energies > 15 eV to transitions into the conduction band from the 5 d band which is filled with electrons and, as confirmed by the X-ray photoelectron spectra, 'buried' beneath the valence band.

7.5.4 Monosulphides of the niccolite group

The niccolite structure of the mineral which gives its name to this structure-type (NiAs) is adopted by the high-temperature forms of many sulphides (e.g. FeS, CoS, NiS), but on cooling these phases distort or transform to other structure-types or even dissociate into mixtures of phases. The niccolite structure also has the ability to accommodate additional metal atoms or to omit metal atoms, either randomly or on planes perpendicular to the *c*-axis. The latter phenomenon is responsible for the compositional and structural complexities of the 'pyrrhotites', metal-deficient iron monosulphides in which vacancy ordering produces superstructures based on the NiAs structure unit cell and compositions centring on Fe_7S_8, Fe_9S_{10}, $Fe_{11}S_{12}$, etc.[3]

Chemical bonding in many of the synthetic sulphides of this group and in the high temperature forms of CoS and NiS has been discussed[47,3] but is of limited mineralogical interest. The bonding models proposed for high temperature FeS, and for the distorted modification stable at room temperature as the mineral troilite, will be considered here; they also provide information relevant to an understanding of the 'pyrrhotites'. In high temperature FeS, iron occurs in regular octahedral coordination to sulphur

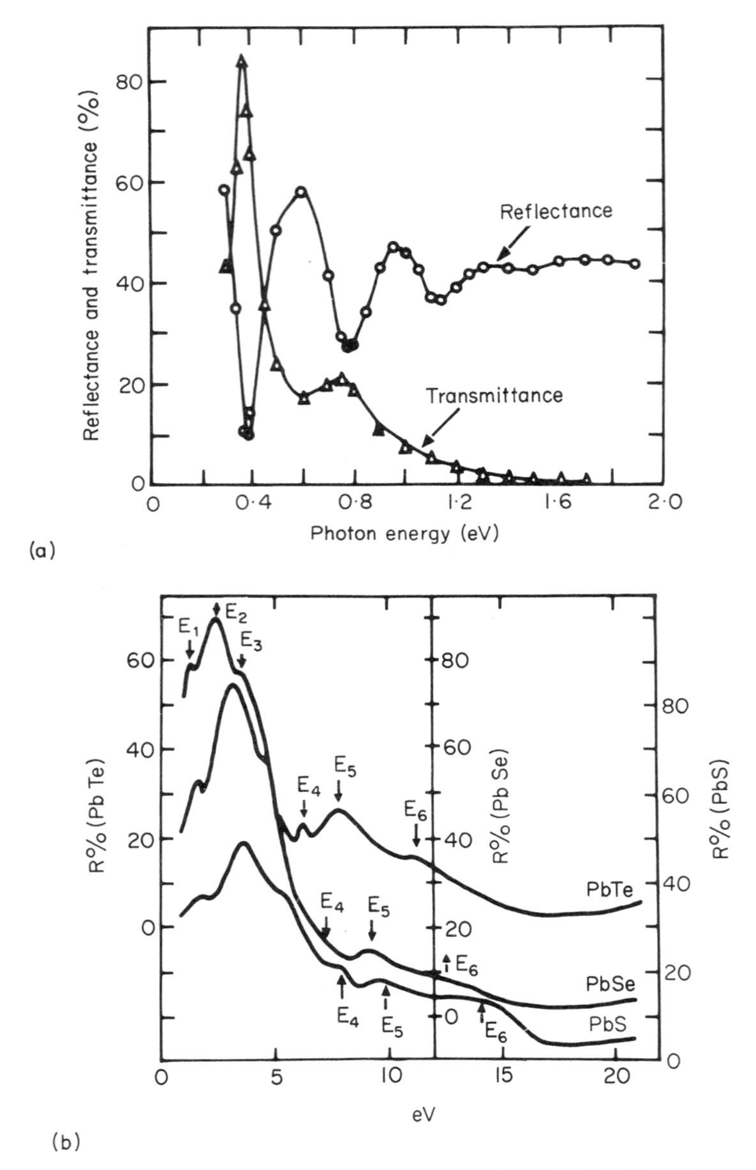

Fig. 7.10 (a) Reflectance and transmittance data for a thin film (0.37 μm thick) of galena on a NaCl substrate (after Schoolar and Dixon[45]). (b) Room temperature reflectance of PbS, PbSe and PbTe (after Cardona and Greenaway[46]) (reproduced from Vaughan and Craig[3] with the publisher's permission).

and the octahedral share faces along the *c*-axis; the sulphurs are in trigonal prismatic coordination. Below 140° C (T_α), FeS distorts from this NiAs-type structure to the troilite modification and this involves triangular clusters of iron atoms forming in the basal plane and some associated movement of sulphur atoms. FeS is a semiconductor which is antiferromagnetically ordered ($T_n = 325°$ C); below T_α there is a sharp change in *c*-axis conductivity and at, or close to, this temperature the magnetic spin direction changes from parallel to the *c*-axis to perpendicular to it.

Schematic MO/band structure models of FeS have been presented by Wilson[47] and by Goodenough[9,10] and X_α calculations for the appropriate FeS_6^{-10} cluster undertaken by Tossell.[48] In Fig. 7.11 are shown the results of these calculations along with schematic diagrams for both high temperature FeS and the troilite modification. The calculated energy levels are labelled according to the irreducible representations of the O_h symmetry group. The $2a_{1g}$, $2t_{1u}$, and $1t_{2g}$ orbitals are the main Fe–S bonding orbitals and are followed by a group of essentially S3p non-bonding orbitals ($2e_g$, $1t_{2u}$, $3t_{1u}$, and $1t_{1g}$). Above these lie the $2t_{2g}$ and $3e_g$ antibonding crystal field orbitals of Fe3d and S3p character. Calculated energies are in reasonable agreement with available data from X-ray emission and absorption spectra[48] and, as Fig. 7.11 shows, with the qualitative models.

Goodenough[10] has discussed the distortion at $T\alpha$ to form the troilite structure in terms of band models (see Fig. 7.11(b)). At high temperature, the majority spin electrons in the t_{2g} orbitals parallel to the *c*-direction (where cation–cation interaction can take place) form a filled band; the remaining t_{2g} and e_g electrons remain localized. The minority spin electrons in t_{2g} orbitals can form bands both perpendicular and parallel to c with the e_g^* orbitals again remaining discrete. Only the Γ_1^* (β) band shown in Fig. 7.11 contains electrons and it is actually half-filled. This narrow band invites a spontaneous distortion that will split the band in two and this is what happens below T_α.

Calculations on FeS_6^{-10} clusters at different internuclear distances have also been used to model the behaviour of FeS at high pressure.[48] These studies confirm the relative ease with which spin-pairing can occur in this material as was suggested by studies of the ^{57}Fe Mössbauer spectrum.[49]

7.5.5 More complex sulphides

Amongst sulphides of more complex composition or structure which have been studied in recent years are chalcopyrite,[50] the copper and silver sulphides[51,52] and the thiospinels.[53]

Chalcopyrite, although having a fairly simple structure based on that of sphalerite[54] contains both Cu^{1+} and Fe^{3+} (as indicated by neutron diffraction and Mössbauer studies[3]) in regular tetrahedral coordination. A fairly complete series of X-ray emission spectra and the X-ray photoelectron

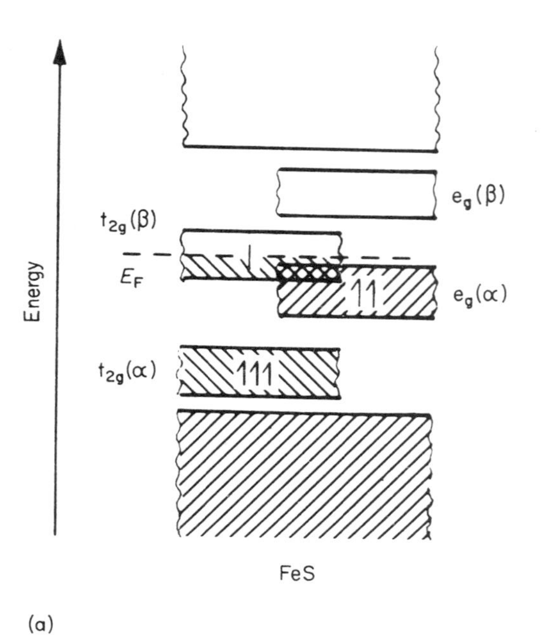

Energy
$e_g(\beta)$
$t_{2g}(\beta)$
E_F
$e_g(\alpha)$
$t_{2g}(\alpha)$
FeS
(a)

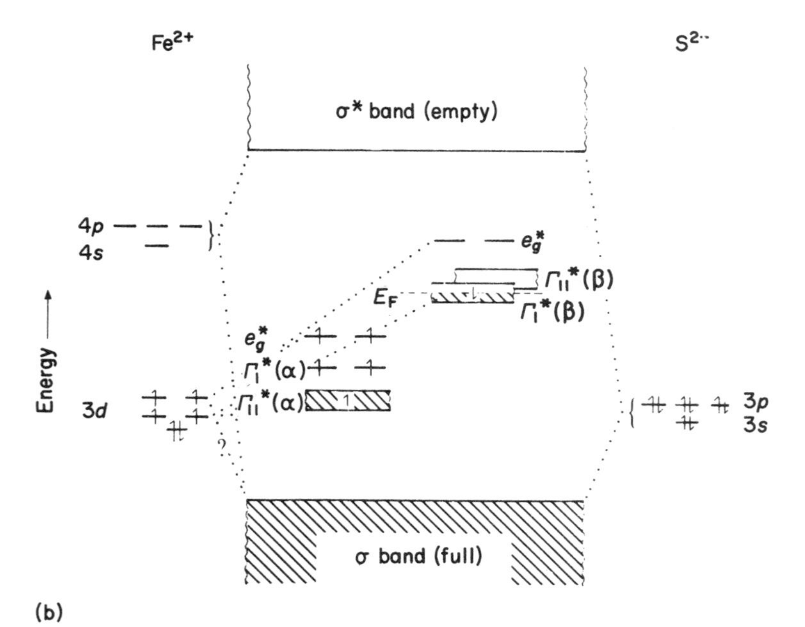

Fe^{2+}
S^{2-}
σ* band (empty)
4p
4s
e_g^*
$\Gamma_{II}^*(\beta)$
E_F
$\Gamma_I^*(\beta)$
Energy
e_g^*
$\Gamma_I^*(\alpha)$
3d
$\Gamma_{II}^*(\alpha)$
3p
3s
σ band (full)
(b)

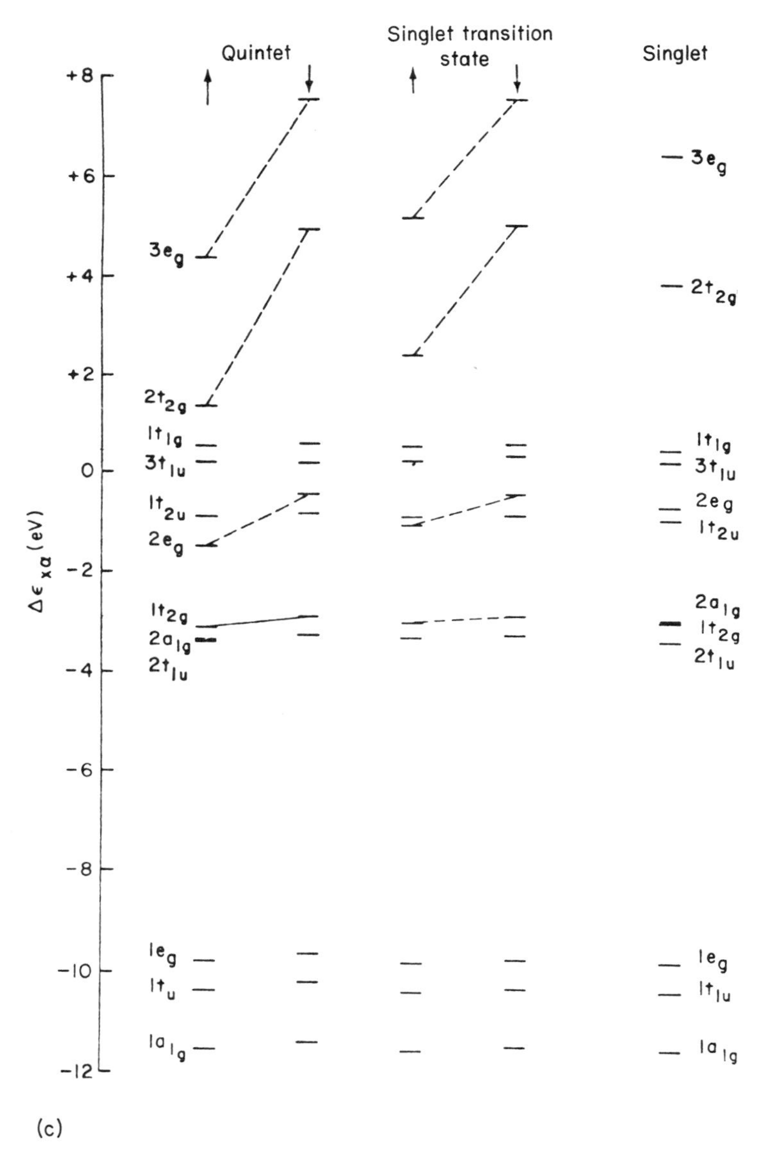

Fig. 7.11 Electronic structure models for FeS. (a) Schematic 'one electron' energy band diagram for NiAs-structured FeS (after Wilson[47]). (b) Schematic energy level diagram for the troilite form of FeS (after Goodenough[10]). (c) Molecular orbital energy levels for the FeS_6^{-10} cluster, representative of the basic building unit in FeS, calculated using the SCF-X_α method (after Tossell[48]).

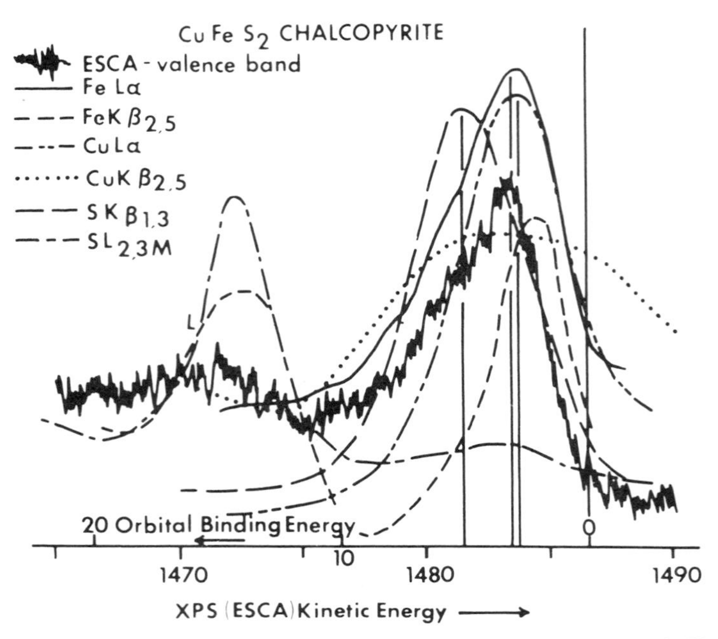

Fig. 7.12 Valence region X-ray photoelectron and FeL_{α}, $FeK_{\beta_{2,5}}$, CuL_{α}, $CuK_{\beta_{2,5}}$ $SK_{\beta_{1,3}}$ and $SL_{2,3}$ M X-ray emission spectra of chalcopyrite (reproduced from Tossell *et al.*[56] with the publisher's permission).

spectrum have been reported and interpreted using X_{α} calculations.[50] The spectra are shown in Fig. 7.12 and these may be qualitatively assigned using calculations for the CuS_4^{7-} and FeS_4^{5-} clusters (Fig. 7.13). The orbitals shown here comprise la_1 and lt_1 which appear to be mainly S3s in character, $2a_1$ and $3t_2$ which are mostly S3p with some metal s or p, the $1t_1$ which is a non-bonding S3p, the $2t_2$ and 1e which are the bonding metal 3d–S3p orbitals and the 2e and $4t_2$ which are the antibonding metal 3d-S3p or 'crystal field' orbitals. In the FeS_4^{5-} cluster, the composition of the 1e, $2t_2$, 2e and $4t_2$ orbitals is highly spin-dependent with 70–80% Fe3d character in the 1e↑, $2t_2$↑, 2e↓ and $4t_2$↑ orbitals. The strong peak in the $SL_{2,3}$ spectrum at about 14 eV arises from the predominantly S3s non-bonding $1a_1$ and $1t_2$ orbitals, and broad features at this energy in the Fe and $CuK_{\beta_{2,5}}$ spectra indicate some mixing of Fe4p and Cu4p into these orbitals. The main $SK_{\beta_{1,3}}$ peak at about 5 eV is assigned to predominantly S3p-type orbitals ($2a_1$, $2t_2$, 1e, $3t_2$ and $1t_1$). The maximum in the FeL_{α} spectrum at ~ 3 eV can be assigned to 2e and $4t_2$ orbitals in the FeS_4^{5-} unit. The CuL_{α} spectrum also peaks at about 3 eV and is assigned to the 2e and $4t_2$ orbitals in the CuS_4^{7-} unit. The shoulder at low binding energy in this spectrum may represent a contribution from orbitals such as the Cu3d $4t_2$.

Comparison of the energies of measured and calculated orbital eigenvalues

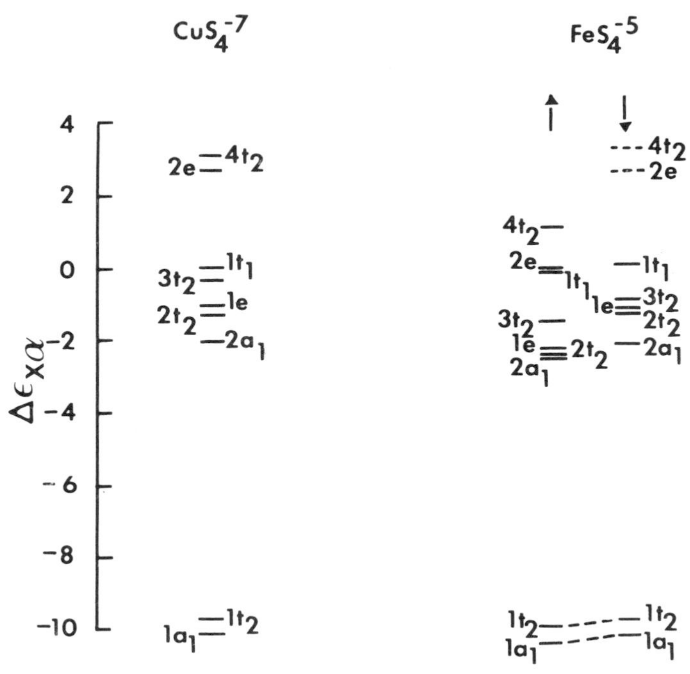

Fig. 7.13 Energy level diagram showing ground state orbital eigenvalues for the clusters CuS_4^{-7} and FeS_4^{-5} calculated using the SCF-X_α method (occupied orbitals are shown as solid lines, unoccupied orbitals as broken lines, energies in eV are relative to the average energy of the lt_1 orbital) (reproduced from Vaughan and Tossell[29] with the publisher's permission).

showed generally good correlation between experiment and calculation, although the calculations do overestimate the binding energies of the Fe3d orbitals. Somewhat better agreement with experiment is found for an X_α band structure study of $CuFeS_2$[55] although even here the Fe3d type orbitals are calculated to be somewhat too stable. It is not apparent why the Cu^{I} cluster results agree well with experiment while the Fe^{III} cluster results are in poor agreement. The poor result may be due to perturbation of the FeS_4^{5-} cluster by neighbouring atoms in the $CuFeS_2$, with the Fe3d orbitals destabilized either through mixing with the Cu3d orbitals or as a result of saturation of the S valences. The high pressure properties of $CuFeS_2$ have been interpreted using these models[50] and explanations offered for the changes which occur in conductivity and magnetic ordering at elevated pressures.

In chalcocite (Cu_2S), copper occurs in two kinds of triangular coordination[56] and Tossell and Vaughan[52] used published SK_β and $L_{2,3}$ X-ray emission spectra and the X-ray photoelectron spectra to construct a diagram

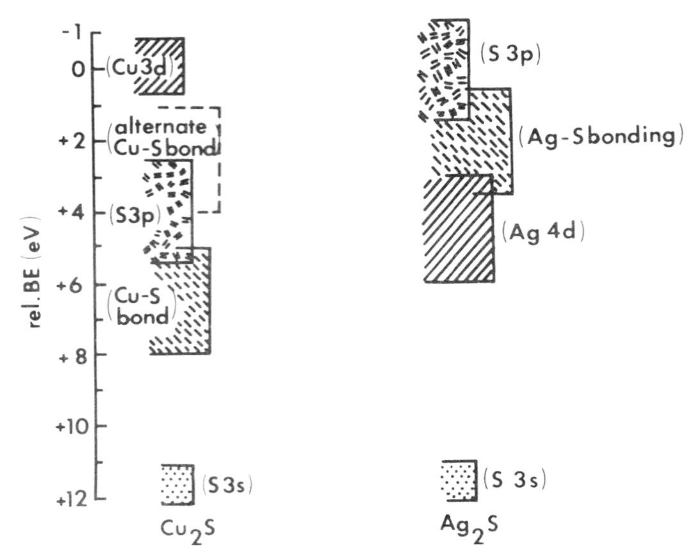

Fig. 7.14 Cu_2S and Ag_2S orbital binding energies estimated from spectra (reproduced from Vaughan and Tossell[29] with the publisher's permission).

of the type shown in Fig. 7.14. Calculations by the X_α method on a CuS_3^{5-} cluster showed good agreement and helped in the interpretation of the spectra. These calculations were also used in an analysis of the electronic structure of covellite (CuS) which contains copper in tetrahedral as well as triangular coordination.[51,54] Here, as illustrated in Fig. 7.15, calculations on the clusters CuS_3^{4-} and CuS_4^{7-} were initially performed on the assumption of formal oxidation states in covellite of Cu_{III}^{2+} $(Cu_{IV}^{+})_2$ $S^{2-}(S_2^{2-})$. However, the metallic conductivity observed in CuS[21] indicates that the highest energy orbitals containing electrons form a collective electron band. Further calculations indicated that to form such a band, charge should flow from the $4t_2$ orbital on the tetrahedral Cu^+ to the 4e orbital on the triangular Cu^{2+} (see Fig. 7.15). The resulting change in electronic structure was represented by performing cluster calculations on CuS_3^{5-} and $CuS_4^{6.5-}$ (Fig. 7.15). As shown by the shaded areas on the diagram, beneath the incompletely filled band of dominantly Cu3d character, lies a non-bonding sulphur band and the metal–sulphur bonding band. The electronic structure of Cu_2S is similar to this (Fig. 7.14) but in marked contrast to the electronic structure proposed for Ag_2S.[52] The energy level diagram, also shown in Fig. 7.14, has been constructed from sulphur K_β and $L_{2,3}$ spectra and from X-ray photoelectron spectra which were probably from the high temperature modification of Ag_2S (argentite) in which silver occurs in tetrahedral and octahedral coordination. The diagram shows a sulphur 3p non-bonding band at the top of the valence region in Ag_2S.

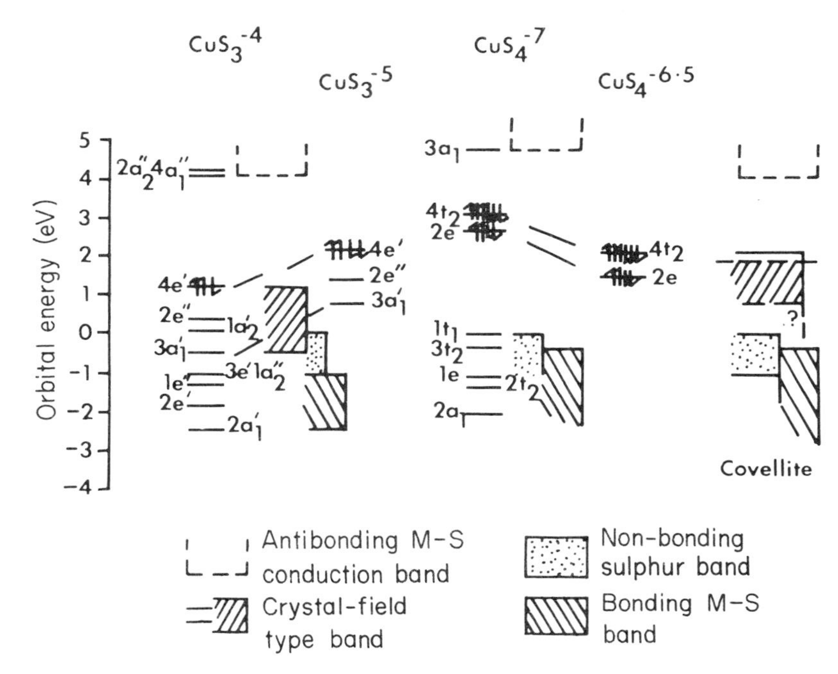

Fig. 7.15 Electronic structure model for covellite (CuS). Discrete energy levels are shown for the clusters CuS_3^{4-}, CuS_3^{5-}, CuS_4^{7-} and $CuS_4^{6.5-}$ and based on SCF-X_α calculations. A composite 'one-electron' band model energy-level diagram is illustrated for the mineral (reproduced from Vaughan and Tossell[29] with the publisher's permission).

Sulphides with the spinel structure have been discussed in terms of qualitative molecular orbital and band theory models.[58,11] It was postulated that in such thiospinels as linnaeite (Co_3S_4), carrollite ($CuCo_2S_4$), polydymite (Ni_3S_4) and violarite ($FeNi_2S_4$) which exhibit metallic conductivity, the highest energy levels containing electrons are a partially-filled σ-antibonding band formed by the overlap of the dominantly metal 3d electron e_g orbitals on octahedral sites and t_2 orbitals on the tetrahedral sites. Calculations on cluster units such as CoS_6^{9-}, CoS_4^{6-} and CuS_4^{7-} appropriate to Co_3S_4 and $CuCo_2S_4$ species have been undertaken using the X_α method.[53] As shown in the results presented in Fig. 7.16 for CoS_6^{9-} and CoS_4^{6-}, the empty e_g orbitals of the octahedral cluster are close in energy to the partially-filled e orbitals of the tetrahedral cluster. Therefore, dispersion of their energies with wave vector could well lead to band overlap. The small energy difference of these orbitals also favours strong interaction and energy splitting as is apparent from simple perturbation theory considerations.[34] Both arguments suggest that band overlap will occur. This model, which is in agreement with the previously

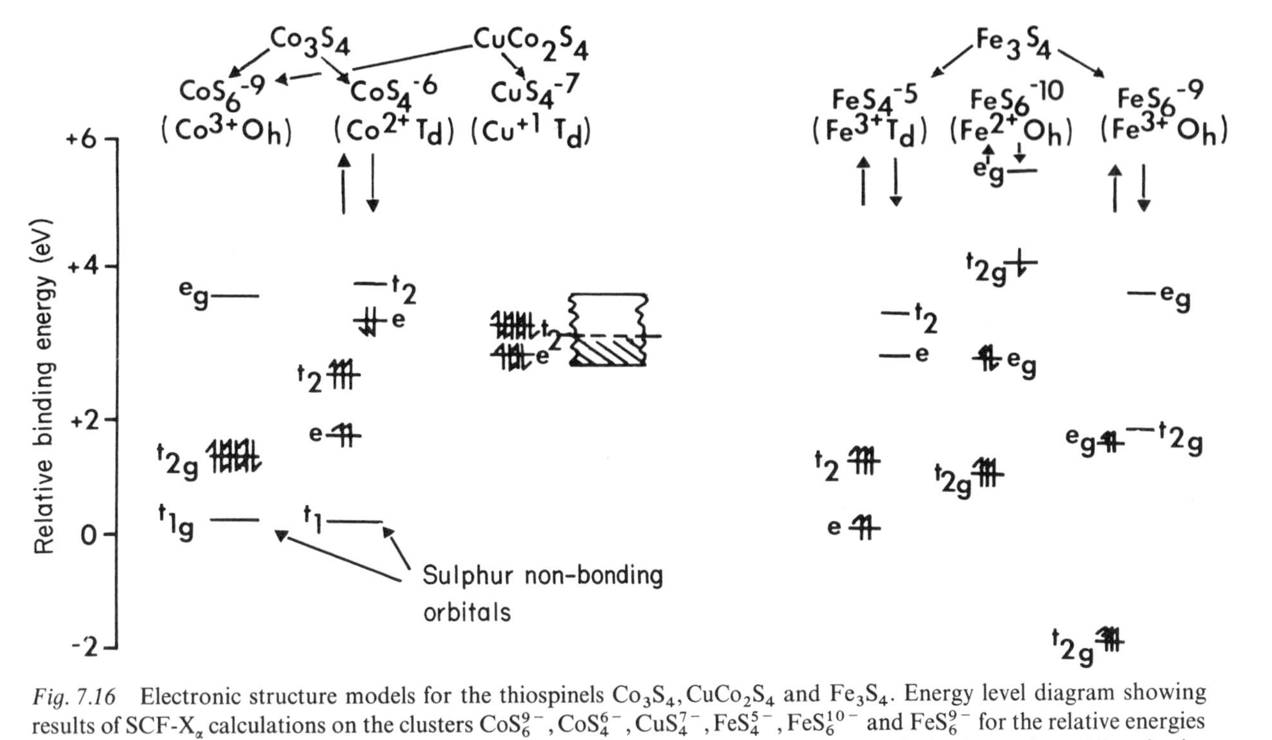

Fig. 7.16 Electronic structure models for the thiospinels Co_3S_4, $CuCo_2S_4$ and Fe_3S_4. Energy level diagram showing results of SCF-X_α calculations on the clusters CoS_6^{9-}, CoS_4^{6-}, CuS_4^{7-}, FeS_4^{5-}, FeS_6^{10-} and FeS_6^{9-} for the relative energies of valence orbitals. For those systems containing unpaired electrons the effects of spin-splitting into spin-up (↑) and spin-down (↓) molecular orbitals is shown (reproduced from Vaughan and Tossell[29] with the publisher's permission).

proposed qualitative theories, can be extended to $CuCo_2S_4$ by incorporating the results of X_α calculations on the CuS_4^{7-} cluster (Fig. 7.16). Here, overlap can again be envisaged between empty e_g orbitals on octahedral site cobalt atoms and e and t_2 orbitals of the tetrahedral site copper to give rise to a partly-filled band.

In contrast to the metallic group of thiospinels, greigite (Fe_3S_4) and daubréelite ($FeCr_2S_4$) show evidence of having localized valence electrons (i.e. they appear to be semiconductors which exhibit ordered magnetism). Calculations on cluster units appropriate to greigite have been performed using the X_α method[53] and again serve to confirm qualitative theories. As shown in the results on FeS_4^{5-}, FeS_6^{10-} and FeS_6^{9-} clusters (Fig. 7.16), large separations in energy occur between t_2 and e_g orbitals on tetrahedral sites as a consequence of spin-splitting.

Such bonding models for the thiospinels have been applied to the interpretation of mineral properties[11] and used to explain the ranges of solid solution observed in these materials, with extensive solid solution between members of the metallic group but not between these and greigite or daubréelite.[3,11,53]

7.5.6 Oxides of the haematite group

Haematite ($\alpha - Fe_2O_3$), ilmenite ($FeTiO_3$) and the rarer mineral eskolaite (α–Cr_2O_3) have crystal structures closely related to that of corundum (α–Al_2O_3). The metals are in octahedral coordination in an approximately hexagonal close-packed array of oxygens. All three minerals are semiconductors and whereas ilmenite is antiferromagnetically ordered at low temperatures ($T_N \approx 60$ K), eskolaite is antiferromagnetic up to a Néel temperature close to room temperature ($T_N \approx 308$ K) and haematite antiferromagnetically ordered up to relatively high temperature ($T_N = 963$ K). The magnetic behaviour of haematite is complex in that it shows a weak 'parasitic' ferromagnetism over a range of temperature ($250\,K < T < T_N$).

As noted in Chapter 1, MO calculations utilizing the X_α method have been performed for metal–oxygen polyhedral clusters containing Fe^{2+}, Fe^{3+}, Ti^{4+} (and also Cr^{3+}) in octahedral coordination to oxygen[59,60] and these can be used to model the electronic structures of haematite, ilmenite and eskolaite. Comparison of the calculated energies with published X-ray and UV photoelectron, X-ray emission and X-ray absorption spectra of haematite, wüstite and rutile were used to support the validity of the calculations which were later used to construct schematic ('one-electron') MO/band theory energy level diagrams.[61] The diagrams for Fe_2O_3, $FeTiO_3$ and Cr_2O_3 are shown in Fig. 7.17. Since Goodenough[9,62] has argued that the outermost electrons in the valence region in these materials may be described using localized models, these (dominantly metal 3d) electrons are represented by discrete levels whereas the other levels are shown as broadening into bands.

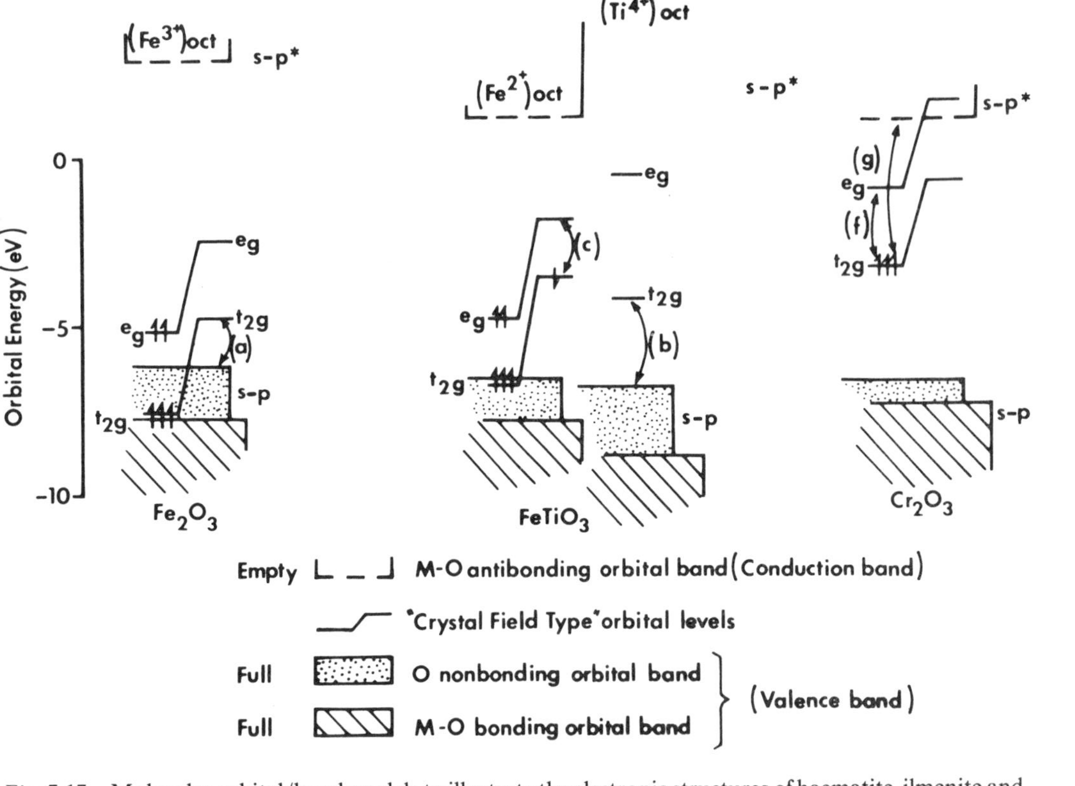

Fig. 7.17 Molecular orbital/band models to illustrate the electronic structures of haematite, ilmenite and eskolaite. The 'crystal-field type' levels are shown as discrete levels split into spin-up and spin-down levels whereas the others are shown as bands (after Vaughan and Tossell[61]).

In the FeO_6^{-9} cluster of haematite, the highest-energy filled orbital is an e_g orbital that is antibonding in character (mainly Fe3d and O2p). The filled t_{2g} orbital (a bonding orbital of Fe3d and O2p character) lies several eV lower in energy, beneath energy levels shown at the top of a delocalized band in Fig. 7.17. The top of this s – p band is made up of O2p nonbonding orbitals. Several eV below the top of this band are the main bonding orbitals of Fe3d, 4s, 4p and O2p character. The vacant orbitals in the valence region are the $e_g\downarrow$ and $t_{2g}\downarrow$ and (about 9 eV above the highest filled orbital) the diffuse antibonding orbitals of the s – p* band with metal and oxygen s and p orbital character.

In Fig. 7.17 a similar model for ilmenite is based on calculations on the FeO_6^{-10} and TiO_6^{-8} clusters. The nature of the s – p (bonding) and s – p* (antibonding) bands is much the same in Fe_2O_3. However, the 3d levels of Fe^{2+} do not overlap the s–p band as much and, along with the vacant Ti3d levels, there is the additional t_{2g} electron of Fe^{2+}. The vacant t_{2g} and e_g levels of TiO_6^{-8} have both Ti3d and O2p character with the e_g level antibonding. The model for eskolaite (Fig. 7.17) based on the CrO_6^{-9} cluster shows a similar overall electronic structure. In this case, both e_g orbitals exhibit appreciable oxygen as well as chromium character and the t_{2g} orbitals rather less oxygen character.

Properties of these materials of mineralogical interest can be considered using these models. So, for example, although both haematite and ilmenite are semiconductors, the models (Fig. 7.17) shows a smaller separation between occupied and unoccupied energy levels in haematite than in ilmenite, in agreement with the lower intrinsic resistivity of haematite (~ 0.3 ohm m) than ilmenite (~ 0.8–1.8 ohm m).[21] The specular reflectance in the visible region is of considerable mineralogical interest and the data of von Gehlen and Piller[63] for haematite and ilmenite are shown in Fig. 7.18. Examination of the electronic structure models (Fig. 7.17) shows that 'allowed' electronic transitions occur in ilmenite from mainly oxygen s – p to titanium t_{2g} orbitals at an

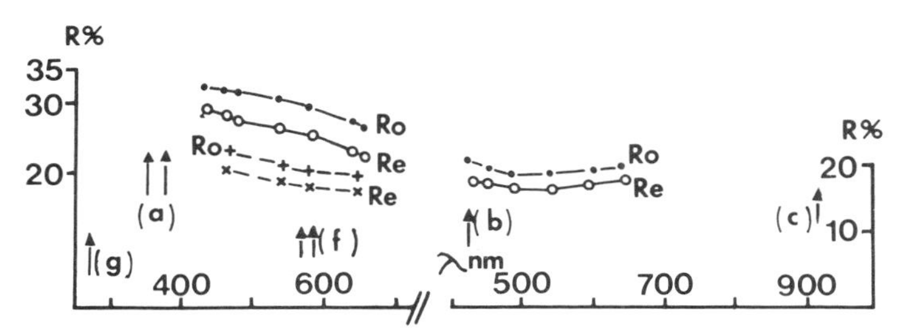

Fig. 7.18 Spectral reflectance data for haematite, ilmenite and eskolaite in the visible light range. Data are shown for both the ordinary (Ro) and extraordinary (Re) vibration directions in plane polarized light. Arrows indicate the energy regions of the major electronic transitions and are labelled to correspond to transitions shown in Fig. 7.17 (modified after von Gehlen and Piller[63] and Vaughan and Tossell[61]).

energy corresponding to ~ 400 nm as shown by the labels in Figs. 7.17 and 7.18. 'Allowed' transitions also occur from iron $t_{2g}\downarrow$ to $e_g\downarrow$ orbitals at energies corresponding to ~ 1000 nm. These transitions, corresponding to the absorption and subsequent re-emission of energy, account for the higher values of reflectance at either end of the 400–700 nm profile. In haematite, the $s-p \rightarrow t_{2g}$ transitions at an energy corresponding to ~ 400 nm account for the decrease in reflectance from 400 to 700 nm and the greater concentration of iron contributing to this transition, for the greater overall reflectance. Spectral reflectance data for eskolaite are also plotted on Fig. 7.18 and here both the spin-allowed $t_{2g} \rightarrow e_g$ transition (f) and the $t_{2g} \rightarrow s-p^*$ conduction band transition (g) contribute to increased reflectance toward 400 nm.

Thermochemical data show considerable differences in the free energies of formation of the three compounds, with ΔG^o_{298} for ilmenite being − 277.1 kcal/gfw, for eskolaite − 253.2 kcal/gfw and for haematite − 177.7 kcal/gfw. In comparing ilmenite with haematite, the crystal-field theory approach would emphasize the stabilization gained by the sixth 3d electron of Fe^{2+} whereas reference to the MO/band model emphasizes another important factor. Although the energies of Fe^{3+}–O and Fe^{2+}–O bonding orbitals are very similar, the Ti^{4+}–O bonding orbital set is stabilized (by ~ 1 eV) relative to the iron–oxygen bonds. This very important contribution to the stability of ilmenite relative to haematite, or as Fig. 7.17 shows, to eskolaite, is not considered in a crystal-field approach. Stabilization of the Cr^{3+}–O bonding orbital set is very similar to that of Fe^{3+}–O but the Cr_2O_3 does not have the destabilizing effect of electrons in the (antibonding) e_g orbitals.

7.5.7 The spinel oxides

The spinel crystal structure, with its one tetrahedral and two octahedral cation sites per formula unit, is well known.[64] Amongst the major opaque spinel oxide minerals, chromite is a *normal* spinel with nominally divalent ions (Fe^{2+}) in the tetrahedral sites and trivalent ions (Cr^{3+}) in the octahedral sites, i.e. $Fe^{2+}_{IV}(Cr^{3+}_{VI}Cr^{3+}_{VI})O_4$, whereas magnetite is an *inverse* spinel, i.e. $Fe^{3+}_{IV}(Fe^{2+}_{VI}Fe^{3+}_{VI})O_4$. Ulvöspinel is unusual in that it contains tetravalent ions, i.e. $Fe^{2+}_{IV}(Fe^{2+}_{VI}Ti^{4+}_{VI})O_4$. Pure endmember chromite is a semiconductor and is ferromagnetically ordered below 90 K. The behaviour of magnetite is extremely complex and still not fully understood. Above a transition temperature of ~ 120 K (the *Verwey* transition) hopping of electrons between the octahedral site iron atoms gives relatively high (metallic) conductivity, whereas below this temperature magnetite is a semiconductor. Magnetite is the classic ferrimagnetic compound.

A similar approach to the bonding in the spinel oxides can be adopted to that just outlined for the oxides of the haematite group. The results of MO calculations on the clusters FeO_6^{-9}, FeO_6^{-10}, FeO_4^{-6}, FeO_4^{-5}, TiO_6^{-8} and

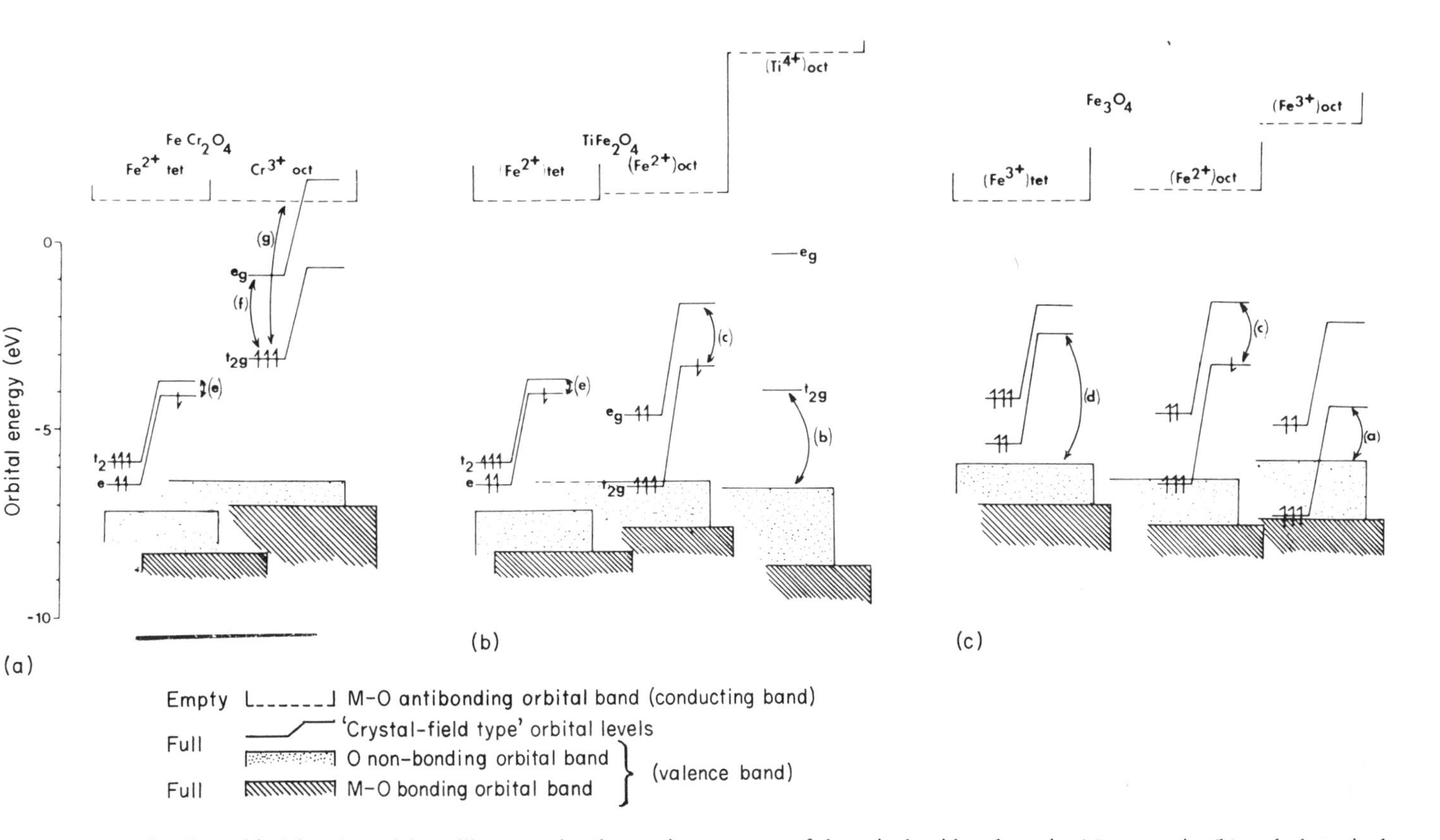

Fig. 7.19 Molecular orbital/band models to illustrate the electronic structures of the spinel oxides chromite (a) magnetite (b) and ulvöspinel (c). The 'crystal-field type' levels are shown as discrete levels split into spin-up and spin-down levels whereas the others are shown as bands (after Vaughan and Tossell[61]).

CrO_6^{-9} can be used to construct schematic ('one-electron') MO/band models as shown in Fig. 7.19. The calculations employed the X_α method and were reported in the publications of Tossell *et al.*,[59,60] Tossell[65,66] and Vaughan and Tossell.[61] Again, in constructing these models, it has been assumed that the outermost (dominantly metal 3d) electrons are localized on the cations as the electrical and magnetic data indicate. The chromite model combines the results of the tetrahedral FeO_4^{-6} cluster and octahedral CrO_6^{-9} cluster calculations. Here, the FeO_4^{-6} empty conduction-band orbitals have metal and oxygen character and the crystal field type e orbital is mostly Fe3d. The empty t_2 orbital is also dominantly Fe3d but with some 4p character, whereas the $t_2\uparrow$ orbital is equally oxygen p and Fe3d with some 4p. As in other related systems, the orbitals immediately beneath these in energy are almost totally oxygen (non-bonding) in character. A similar overall structure is indicated by the calculations on CrO_6^{-9}. Here, both e_g orbitals exhibit appreciable oxygen as well as chromium character and the t_{2g} orbitals rather less oxygen character. Magnetite can be modelled using the results of cluster calculations on FeO_4^{-5}, FeO_6^{-10} and FeO_6^{-9}. The valence region X-ray and UV photoelectron spectra of magnetite can be correlated with the results of these calculations.[66]

The atomic orbital compositions of the molecular orbitals for the octahedral-site ions can be regarded as similar to those already discussed for haematite and ilmenite. The tetrahedrally coordinated Fe^{3+} is similar to tetrahedral Fe^{2+} in having the crystal-field type levels above and separate from the oxygen-non-bonding set and the Fe–O bonding orbitals of the system. For ulvöspinel, the FeO_6^{-10} and TiO_6^{-8} calculations are used in combination with FeO_4^{-6}.

The spectral reflectance profiles of these spinels (Fig. 7.20) can be interpreted in a similar way to those of the haematite group oxides. In chromite, the transition chiefly contributing to the reflectance is the 'spin-allowed' crystal field transition in the octahedral Cr^{3+} ions (f) at ~ 530 nm. The ulvöspinel reflectance profile resembles that of ilmenite since the same 'allowed' transitions in octahedral Fe^{2+} and Ti^{4+} ions contribute to the higher reflectance values toward either end of the visible range. Magnetite has contributions from oxygen → metal charge transfer transitions ((a) and (d)) and the spin-allowed crystal field transition in octahedral Fe^{2+}. Data on the hardness properties of these minerals are incomplete but available information suggests that chromite is the hardest of the three which correlates with the absence of electrons in the e_g orbital levels in octahedral sites. Ulvöspinel is predicted to be the softest of the three phases with the e_g electron of the octahedral Fe^{2+} and $t_2\downarrow$ electron of tetrahedral Fe^{2+}. Thermochemical data are also incomplete but chromite probably has a significantly greater value for $-\Delta G^o_{298}$ than magnetite. In terms of crystal-field theory, these data could be interpreted as due to the considerable stabilization of Cr^{3+} in octahedral sites. The MO model suggests that this may be partly compensated for by lesser

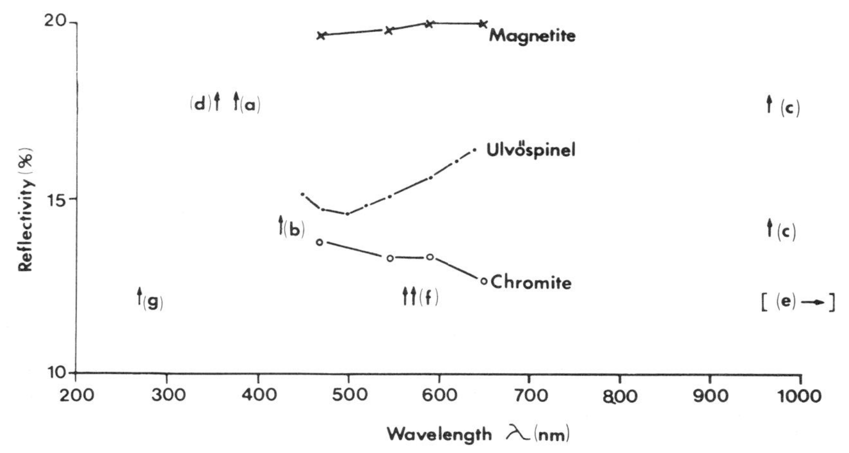

Fig. 7.20 Spectral reflectance data for chromite, magnetite and ulvöspinel in the visible light range. Arrows indicate the energy regions of the major electronic transitions and are labelled to correspond to transitions shown in Fig. 7.19 (after Vaughan and Tossell[61]).

stabilization of metal–oxygen bonding interactions than in octahedrally coordinated Fe^{2+}, Fe^{3+}, or Ti^{4+}. Although ulvöspinel has an additional $e{\downarrow}$electron which is antibonding in character, the metal–oxygen bonding orbitals are all stabilized relative to those in magnetite so that the value of $-\Delta G^{\circ}_{298}$ may well be greater than that of magnetite.

7.6 CONCLUDING REMARKS

The opaque minerals present special problems as regards an understanding of their chemical bonding. The delocalization of valence electrons, which is often the cause of their opacity, is commonly linked with interesting electrical and magnetic behaviour and has often led to detailed study of these materials by solid state physicists. However, in common with other mineral groups, the most satisfactory bonding models for the mineralogist derive from a molecular cluster approach, although it is also important to relate such models to those derived from qualitative and quantitative band theory.

Using such approaches, it has proved possible to assign and to interpret spectroscopic data from a wide range of techniques, notably from X-ray emission and photoelectron spectroscopy and from specular or diffuse reflectance spectroscopy. This has, led, in turn, to interpretation of the optical properties of interest to the mineralogist in terms of bonding models. In certain cases, structural relationships have been explained as well as trends in bond length and cell parameter variation, and also systematic variations in hardness in some mineral systems. It has also been possible to provide explanations for general trends in overall stability, for the occurrence of metastable species, and for the limits of solid solution in certain systems. Modelling and prediction of the behaviour of opaque minerals at elevated pressures has also been possible in some materials.

In the future, the further development of quantum mechanical models combined with careful spectroscopic studies should enable a much greater understanding of all the above mentioned properties, as well as others still poorly understood such as the magnetic and electrical behaviour. There are also very wide areas of potential application involving the surface properties of these minerals and their dissolution characteristics which are relevant to mineral processing and hydrometallurgy[67] as well as the geological sciences. It should also be possible to model behaviour in melts and solutions and to study reaction mechanisms in order to introduce new theoretical concepts into discussions of the genesis of metalliferous ores.

REFERENCES

1. Wells, A.F. (1961) *Structural Inorganic Chemistry*, Oxford University Press Oxford.
2. Dana, J.W. and Dana, E.S. (1951) *The System of Mineralogy*, Vol II (revised by C. Palache, H. Berman and C. Frondel), John Wiley, New York.
3. Vaughan, D.J. and Craig, J.R. (1978) *Mineral Chemistry of Metal Sulphides*, Cambridge University Press, Cambridge.
4. Rumble, D. (ed.) (1976) *Oxide Minerals*, Mineral. Soc. Amer. Short Course Notes Vol. 3, Min. Soc. Amer.
5. Pauling, L. (1970) *Mineral. Soc. Amer. Spec. Paper*, **3**, 125.
6. Burns, R.G. (1970) *Mineralogical Applications of Crystal Field Theory*, Cambridge University Press, Cambridge.
7. Nickel, E.H. (1968) *Can. Mineral.*, **9**, 311.
8. Nickel, E.H. (1970) *Chem. Geol.*, **5**, 233.
9. Goodenough, J.B. (1963) *Magnetism and the Chemical Bond*, Wiley, New York.
10. Goodenough, J.B. (1967) in *Propriétés Thermodynamiques Physiques et Structurales des Dérivés Semi-Metalliques.* Centre National de la Recherche Scientifique, Paris.
11. Vaughan, D.J., Burns, R.G. and Burns, V.M. (1971) *Geochim, Cosmochim. Acta*, **35**, 365.
12. Gibbs, G.V. (1982) *Am. Mineral.*, **67**, 421.
13. Wendlandt, W.W. and Hecht, H.G. (1966) *Reflectance Spectroscopy*, Wiley-Interscience, New York.
14. Bither, T.A., Bouchard, R.J., Cloud, W.H. *et al.* (1968) *Inorg. Chem.*, **7**, 2208.
15. Hedelman, S. and Mitchell, W.N. (1968) in *Modern Aspects of Reflectance Spectroscopy*, (ed. W. Wendlandt) Plenum Press, New York.

16. Craig, J.R. and Vaughan, D.J. (1981) *Ore Microscopy and Ore Petrography*, Wiley-Interscience, New York.
17. Piller, H. (1977) *Microscope Photometry*, Springer-Verlag, Berlin.
18. Embrey, P.G. and Criddle, A. (1978) *Am. Mineral.* **63**, 853.
19. Clark, R.J.H. (1964) *J. chem. Educ.*, **41**, 488.
20. Kubelka, P. and Munk, F. (1931) *Z. Tech. Phys.*, **12**, 593.
21. Shuey, R.T. (1975) *Semiconducting Ore Minerals*, Elsevier, Amsterdam.
22. Baleshta, T.M. and Dibbs, H.P. (1969) *An Introduction to the Theory, Measurement and Application of Semiconductor Transport Properties of Minerals*, Mines Branch Tech. Bull TB 106, Ottawa, Canada.
23. Burns, R.G. and Vaughan, D.J. (1970) *Am. Mineral.*, **55**, 1576.
24. Goodenough, J.B. (1972) *J. Solid State Chem.*, **5**, 144.
25. Vaughan, D.J. (1971) D. Phil. thesis, University of Oxford.
26. Wood, B.J. and Strens, R.G.J. (1979) *Mineral. Mag.*, **43**, 509.
27. Slater, J.C. and Johnson, K.H. (1972) *Phys. Rev.*, **5B**, 844.
28. Johnson, K.H. (1973) *Adv. Quant. Chem.*, **7**, 143.
29. Vaughan, D.J. and Tossell, J.A. (1983) *Phys. Chem. Min.*, **9**, 253.
30. Vaughan, D.J. (1973) *Bull. Mineral.*, **101**, 484.
31. Tossell, J.A., Vaughan, D.J. and Burdett, J. (1981) *Phys. Chem. Mineral.*, **7**, 177.
32. Tossell, J.A. (1983) *Phys. Chem. Mineral.*, **9**, 115.
33. Burdett, J. (1979) *Nature, Lond.*, **279**, 12.
34. Burdett, J.K. (1980) *J. Am. chem. Soc.*, **102**, 450.
35. Hulliger, F. and Mooser, E. (1965) *Prog. Solid State Chem.*, **2**, 330.
36. Hullinger, F. (1968) *Struct. Bonding*, **4**, 83.
37. Nickel, E.H. (1968) *Can. Mineral.*, **9**, 311.
38. Goodenough, J.B. (1972) *J. Solid State Chem.*, **5**, 144.
39. Dalven, R. (1969) *Infrared Phys.*, **9**, 141.
40. Zemel, J.N., Jensen, J.D. and Schoolar, R.B. (1965) *Phys. Rev.*, **140**, A330.
41. McFeely, F.R., Kowalczyk, S., Ley, L. *et al.* (1973) *Phys. Rev.*, **7B**, 5228.
42. Tung, Y.W. and Cohen, M.L. (1969) *Phys. Rev.*, **180**, 823.
43. Herman, F., Kortum, R.L., Outenburger, I.B. and Van Dyke, J.P. (1968) *J. Phys., Paris*, **29**, C4.
44. Hemstreet, L.A. Jr. (1975) *Phys. Rev.*, **11B**, 2260.
45. Schoolar, R.B. and Dixon, J.R. (1965) *Phys. Rev.*, **137**, A667.
46. Cardona, M. and Greenaway, D.L. (1964) *Phys. Rev.*, **133**, A1685.
47. Wilson, J.A. (1972) *Adv. Phys.*, **21**, 143.
48. Tossell, J.A. (1977) *J. Chem. Phys.*, **66**, 5712.
49. Vaughan, D.J. and Tossell, J.A. (1973) *Science*, **179**, 375.
50. Tossell, J.A., Urch, D.S., Vaughan, D.J. and Weich, G. (1982) *J. Chem. Phys.*, **77**, 77.
51. Tossell, J.A. (1978) *Phys. Chem. Mineral*, **2**, 225.
52. Tossell, J.A. and Vaughan, D.J. (1981) *Inorg. Chem.*, **20**, 3333.
53. Vaughan, D.J. and Tossell, J.A. (1981) *Am. Mineral.*, **66**, 1250.

54. Hall, S.R. and Stewart, J.M. (1973) *Acta Cryst.*, **29B**, 579.
55. Hamajima, T., Kambara, T., Gondaira, K.I. and Oguchi, T. (1981) *Phys. Rev.*, **B24**, 3349.
56. Evans, H.T. Jr. (1971) *Nature, Lond.*, **232**, 69.
57. Vaughan, D.J. and Tossell, J.A. (1980) *Can. Mineral.*, **18**, 157.
58. Goodenough, J.B. (1969) *J. Phys. Chem. Solids*, **30**, 261.
59. Tossell, J.A., Vaughan, D.J. and Johnson, K.H. (1973) *Nature, Phys. Sci.*, **244**, 42.
60. Tossell, J.A., Vaughan, D.J. and Johnson, K.H. (1974) *Am. Mineral.*, **59**, 319.
61. Vaughan, D.J. and Tossell, J.A. (1978) *Can. Mineral.*, **16**, 159.
62. Goodenough, J.B. (1971) in *Progress in Solid State Chemistry.*, Vol. 5 (ed. H. Reiss), Pergamon Press, Oxford.
63. von Gehlen K. and Piller, H. (1965) *Neues Jahrb. Mineral. Monatsh.*, 97.
64. Grimes, N.W. (1975) *Physics in Technology*, Jan., 22.
65. Tossell, J.A. (1976) *Am. Mineral.*, **61**, 130.
66. Tossell, J.A. (1978) *Phys. Rev.*, **B17**, 484.
67. Vaughan, D.J. (1984) in *Hydrometallurgical Process Fundamentals* (ed. R. Bautista), Plenum Press, New York.
68. Li, E.K., Johnson, K.H., Eastman, D.E. and Freeouf, J.L. (1974) *Phys. Rev. Lett.*, **32**, 470.

8

Mineral Surfaces and the Chemical Bond

Frank J. Berry

8.1 INTRODUCTION

Current interests in the nature of solid surfaces are, to some extent, a reflection of recent developments in techniques with a capacity to examine those superficial regions of solids which have hitherto resisted investigation. Such developments have enabled more accurate assessments to be made of the bonding characteristics, structural properties, and elemental compositions of solid surfaces. The subject has therefore adopted a new perspective and, for the first time, fundamental properties of solid surfaces and molecular processes at the solid–gas interface are amenable to rigorous examination.

Such matters are clearly relevant when considering the structure and chemistry of naturally occurring solids. However, despite the determination of the bulk structural properties of many minerals by techniques such as X-ray diffraction there is a sparsity of data by which the surface properties of these naturally occurring materials may be described. Hence studies of the chemical and structural properties of mineral surfaces are not only timely, but are also relevant to the development of an improved understanding of natural processes such as weathering and the role of surface phenomena in more applied fields such as mineral extraction. This chapter therefore attempts to provide the solid state chemist or mineralogist with a guide to some of the more common techniques which may be used for surface investigation and their potential value for the study of natural materials.

8.2 SPECTROSCOPIC TECHNIQUES

Of the various spectroscopic techniques described in this book those referred to as 'surface techniques' have been the most recent to have developed.

Table 8.1 Glossary of surface spectroscopic techniques

AES	Auger electron spectroscopy
CEMS	Conversion electron Mössbauer spectroscopy
EELS	Electron energy loss spectroscopy
EXAFS	Extended X-ray absorption fine structures
EXELFS	Extended electron loss fine structure
ISS	Ion scattering spectroscopy
LEED	Low energy electron diffraction
PES	Photoelectron spectroscopy
SIMS	Secondary ion mass spectroscopy
UPS	Ultra-violet photoelectron spectroscopy
XPS	X-ray photoelectron spectroscopy

However, it must be acknowledged that the word surface means different things to different scientists. For example, the surface physicist concerned with atomic construction on single crystal surfaces seeks to examine the outermost atomic layer whilst the materials scientist interested in alloy corrosion processes is frequently interested in a complex layer several microns thick. The actual thickness of the surface layer which is examined by a particular technique is usually a characteristic feature of that specific method of investigation and appropriate comments on sampling depth are made at pertinent places in this chapter. Many of the techniques which have been developed for the examination of the superficial regions of solids are commonly known by an acronym and although a glossary of some of the more common terms is given in Table 8.1 the reader is referred to the literature[1] for a comprehensive survey of the principles and applications of a wide range of chemical and physical techniques which can be used for methods of surface characterization.

8.2.1 Electron spectroscopy

Although there has been a proliferation of 'surface spectroscopies' in recent years only a few have developed as techniques capable of giving information which can be used in the true sense of surface analysis and characterization. Of these, the electron spectroscopic techniques which involve the ejection from the solid of bound electrons by bombardment with X-rays, ultra-violet protons, or electrons, have been developed with most vigour. The principles of these techniques, and the rigorous interpretation of the results in terms of surface phenomena, involve a consideration of complex matters such as the electronic properties of solid surfaces and their interaction with electromagnetic fields. Such matters are outside the scope of this chapter but have been excellently described elsewhere.[2]

The natural division of the electron spectroscopies into two classes provides a convenient means by which their application to the study of solid surfaces may be considered. It will be recalled that the *photoelectron* spectroscopies are techniques which involve the direct ejection of electrons from bound electronic energy levels, whilst *Auger* electron spectroscopy (AES) entails electron ejection from an inner shell with the vacancy being filled by an outer electron in a process involving the transfer of energy to another electron (the Auger electron) which leaves the atom. The general theory and practice associated with these techniques has been described in Chapter 2.

It is important to record that of the photoelectron spectroscopic techniques X-ray photoelectron spectroscopy (XPS) which is sometimes referred to as ESCA (Electron Spectroscopy for Chemical Analysis) and which involves the ejection of core electrons by X-rays has been more commonly applied to the study of mineral surfaces than ultra-violet photoelectron spectroscopy (UPS). Indeed XPS, the principles of which have been described in detail in Chapter 2, is one of the most powerful surface sensitive techniques to have developed in recent years with the great advantages of being sensitive to nearly every element and thereby providing a means of qualitative and quantitative surface elemental analysis. However it is possible that UPS, which has not been covered in any detail in Chapter 2, and which involves the ejection of less tightly bound valence electrons from the solid by bombardment with ultra-violet photons may develop as an important method of mineral surface examination in the future, and therefore deserves a few introductory comments. The UPS photons are generally provided by helium gas discharge lamps and the experimental methods resemble those of XPS, indeed XPS and UPS sources are often available in a single instrument. The UPS spectra give electron energy distributions which may be related to the electron density of states of the solids and usually require accurate band structure calculations for meaningful interpretation. Elemental analysis is not a direct manifestation of the UPS spectrum; its chief power lies in its application to studies of adsorption at solid surfaces where features of the spectra may be interpreted in terms of modified molecular orbitals of the gaseous adsorbate when bonded to the solid surface. However, this power is not without limitations since, although the adsorption of a gas such as oxygen will usually give rise to features in a region characteristic of oxygen, the small energy range within which the valence levels of all species fall coupled with the common broadness of such features frequently results in inconclusive spectra. The main application of the technique to date has been for the acquisition of data relating to bulk band structure and for developing theories of the photoemission process within solids. Such studies, even from pure metals, have often given complex data which has resulted in the slow development of the technique for the study of surfaces of complex solids such as those encountered in mineral chemistry. It is worth noting that the approaches by which both UPS and XPS data may be interpreted in terms of surface phenomena are well illustrated in a review[3] of the application of electron spectroscopy to the study of reactivity of solid surfaces.

The sensitivity of XPS, UPS, and AES to surface effects derives from the strong interactions between electrons and matter such, for example, that electrons with energies less than 1000 eV usually travel a distance of less than 1 nm in a solid before suffering an inelastic collision. Hence the spectra relate to atoms within the first few layers of the solid surface and are sensitive to the presence of adsorbed layers. The major factor which determines the probing depth of electron spectroscopic techniques is the energy of the escaping electron and the areas of the peaks in a photoelectron spectrum may, with appropriate calibration and cautious awareness of the complex factors which affect signal intensity, be used to assess the relative concentrations of different atoms in regions between 20 and 100 Å of the surface. The different forms of electron spectroscopy which are currently available reflect the different bombarding species which may be used to induce electron ejection. The probing level of Auger spectroscopy commonly permits sampling of less than a few hundred Angstroms, indeed, low energy Auger electrons only escape from the first atomic layers of a surface and the technique predates XPS for the study of surface phenomena. Although Auger electron spectroscopy is a powerful means of elemental detection and surface analysis it must be remembered that the bombarding electron beam can damage the surface layer and cause structural rearrangements or, when adsorbed species are under investigation, may cause decomposition or desorption effects. Furthermore, although chemical shifts in Auger electron spectra may be considered in terms of changes in the atomic environments, such effects are frequently more difficult to interpret than changes in XPS spectra. Indeed changes in AES peak shape are as common as clearly observable shifts and the shift mechanism is more complex than that operating in XPS since it is a function of more than one energy level. The problem is particularly acute if the upper levels are valence shell-type and in such cases no simple correlation between Auger shifts and, for example, atomic charge can be expected. However, in other ways, Auger electron spectroscopy is a useful technique as compared with photoelectron spectroscopy. The facility to rapidly record satisfactory signal strengths from small areas is a powerful feature of the technique and is of particular relevance to the study of minerals where only small samples are available or where specific regions of the surface require special attention. It is clear from a survey of the literature that most of the reported surface studies of minerals have involved the use of either X-ray photoelectron spectroscopy or Auger electron spectroscopy. Although this observation is a reflection of the pre-eminence of these techniques for the study of solid surfaces in general, it is important to comment on some other methods which, in principle at least, are well suited for the examination of mineral surfaces.

8.2.2 Conversion electron Mössbauer spectroscopy

The use in mineral chemistry of conventional transmission mode Mössbauer spectroscopy, which involves the detection of γ-rays transmitted through thin samples, has been reviewed in Chapter 5. The technique is clearly a powerful means by which the electronic environment of nuclei may be investigated. In particular, the Mössbauer chemical isomer shift δ, which is dependent on the electron density at the nucleus, and the quadrupole splitting Δ, which increases according to the degree of distortion of the electronic and lattice environment about the Mössbauer nucleus, are important parameters by which the nature of chemical bonding and structure may be investigated.

Transmission mode Mössbauer spectroscopy does have some application in the study of surfaces of minerals, for example, materials with high surface areas such as the sheet silicates, or those with low surface/bulk ratios which may be studied by incorporating ^{57}Co into the surface and using the samples as a source in conjunction with a single line absorber. Such methods, which have been used to study corrosion and passivation, are equally well suited for the study of the weathering of minerals. However, for the majority of materials the transmission mode Mössbauer measurements sample the bulk of the solid and therefore the interpretation of the data is mainly applicable to internal structural and chemical properties.

The application of Mössbauer spectroscopy to the study of solid surfaces is best achieved by using the technique in alternative modes of operation which involve the detection of radiation emanating from the front surface of the solid material. In such experiments the γ and X-rays which result from internal conversion processes and which are resonantly emitted from the surface may be counted with apparatus similar to that normally used in transmission mode Mössbauer spectroscopy (Chapter 5). Since the escape depth of the γ- and X-radiation is quite large it is possible to study surfaces of several μm thickness and grossly modified surfaces are therefore amenable to examination by such methods.

However, a significant advance in the use of Mössbauer spectroscopy for the study of solid surfaces has involved the development of conversion electron Mössbauer spectroscopy (CEMS), the detailed principles and applications of which have been described in other articles.[4–7] Iron-57 CEMS has been the most commonly applied form of this technique and, in brief, involves the absorption of γ-rays emitted from the Mössbauer source by ^{57}Fe nuclei in the sample so as to give excited nuclei the majority (*c.* 91%) of which decay by emitting internally converted electrons whilst the remainder decay by re-emitting 14.4 keV γ-rays. The ability of CEMS to investigate the superficial regions of solids is vested in the shallow escape depth of the internally converted electrons, indeed 95% of the 7.3 keV electrons ejected from the K-shell originate from within 300 nm of the surface and 60% emanate from within 54 nm. Electrons generated at different depths within the sample emerge with different energies, hence it is possible to energy analyse these electrons so as to study discrete layers of the sample surface. The sensitivity of the method is largely dependent on the concentration of iron in the surface regions and when the surface is enriched in ^{57}Fe the sensitivity is dramatically increased and spectra can be accumulated within minutes.

The greatest restriction on the general application of CEMS results from the relatively few elements which can be sensed by the method. Most CEMS experiments have involved studies of iron-bearing solids and, although investigations of tin compounds are now being successfully conducted, the total number of isotopes which may be developed for use in CEMS in the future is unlikely to be large.

The experimental features of CEMS are similar to those used in conventional transmission mode Mössbauer experiments. Although the detection of internally converted electrons instead of γ-rays is, in principle, a reason why CEMS should be a more efficient method of Mössbauer spectroscopy, the low energy of the electrons inhibits their ability to penetrate a detector window. Hence, the sample needs to be inside the detector which is called a resonance counter. Instruments similar to that depicted in Fig. 8.1 have been successfully developed by several workers including the author of this chapter. The apparatus is essentially an 8 mm thick flat cylindrical flow counter with Mylar windows. The sample is fixed onto one of these windows and the pair are set

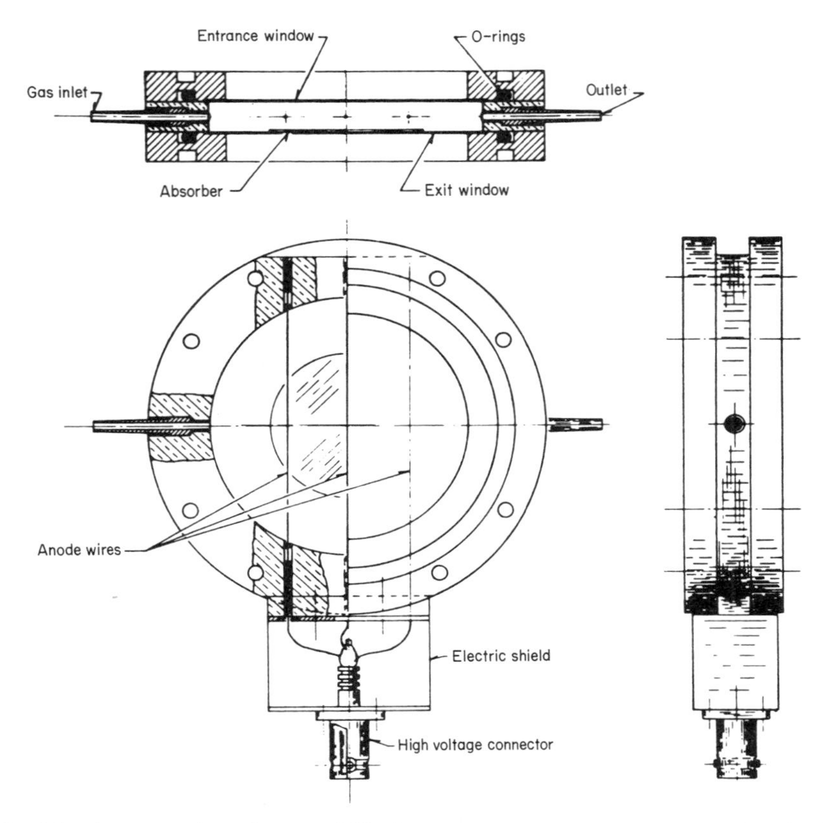

Fig. 8.1 A conversion electron Mössbauer resonance counter (reproduced with permission from Fenger, J. Danish Atomic Energy Commission, Riso, Denmark).

into aluminium rings which are screwed onto a Perspex block containing three parallel 0.1 mm thick stainless steel anode wires. The inlet and outlet tubes permit the flow of a counting gas composed of 90% helium and 10% methane. The back scattered internally converted electrons are detected as a function of the Doppler modulated incident γ-ray energy and, in contrast to transmission mode Mössbauer spectra, maxima in electron counts occur at velocities when resonant absorption occurs (Fig. 8.2).

During recent years considerable effort has been made to improve the basic resonance counters and develop instrumentation for operation under more informative conditions. Reports of the construction of resonance counters for operation at 77 K which would be expected to give superior spectra, and thereby enhanced applicability to problems which require high spectral resolution, have been reviewed.[6,7]

It is also relevant to comment on the use of resonance counters for depth resolved CEMS studies.[4–7] One method involves the variation of the incident γ-ray angle through 90° from normal to give both increased count rates and the detection of effects from different surface regions in materials enriched in iron-57. A recent development of this technique has used the variation of the γ-ray angle and the sample temperature. Attempts to limit the probing depth by either altering the distance between the counter wires and the sample or by exploiting the poor resolution of the helium–methane counter illustrate other methods by which the sensitivity of the basic instrument to surface effects in bulk solids may be refined. However, many of the most successful depth resolved CEMS studies have involved the use of β-ray spectrometers.[4,5] The resolution of effects originating from different surface layers using these devices results from the attenuation of the conversion electrons by inelastic collisions as they escape from the deeper regions. By using such methods the

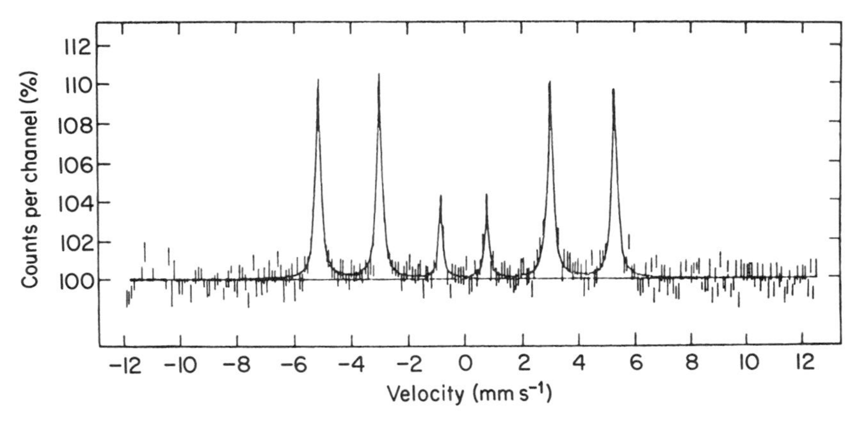

Fig. 8.2 Conversion electron Mössbauer spectrum of natural iron (reproduced with permission from Fenger, J. (1973) *Nucl. Instrum. Methods*, **106**, 203).

resolution of thin layers (*c.* 40 nm) of iron-bearing solids into differing regions, perhaps to within 5 nm of the surface, has been suggested.

8.2.3 Other techniques

It is pertinent in this chapter to comment on a few of the other techniques mentioned in Table 8.1 which, although yet to be extensively applied in mineral chemistry, may develop in the future as important methods for the investigation of mineral surfaces.

The first of these techniques, secondary ion mass spectrometry (SIMS), involves the bombardment of the solid sample with high energy argon ions so that surface atoms or layers of several atoms are eroded by sputtering. The technique, involving high beam current densities, is currently widely used in the microelectronics industry for the sensitive detection and analysis of dopants in semiconductors and, under these circumstances, is not a surface sensitive technique. However, the use of lower current densities and a focused beam rastered over an area to produce low erosion rates as compared with the analysis time should, in principle, enable the examination of monolayers. Recent developments have overcome the charging problems which have previously inhibited the study of insulator surfaces and new techniques for uniform sputtering should enable reliable depth-profiling of thick surface layers in the future.

Extended X-ray absorption fine structure (EXAFS) is another technique which, although only recently developed, seems to be potentially useful for mineral surface examination. The technique concerns situations when the photon energy becomes just sufficient to cause the emission of photoelectrons from an electron shell so that a sharp increase in the cross-section for photon absorption, commonly known as the absorption edge, occurs. Above this edge an oscillatory variation in the absorption cross-section is often observed which reflects the local structure surrounding the atom under study. The greater ease of interpretation of data in the range > 100 eV above the absorption edge has given rise to the name extended X-ray absorption fine structure. The structural information is contained in the oscillations of the absorption cross-section and from analysis of the data, especially that near absorption edges, accurate estimates of interatomic distances and coordination numbers of atoms can be obtained. The technique is applicable to the study of amorphous as well as crystalline materials and the surface sensitivity enhanced by measuring Auger electrons generated by core hole filling.

Another technique which is worthy of mention is ion scattering spectroscopy (ISS). In this method a beam of ions is rebounded from the surface atoms by elastic collisions so that the loss of energy of the primary ions may be related to the masses of the atoms in the solid surface. The technique is similar to Rutherford scattering but involves the use of low energy ions (< 5 keV) which greatly reduces the depth of interaction and thereby renders surface sensitivity to the technique. Although cross-section measurements remain ill-defined for low energy ions and much remains to be done in terms of the interpretation of spectra, especially from a quantitative aspect, the technique does seem to have considerable potential for surface investigation since it avoids the weighting effects of top-layer plus subsurface-layer composition which affect XPS, AES and, to a lesser extent, some SIMS data.

It is also pertinent in this chapter to acknowledge electron diffraction as one of the most important methods for determining the structure and chemical compositions of solid surfaces. Although the large number of diffraction techniques which are so important in solid state structural studies have not been included in this book since adequate coverage would require a text in its own right, it is proper that the power of reflection high energy electron diffraction (RHEED) and low-energy electron diffraction (LEED) for surface examination be noted. Indeed, electron beam techniques which enable electron energy loss spectroscopy (EELS) are another potentially powerful means of surface analysis.

Given that XPS, AES and CEMS constitute the most commonly applied spectroscopic methods for the study of mineral surfaces the remainder of this chapter is devoted to results obtained from these techniques.

8.3 APPLICATIONS IN MINERAL CHEMISTRY

8.3.1 Photoelectron spectroscopy

The applications of XPS in mineralogy and geochemistry were excellently reviewed[8] in 1979 and the matter is given further consideration here to illustrate its particular power and versatility for the examination of mineral surfaces. This application is well illustrated by a recent study[9] of gold deposition at low temperatures on sulphide minerals. The work is important because gold deposits are commonly associated with quartz veins containing variable amounts of pyrite and other sulphides. Gold is thought to be transported as $AuCl_2^-$ or $AuCl_4^-$ in strongly acidic and saline solutions and as AuS^- or $Au(S_2O_3)_2^{3-}$ in weakly acidic and alkaline media by hydrothermal solutions rising from great depths. In some cases the gold occurs as finely divided particles intimately intergrown with pyrite suggesting that gold, quartz and pyrite are precipitated in response to changes in temperature. During the study[9] XPS was used to investigate the reaction between $KAuCl_4$ in solution and various sulphide minerals (Fig. 8.3). The results enabled the formulation of a new mechanism for the deposition of gold at low solution concentration and temperature which involves the initial adsorption of gold(III) as the hydrated or hydrolysed species and its subsequent rapid reduction to metallic gold on the sulphide mineral surface. The results, which showed gold deposition to continue even after the surface was entirely covered with the first layer of gold(O), suggest that the electrons required for the

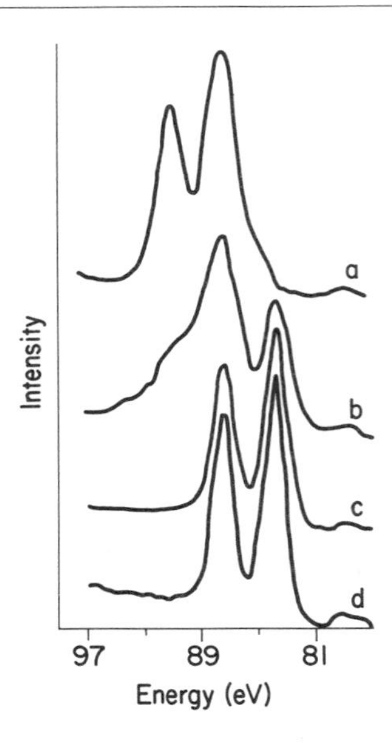

Fig. 8.3 Gold 4f XPS spectra for (a) $KAuCl_4$ solution (10 ppm) evaporated onto Teflon; (b) sphalerite (ZnS) plate reacted with this $KAuCl_4$ solution; (c) pyrite (FeS_2) plate reacted with this $KAuCl_4$ solution; (d) metallic gold (reproduced with permission from Bancroft, G.M. and Jean, G. (1982) *Nature, Lond.*, **298**, 730).

subsequent reduction processes emanate from within the interior of the crystal. Other aspects of the chemistry of sulphur-containing minerals are given further consideration in Chapter 7 which deals with the opaque minerals.

Although Mössbauer spectroscopy is a valuable technique for distinguishing between Fe^{2+} and Fe^{3+} in minerals (Chapter 5) there are several illustrations[8] of the successful application of XPS to such problems although many of these have been mainly concerned with iron in bulk lattice sites. However, it is pertinent to record that XPS studies[10] of well characterized silicate minerals containing Fe^{2+} and Fe^{3+} have been unable to distinguish between the oxidation states of iron in these materials and the possibility that this difficulty might involve a superficial iron species is a matter which is well suited for further investigation by conversion electron Mössbauer spectroscopy (CEMS).

Other applications of XPS to the chemistry of mineral surfaces have derived from its ability to examine phenomena at the solid–gas[11] and solid–liquid[12,13] interfaces. Such matters are of contemporary relevance since there has been much recent interest in the factors which may influence the sorption of heavy metals and radioactive wastes on mineral surfaces. The potential of XPS for these types of investigations is well illustrated by studies[8,14–16] of the adsorption of Ba^{2+} and Pb^{2+} on powdered and crystalline calcite and of

Hg^{2+} on iron sulphides. The investigation of the sorption of Ba^{2+} from solution involved the use of atomic-absorption spectroscopy to follow solution concentrations and of XPS to detect the Ba^{2+} sorbed on the solid calcite surfaces. The study demonstrated the power of XPS for quantitative determinations of trace levels, *c.* 10^{-8}g, and, when combined with the atomic absorption data, clearly showed the dependence of barium ion sorption on the initial Ba^{2+} concentration and the surface area of the calcite. The results of both methods were consistent with monolayer coverage by the barium ions of the calcite surfaces. These studies[14–16] demonstrated that mineral surfaces have the capacity to act as heavy metal sinks and that low concentrations of metals such as mercury (even in the ppb range) may be rapidly adsorbed by iron sulphides. Investigations of this type have clear environmental implications, for example, from the results reviewed above it would seem feasible to consider the use of sulphides for the removal of mercury from polluted waters or industrial pulp and paper effluent.

It is also pertinent to record a related and recent surface study by XPS of a leached sphene glass.[17] The study is important because of the current interest in titanates and titanate- or titanosilicate-based ceramics as possible nuclear waste hosts. This particular application results from their excellent resistance to leaching which has been attributed to a largely insoluble titanium dioxide rich surface layer which is formed by the selective leaching of univalent and divalent cations. The XPS study, complemented by SIMS (secondary ion mass spectrometry), examined the leaching of the sphene ($CaTiSiO_5$) in the form of natural samples, sintered ceramics, and glasses of stoichiometric sphene composition. The results enabled the characterization of the leached surface layers of the titanosilicate and showed how the leaching of sphene glasses in deionized water parallels leaching mechanisms which have been proposed for the titanates.

The applications of XPS in other matters of geochemical significance are illustrated by an investigation of the adsorption of metal ions by hydrous manganese oxide.[18] In these studies aluminium plates were coated with a rather smooth thin hard film of manganese oxide by repeated alternate dipping of the plates into dilute manganese sulphate and potassium permanganate solutions. The uptake of Ni^{2+}, La^{3+} and Ba^{2+} ions on the manganese oxide films was followed directly by XPS and plots of metal uptake on the surface against time for various initial concentrations were found to be characteristic of sorption processes. The results enabled the sequence of cation selectivity to be determined as $Ba^{2+} \gg Ni^{2+} > La^{3+}$. This demonstration of the control of trace element build up in manganese oxide by surface chemical and adsorption processes illustrates the immense potential economic importance of marine manganese nodules as sinks for environmentally important elements. Indeed, more than sixty elements have now been detected in nodules at concentrations many orders of magnitude greater than are present in seawater.

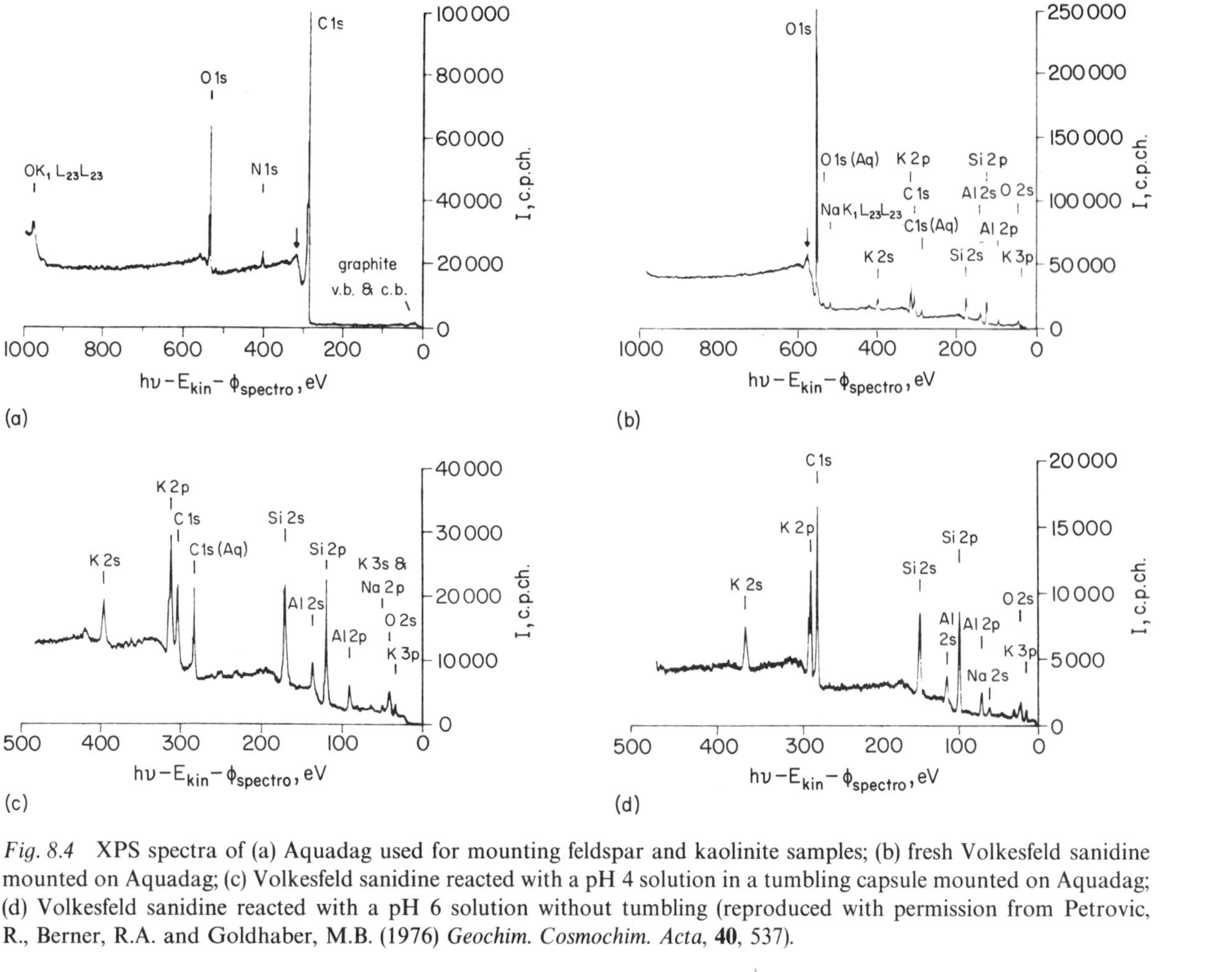

Fig. 8.4 XPS spectra of (a) Aquadag used for mounting feldspar and kaolinite samples; (b) fresh Volkesfeld sanidine mounted on Aquadag; (c) Volkesfeld sanidine reacted with a pH 4 solution in a tumbling capsule mounted on Aquadag; (d) Volkesfeld sanidine reacted with a pH 6 solution without tumbling (reproduced with permission from Petrovic, R., Berner, R.A. and Goldhaber, M.B. (1976) *Geochim. Cosmochim. Acta*, **40**, 537).

Further illustrations of the use of XPS to study the kinetics of surface chemical reactions resulting from diffusion, dissolution, exchange and adsorption processes may be cited and the technique, which can directly sense minute changes in surface ion concentration, has certain advantages over other methods such as atomic absorption which follow the processes by monitoring changes in solution concentration. Such advantages are evident in the investigation[19] of Ba^{2+}–Sr^{2+} exchange at the mineral–solution interface of strontium sulphate in barium dichloride solution and which involved the monitoring of the Ba $4d_{5/2}$/Sr $3d_{5/2}$ peak area ratios in the XPS spectra as a function of time. In other studies[20–22] changes in the XPS Al 2p/Si 2p and Mg 2p/Si 2p peak area ratios in aluminium and magnesium silicates have been used to identify congruent or incongruent dissolution processes and to estimate the Mg^{2+} diffusion coefficient and diffusion rates at higher temperatures. It is also pertinent to cite XPS investigations[23] of oxide and silicate minerals during which sanidine grains of 100–600 μm diameter were treated at 82° C in aqueous electrolyte solutions (pH ranging between 4 and 8) for 193 or 377 hours. Dissolution equivalent to the removal of silica from the outer 300–900 Å of the grains was achieved and the shallow subsurfaces of the feldspar particles analysed for potassium, aluminium and silicon by XPS (Fig. 8.4). The results showed that any alkali-depleted subsurface zone (leached layer) in the feldspar was less than *c.* 17 Å thick. It was concluded that, in the absence of a compact precipitate layer, dissolution of akali feldspar minerals in the temperature range corresponding to deep diagenesis is controlled by processes at the feldspar–solution interfaces without the formation of a leached layer which exceeds the thickness of one feldspar unit cell. Whether the same applies at the temperatures of shallow diagenesis and weathering was not evaluated with certainty, but considerations of leached layers on alkali silicate glasses suggested that similar processes were possible. It is also relevant in this chapter to note how studies[24] of the surface Si/Al ratios of zeolites and of cation distributions in Ag^{+} and Cu^{2+} exchanged Na A-zeolites have shown, despite low accuracy measurements of the Si 2p and Al 2p peak intensities, how XPS can provide information on these important types of materials which is not readily obtained from other techniques.

XPS has also been found to be a useful technique for elucidating the surface properties of solids which may be relevant in applied aspects of mineralogy such as mineral processing. For example, XPS studies[25] of lead monoxide have demonstrated that oxidized surfaces give spectra similar to those obtained from surfaces formed by the chemisorption of oxygen on lead, and of rhombic and tetragonal lead(II) oxide. It is especially interesting that the XPS O 1s spectrum was sufficiently sensitive to the different crystal modifications to allow the identification of a polymorphic transformation on a monolayer of the oxide film. However, it is also relevant to record a study[26] of a range of transition metal films which reported the failure of both XPS and Auger spectroscopy to give clear shifts on adsorption of a monolayer equivalent of oxygen or carbon monoxide. These latter results can be compared with the large shifts which may be observed when oxidation of many atomic layers of the bulk occurs and it therefore seems that studies of sub-monolayer chemisorption processes using chemical shift data from the substrate must be pursued with caution and that under these circumstances more reliable data are likely to be obtained by recording the shifts in the adsorbate peaks. It is possible that UPS may be a powerful technique in investigations of this type. Another relevant study[27] has reported that rutile-type titanium dioxide is always covered with chemisorbed and physisorbed moisture. The O 1s and Ti 2p spectra from the clean titanium dioxide surfaces showed the oxygen ions to terminate the surface and the Ti^{4+} ions to be screened beneath the superficial oxygen species. The interpretation of the oxygen and titanium binding energies in terms of the formation of ˙OH radicals from the singly bonded Ti–OH groups, and which may be related to the surface photocatalytic behaviour of titanium dioxide, illustrates how XPS data may be associated with phenomena of potential technological significance.

The final example of the versatility of XPS and its applicability to studies of surface chemistry is provided by a study[28] of changes in iron oxidation state which are induced by argon ion bombardment of different iron oxides. The surfaces were found to suffer reduction and to consist predominantly of metallic iron. Further work showed that argon ion sputtering of silicates and oxides produced a surface darkening in the iron-bearing materials which could be likened to the darkening of the lunar surface which results from solar wind bombardment.

XPS has therefore shown itself to be a viable and powerful technique for the investigation of the surface properties of a wide range of minerals and is amenable to exploitation in several areas of mineral chemistry. It would seem that the full potential of the technique in this field of study has yet to be realized and, as experimental procedures and instrumental techniques develop, it is quite reasonable to assume that new applications will emerge.

8.3.2 Auger electron spectroscopy

Although there is a sparsity of data relating to the use of Auger spectroscopy for the investigation of mineral surfaces, studies[29] of specially treated silicon and silicon dioxide illustrate the potential value of the technique to mineralogists. The Auger spectrum (Fig. 8.5(a)) from the silicon surface cleaned by neon ion bombardment and probably somewhat amorphous, changes, particularly in the region below ~ 90 eV (Fig. 8.5(b)), when annealed and subjected to adsorption of oxygen. Indeed the shape of the oxygen peak in Fig. 8.5(b) is quite characteristic of an ideal Auger spectrum. The Auger spectrum of SiO_2 (Fig. 8.6) shows the complete disappearance of the 92- and 107-eV peaks of pure silicon as might be expected from a sample of the pure oxide.

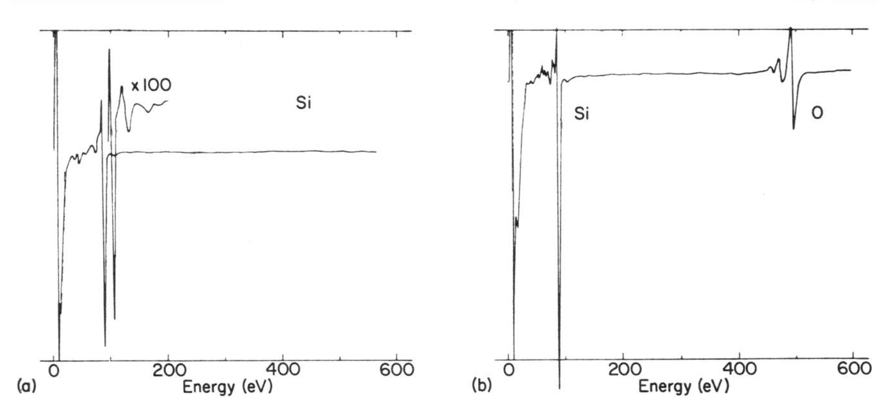

Fig. 8.5 Auger spectrum from (a) single crystal Si(III) after sputter-cleaning with 1 keV Ne ions, (b) similar surface after annealing and adsorption of ~ 0.5 monolayer of oxygen atoms (reproduced with permission from Chang, C.C. (1974) in *Characterisation of Solid Surfaces* (eds P.F. Kane and G.B. Larrabee), Plenum Press, New York, p. 509.

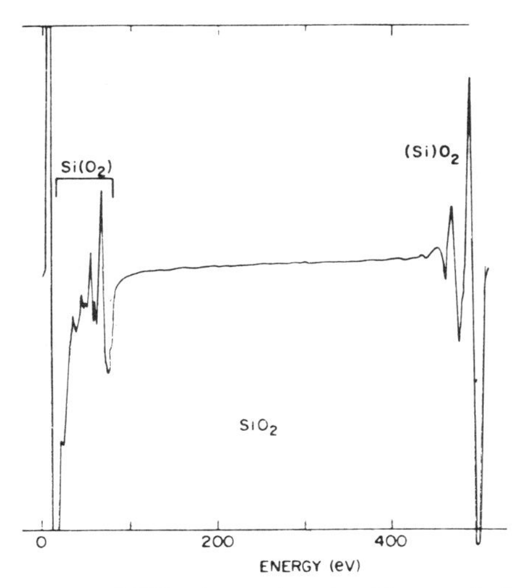

Fig. 8.6 Auger spectrum of SiO_2 (reproduced with permission from Chang, C.C. (1984) in *Characterisation of Solid Surfaces* (eds P.F. Kane and G.B. Larrabee), Plenum Press, New York, p. 509).

The use of Auger electron spectroscopy for the study of the composition of mineral fracture surfaces was demonstrated some years ago[30] by a series of measurements on kidney haematite. Spectra were obtained from massive pieces of the haematite as well as from round, pelletized specimens. The results revealed the presence of *c.* 1–10 atom per cent of calcium and potassium in some of the surfaces as compared with corresponding cationic concentrations

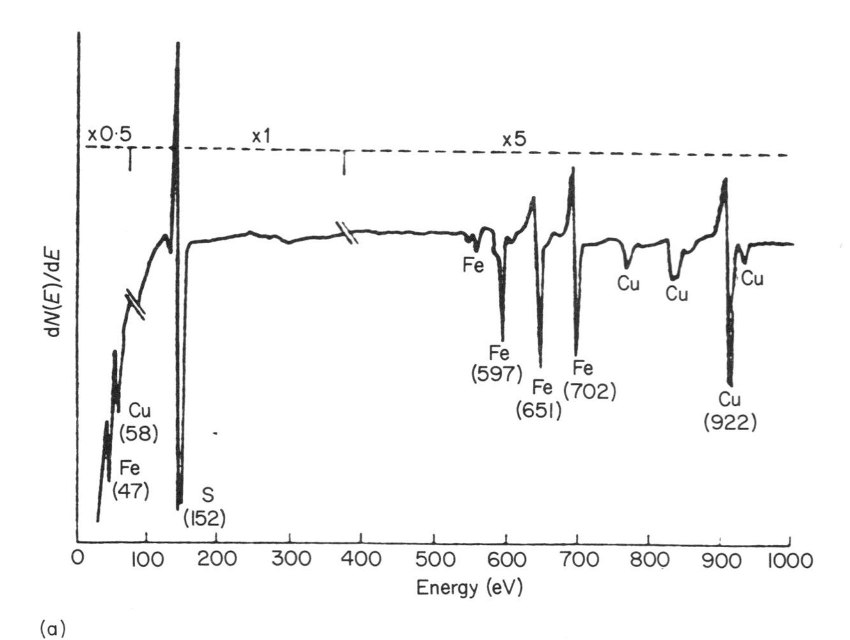

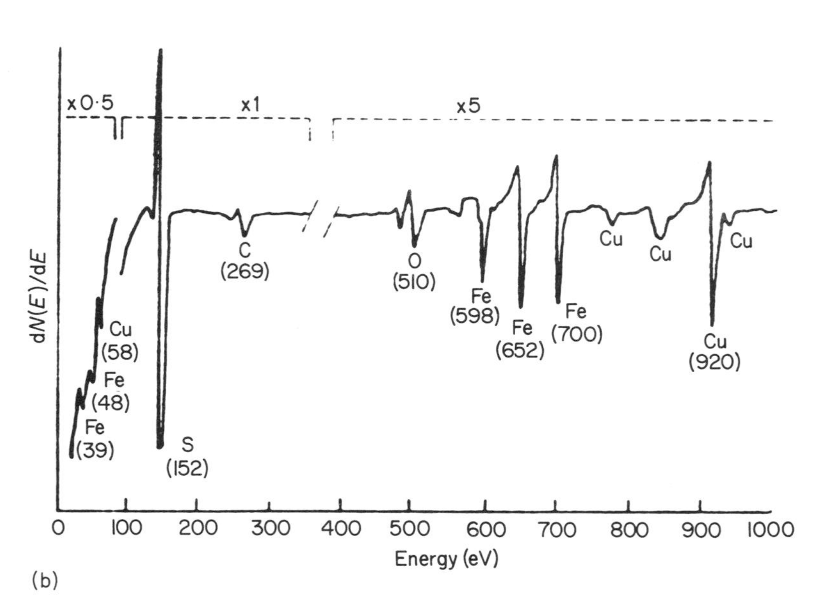

Fig. 8.7 Auger electron spectra from (a) unoxidized fracture surface of chalcopyrite and (b) the fracture surface after exposure to oxygen pressure of 1 torr for 30 s (reproduced with permission from Eadington, P. (1977) *Trans. Inst. Min. Met.*, **86c**, 186).

of *c.* 1–15 ppm. The compositions of the fracture surfaces were found to vary with roasting, washing and wet-grinding.

An Auger electron spectroscopy study[31] of oxidation layers on surfaces of chalcopyrite ($CuFeS_2$) is worthy of note since it illustrates the potential value of the technique in studies which are relevant to more applied aspects of mineralogy. The investigation is important since, in recent years, it has been established that oxygen plays an important role in the attachment of floatation reagents to semiconducting sulphide minerals such as chalcopyrite and galena during their separation from gangue. In these processes small quantities of adsorbed oxygen can control the semiconducting properties of the surface layer and thus influence the chemisorption of the collector whilst extensive oxidation can result in the formation of layers of oxidation products that may participate in exchange reactions with the collector. The element profiles obtained from the Auger electron spectra (Fig. 8.7) recorded across oxidation layers on the surfaces of wet-fractured chalcopyrite showed enhancement of the cooper : iron ratio. The quantity of oxygen adsorbed on the surface of dry fractured chalcopyrite after exposure to 1 torr of oxygen pressure for 30 s (Fig. 8.7(b)) was shown to depend on the conduction properties of the bulk and was found to be greater for '*n*-type' than for '*p*-type' conductors. It is also interesting that the rate of build up of the oxidation layer was found to be some ten times greater for wet-fractured chalcopyrite than for the dry specimen.

Hence Auger spectroscopy has much potential for the study of a variety of aspects of surface mineral chemistry and the technique undoubtedly warrants serious consideration in future attempts to elucidate surface composition and structure.

8.3.3 Conversion electron Mössbauer spectroscopy

The sensitivity of CEMS to the formation of surface oxides on iron-bearing solids was well illustrated by an investigation[32] of the corrosion of a steel plate. The spectrum in Fig. 8.8(a) shows a doublet characteristic of an approximately 2×10^{-3} cm thick layer of β-FeOOH on a small six-line pattern typical of metallic iron whilst Fig. 8.8(b) shows that removal of some superficial rust produces an enhanced contribution from the metallic iron substrate and Fig. 8.8(c) indicates that complete removal of the β-FeOOH corrosion product results in the clean six-line spectrum characteristic of metallic iron. Similar principles were demonstrated in a subsequent CEMS study[33] of oxides of several hundred Angstroms thickness which were formed on iron by heating in air. Non-stoichiometric Fe_3O_4 was observed after oxidation at 225° C whilst short period treatment at 350° C gave a duplex film consisting of Fe_3O_4 and α-Fe_2O_3. Oxidation at 450° C resulted in the formation of nearly stoichiometric Fe_3O_4. The work illustrates the sensitivity of CEMS to the detection of different oxides of iron.

The use of CEMS for the study of natural materials was one of the first

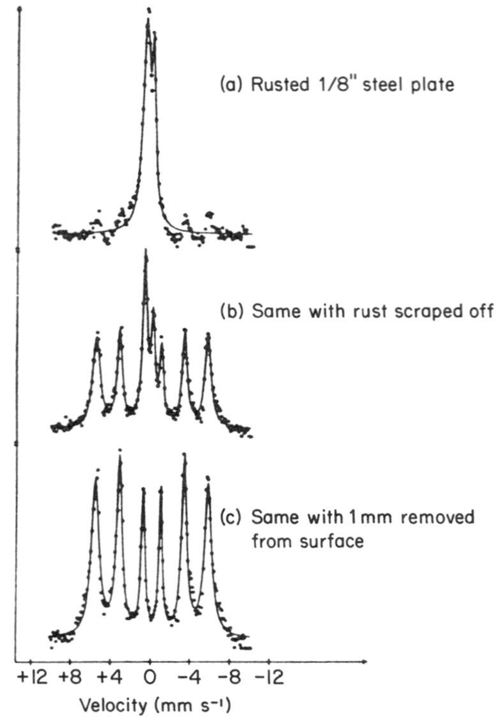

Fig. 8.8 Conversion electron Mössbauer spectra of corroded steel plate (reproduced with permission from Terrell, J.H. and Spijkerman, J.J. (1968) *Appl. Phys. Lett.*, **13**, 11).

applications of the technique. For example, a Mössbauer investigation[34] of lunar fines designed to evaluate the presence of iron(III) oxide in lunar soils also included a CEMS study which revealed evidence for Fe^{3+} and metallic iron within a few thousand ångströms of the particulate surfaces. A study[35] of a Canadian biotite crystal demonstrated the validity of CEMS as a method for determining quadrupole doublet peak area ratios and illustrated the suitability of the technique for the study of natural crystalline materials. A subsequent CEMS study[36] of the initial stages of oxidation of biotite is also important since it illustrates the power of the technique for sensing the early steps in solid state reactions. Although the transmission mode Mössbauer spectrum (Fig. 8.9) of biotite was unchanged by heating until rather severe conditions were used (550 K for 700 h) the CEMS spectra (Fig. 8.10) showed a thermally induced increase in the iron(III) content after only 100 h. The results illustrate the power of CEMS for detecting the initial stages of mild oxidation and their confinement to regions within 300 nm of the biotite surface. The sensitivity of CEMS to the detection of surface oxidation was confirmed by subsequent transmission mode measurements which only detected small increases in the bulk iron(III) content when the material was subjected to vigorous oxidation.

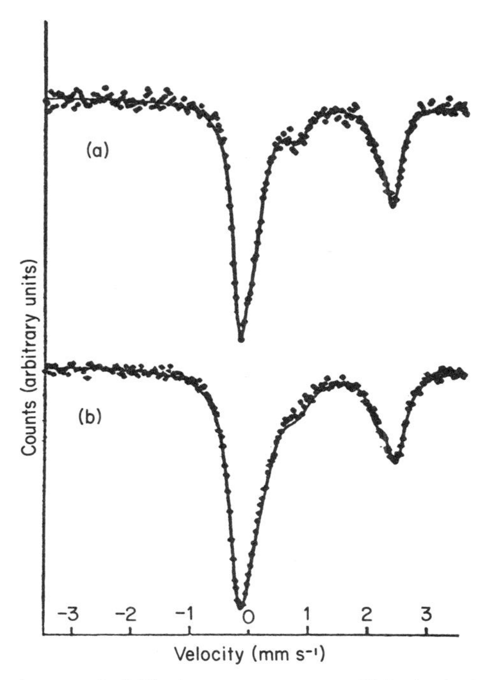

Fig. 8.9 Transmission mode Mössbauer spectrum of biotite before (a) and after (b) heat treatment at 550 K for 700 h (reproduced with permission from Tricker, M.J., Winterbottom, A.P. and Freeman, A.G. (1976) *J. chem. Soc. Dalton Trans.* **1976** 1289).

The results enabled a consideration of the mechanism of superficial oxidation in terms of the surface structure and the reaction temperatures.

Another study[4] of the different surface regions of siderite crystals is worthy of note because it illustrates the potential of CEMS in applied mineralogy and geochemistry. The siderite gave the expected wide doublet CEMS spectrum characteristic of high-spin Fe^{2+}. However, spectra recorded from certain regions of the solid showed doublets corresponding to Fe^{3+} which were ascribed to the presence of superparamagnetic goethite and associated with the effects of weathering of the mineral surface. Such effects have been difficult to monitor in the past and it seems that CEMS may be a valuable technique in this particular application. It is also relevant to cite a study[37] of vivianite which is a naturally occurring hydrated iron(II) phosphate of formula $Fe_3(PO_4)_2 \cdot 8H_2O$. The transmission mode Mössbauer spectra (Fig. 8.11(a)) recorded from a single crystal of the mineral showed the superposition of two wide quadrupole split absorptions corresponding to two Fe^{2+} ions in non-equivalent sites and a narrow doublet characteristic of Fe^{3+} and indicative of partial oxidation. In marked contrast, the CEMS spectrum (Fig. 8.11(b)) showed the absence of Fe^{3+} in the surface regions. Moreover, moderate heating of the sample gave a product for which the transmission mode spectrum

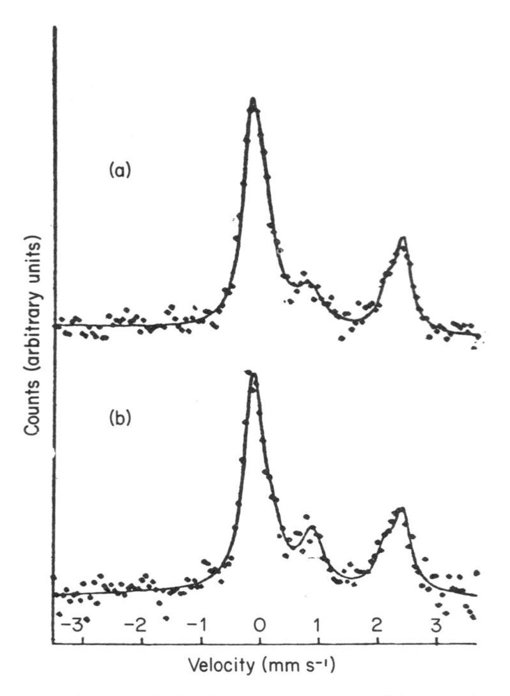

Fig. 8.10 Conversion electron Mössbauer spectrum of biotite before (a) and after (b) heat treatment at 550 K for 100 h. The increase of the Fe(III) content in spectrum (b) is to be noted (reproduced with permission from Tricker, M.J., Winterbottom, A.P. and Freeman, A.G. (1976) *J. chem. Soc. Dalton Trans.*, **1976**, 1289).

(Fig. 8.11(c)) showed facile conversion of Fe^{2+} to Fe^{3+} whilst CEMS (Fig. 8.11(d)) showed that Fe^{3+} was only formed in the surface after prolonged calcination. The results are indicative of an unexpected stability of the surface regions of vivianite to oxidation as compared with the bulk. In this respect it is relevant to note independent contemporary CEMS studies[38,39] which showed that synthetic iron(II) phosphate octahydrate formed on the surface of metallic iron by treatment with phosphoric acid or metal phosphate solutions is also resistant to oxidation. The observations are curious given that bulk iron(II) phosphate octahydrate is easily converted to a high-spin iron(III) species. The apparent resistance of the surface regions of iron(II) phosphate octahydrate and vivianite to oxidation might[37] arise from a preferential dehydration of these regions to some oxidatively more stable lower hydrate of formula $Fe_3(PO_4)_2 nH_2O (n < 8)$.

The CEMS studies of mineral surfaces which have been briefly outlined above clearly show that the technique has much promise for elucidating phenomena which are of interest to mineralogists as well as being of fundamental chemical significance.

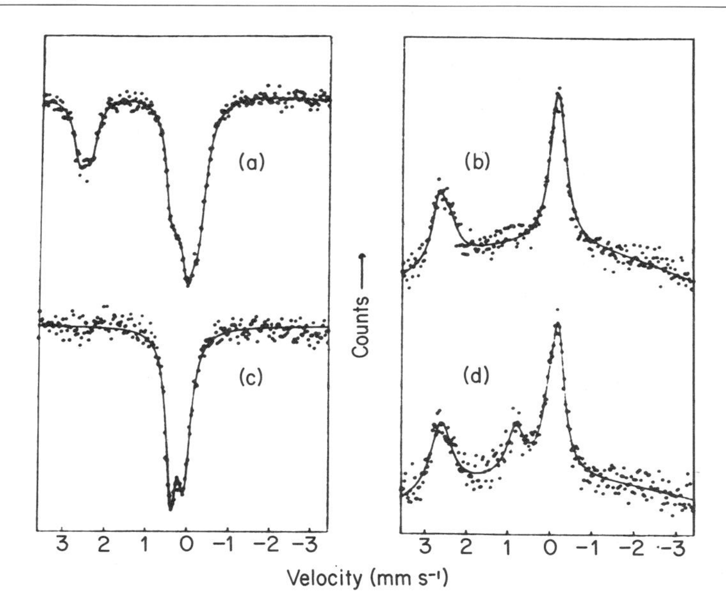

Fig. 8.11 (a) Transmission and (b) conversion electron Mössbauer spectra of a vivianite single crystal before heating. The resulting spectra after heating for 1 h at 120° C indicate conversion of the bulk (c) to a mainly iron(III) species whilst CEMS (d) suggests that the stable surface is mainly iron(II) (reproduced with permission from Tricker, M.J., Ash, L.A. and Jones, W. (1979) *J. inorg. nucl. Chem.*, **41**, 891).

8.4 CONCLUDING REMARKS

The application of recently developed techniques to the examination of mineral surfaces is a viable pursuit and several studies cited in this chapter illustrate that many minerals have surface properties which differ from those of the bulk. Phenomena such as these, together with assessments of changes induced in the surface layers by chemical treatment, will become increasingly amenable to examination as instrumental techniques develop. Clearly this area of science holds much scope for future development, with many experimental limitations now being successfully eroded so that the experimentalists' imagination and resourcefulness may be fully exploited.

REFERENCES

1. Kane, P.F. and Larrabee, G.R. (1974) *Characterisation of Solid Surfaces*, Plenum Press, New York.
2. Feuerbacher, B., Fitton, B. and Willis, R.F. (1978) *Photoemission and the Electronic Properties of Surfaces*, Wiley-Interscience, Chichester.

3. Joyner, R.W. (1977) *Surface Sci.*, **63**, 291.
4. Tricker, M.J. (1977) in *Surface and Defect Properties of Solids*, Vol. 6 (eds M.W. Roberts and J.M. Thomas), Specialist Periodical Reports, The Chemical Society, London, p. 106.
5. Jones, W., Thomas, J.M., Thorpe, R.K. and Tricker, M.J. (1978) *Appl. Surf. Sci.*, **1**, 388.
6. Berry, F.J. (1979) *Transition Met. Chem.*, **4**, 209.
7. Tricker, M.J. (1981) in *Mössbauer Spectroscopy and its Chemical Applications* (eds J.G. Stevens and G.K. Shenoy), Advances in Chemistry Series, American Chemical Society, Washington, p. 63.
8. Bancroft, G.M., Brown, J.R. and Fyfe, W.S. (1979) *Chem. Geol.*, **25**, 227.
9. Bancroft, G.M. and Jeans, G. (1982) *Nature, Lond.*, **298**, 730.
10. Adams, I., Thomas, J.M. and Bancroft, G.M. (1972) *Earth Planet. Sci. Lett.*, **16**, 429.
11. Norton, P.R. (1975) *J. Catal.*, **36**, 211.
12. Hercules, D.M., Cox, L.E., Onisick, S. *et al.* (1973) *Anal. Chem.*, **45**, 1973.
13. Czuha, M. and Riggs, W.M. (1975) *Anal. Chem.*, **47**, 1836.
14. Bancroft, G.M., Brown, J.R. and Fyfe, W.S. (1977) *Chem. Geol.*, **19**, 131.
15. Bancroft, G.M., Brown, J.R. and Fyfe, W.S. (1977) *Anal. Chem.*, **49**, 1044.
16. Brown, J.R. Ph.D thesis, University of Western Ontario (ref. 8).
17. Bancroft, G.M., Metson, J.B., Kanetkar, S.M. and Brown, J.D. (1982) *Nature, Lond.*, **299**, 708.
18. Brule, D.G., Brown. J.R., Bancroft, G.M. and Fyfe, W.S. (1980) *Chem. Geol.*, **28**, 331.
19. Thomassin, J.H., Baillif, P., Calapkuln, F. *et al.* (1975) *C.R. Acad. Sci., Paris, Ser. D*, **281**, 1067.
20. Thomassin, J.H., Touray, J.C. and Tricket, J. (1976) *C.R. Acad. Sci., Paris, Ser. D*, **282**, 1229.
21. Thomassin, J.H., Goni, J., Baillif, P. and Touray, J.C. (1976) *C.R. Acad. Sci., Paris, Ser. D*, **283**, 131.
22. Thomassin, J.H., Goni, J., Baillif, P. *et al.* (1977) *Phys. Chem. Miner.*, **1**, 385.
23. Petrovic, R., Berner, R.A. and Goldhaber, M.B. (1976) *Geochim. Cosmochim. Acta*, **40**, 537.
24. Finster, J. and Lorenz, P. (1977) *Chem. Phys. Lett.*, **50**, 223.
25. Kim, K.S. and Winograd, N. (1973) *Chem. Phys. Lett.*, **19**, 209.
26. Brundle, C.R. and Carley, A.F. (1975) *Chem. Phys. Lett.*, **33**, 41.
27. Sham, T.K. and Lazarus, M.S. (1979) *Chem Phys. Lett.*, **68**, 426.
28. Yin, L.I., Ghose, S. and Adler, I. (1972) *J. geophys. Res.*, **77**, 1360.
29. Chang, C.C. (1974) in *Characterization of Solid Surfaces* (eds P.F. Kane and G.R. Larrabee), Plenum Press, New York, p. 509.
30. Eadington, P. (1974) *Trans. Inst. Min. Met.*, **83c**, 223.
31. Eadington, P. (1977) *Trans. Inst. Min. Met.*, **86c**, 186.
32. Terrell, J.H. and Spijkerman, J.J. (1968) *Appl. Phys. Lett.*, **13**, 11.

33. Simmons, G.W., Kellerman, E. and Leidheiser, H. (1973) *Corrosion*, **29**, 227.
34. Forester, D.W. (1973) in *Proc. 4th Lunar Sci. Conf., Suppl. 4, Geochim. Cosmochim. Acta*, **3**, 2697.
35. Tricker, M.J. and Freeman, A.G. (1975) *Surface Sci.*, **52**, 549.
36. Tricker, M.J., Winterbottom, A.P. and Freeman, A.G. (1976) *J. chem. Soc., Dalton Trans.*, 1289.
37. Tricker, M.J., Ash, L.A. and Jones, W. (1979) *J. inorg. nucl. Chem.*, **41**, 891.
38. Berry, F.J. and Maddock, A.G. (1978) *J. chem. Soc. chem. Commun.*, 308.
39. Berry, F.J. (1979) *J. chem. Soc., Dalton Trans.*, 1736.

Index

Ab initio SCF calculations, applied to Si-O-Si and Si-S-Si linkages, 11
Absorption coefficient, 96, 257
Acanthite (Ag_2S) structure, 253
  molecular orbital energy level diagram, 280
Ag_2S *see* Acanthite
Al_2O_3 *see* Corundum, Ruby, Sapphire, Alumina
Alabandite (α-MnS) structure, 252
Albite, X-ray emission spectra of, 42
Alloys, 251
Altaite (PbTe)
  band structure calculations, 271
  electrical properties, 271
  optical properties, 273
  structure, 252
  X-ray photoelectron spectrum, 271
Alumina, X-ray emission spectra of, 42
Alunite, X-ray emission spectra of, 42
Amphiboles, optical spectra of glaucophane-riebeckite series, 86
Andradite, *see* Garnet
Anthophyllites, Mössbauer spectra of, 178
Apatite, luminescence in, 119
Aragonite, luminescence spectra of, 132, 138
Argentite (Ag_2S), structure type, 253
  see also Ag_2S
Arsenic group, 252
Arsenopyrite (FeAsS)
  application of ligand field theory to, 268
  application of MO theory to, 270
  relationship to other disulphides, 268
  structure type, 252
As_2^{2-}, cluster molecular orbital diagram for, 25
As_4^{4-}, cluster molecular orbital diagram for, 25
Auger spectroscopy, 33, 39
  application to surface studies, 294, 305, 306
  chalcopyrite, 309
  silica, 307
Augite, Mössbauer spectra of, 188

B_2O_3, cluster calculations on, 8
Band theory
  applications to metal oxides and chalcogenides, 256
  applied to the luminescence process, 110
  calculations on MgO, 5
  calculations on PbS, 271
  calculations on SiO_2, 14
Basis set, 4
Berndtite (SnS_2) structure, 253
Biotite, conversion electron Mössbauer spectra of, 310
  Mössbauer spectra of, 188
Bismuthinite (Bi_2S_3) structure, 254
Boehmite (γ-AlO(OH)), X-ray emission, spectrum of, 48
BO_3^{3-}, cluster calculations on, 8
$B(OH)_3$, cluster calculations on, 9
Bond angles in silicates and their calculation, 11
Bornite (Cu_5FeS_4) structure, 253

Breithauptite (NiSb) structure, 253
Brucite ($Mg(OH)_2$),
X-ray emission spectrum of, 46
X-ray photoelectron spectrum of, 46

$CaCO_3$ cluster calculations on, 7
see also Calcite
Calcite
ESR spectra of, 219
ligand field parameters for Mn^{2+} in, 117
luminescence spectra of, 115, 132
X-ray photoelectron spectra of, 302
CaO electronic structure at high pressures, 6
luminescence from F-centres in, 120
Carrier concentration and mobility, 259
Carrollite ($CuCo_2S_4$) chemical bonding in, 281
Cassiterite (SnO_2) structure, 254
Cathodoluminescence, 103, 104, 112
techniques for study of, 121
Cation ordering, determination by luminescence excitation spectroscopy, 137
determination by Mössbauer spectroscopy, 179
Cattierite (CoS_2)
diffuse reflectance spectrum, 265
molecular orbital energy diagram, 266
optical constants, 265
properties, 264, 267
structure, 252
Chalcocite (Cu_2S)
molecular orbital binding energies, 280
structure type, 253, 279
X-ray emission spectrum, 279
X-ray photoelectron spectrum, 279
Chalcogenides, 251
Chalcopyrite ($CuFeS_2$)
Auger spectrum of, 309
chemical bonding in, 278
discrepancies between calculation and experiment, 19
high pressure properties of, 279
Mössbauer spectra of, 171
oxidation state of cations in, 275
structure, 253, 255
X-ray emission spectra of, 278
X-ray photoelectron spectrum of, 278
Charge transfer
$Fe^{2+} \rightarrow Fe^{3+}$, 86
$Fe^{3+} \rightarrow Ti^{4+}$, 87
origin of blue colouring sapphire, 89
oxygen→metal, 72
$Ti^{3+} \rightarrow Ti^{4+}$, 87
Chemical isomer shift in Mössbauer spectroscopy, 144
Chromia *see* Eskolaite
Chromite ($FeCr_2O_4$)
chemical bonding in, 286
properties of, 286, 288
structure, 254, 286
Cinnabar (HgS) structure type, 254
Clausthalite (PbSe)
bond structure calculations, 271
electrical properties, 271
optical properties, 273
structure, 252
X-ray photoelectron spectra, 271
Clays Mössbauer spectra of, 176, 182
Clinopyroxenes, Mössbauer spectra of, 181
ClO_4^- anion, properties of, 16
X-ray emission spectra, 15
Cluster models, 6
CNDO method applied to bond angles in silicates, 11
CO_2 structure calculated by DLS method, 13
CO_3^{2-} cluster
calculations on, 7
C—O distances in, 7
vibrational frequencies of, 7
$CoAs_6^{3-}$ cluster, molecular orbital diagram for, 25
Configurational co-ordinate diagram, 107
Corundum
Mössbauer spectra of Fe-Ti doped, 91
optical absorption spectra of Fe^{3+} in, 82
optical absorption spectra of Ti^{3+} in, 84
CoS_4^{6-} cluster, molecular orbital diagram (valence region), 281
CoS_6^{9-} cluster, molecular orbital diagram (valence region), 281
Co_3S_4, *see* Linnaeite
Cobalt pentlandite (Co_9S_8) structure, 253

Coesite, *see* SiO_2
Columbite-tantalite ($Fe_1Mn(Nb_1Ta)_2O_6$) structure, 254
Conductivity (electrical), 259
  see also Resistivity
Configuration interaction, 4
Conversion electron Mössbauer spectroscopy, 296
  study of biotite, 310
  study of iron oxides, 309
  study of lunar rocks, 310
  study of siderite, 311
  study of vivianite, 311
Cordierite, Mössbauer spectra of, 188
Correlation of electrons, 3
CoS, 273
CoS_2 *see* Cattierite
Covellite (CuS), electronic structure of, 280
  structure type, 253
Crocidolite, Mössbauer spectra of, 188
Cr_2O_3, *see* Eskolaite
CrO_6^{9-} cluster, molecular orbitals and application to eskolaite, 285
  molecular orbitals and application to spinel oxides, 286
Cristobalite, *see* SiO_2
Crookesite (Cu_2Se) structure, 253
Cryolite, X-ray emission spectra of, 42
Crystal field spectra, 64, 74
Crystal field splitting parameter, 65
Crystal field theory, 256
Cummingtonite, Mössbauer spectra of, 178
Cu^{2+}, ESR spectra, 223
Cu_2S
  X-ray emission spectra, 17
  X-ray photoelectron spectra, 17
  see also chalcocite
Cubanite ($CuFe_2S_3$) structure, 253
$CuCo_2S_4$, *see* Carrollite
$CuFeS_2$, *see* Chalcopyrite
CuS, *see* Covellite
CuS_2
  molecular orbital energy level diagram, 266
  optical constants, 265
  properties of, 264, 267
CuS_3^{4-} cluster, molecular orbital energy level diagram, 280
CuS_3^{5-}cluster, molecular orbital energy level diagram, 280
$CuS_4^{6.5-}$ cluster, molecular orbital energy level diagram, 280
CuS_4^{7-} cluster, comparison with ZnS_4^{6-} cluster, 17
  molecular orbital diagram, 18

DLS, *see* Distance-least-squares
Dating of minerals, application of Mössbauer spectroscopy, 183
Daubréelite ($FeCr_2S_4$) chemical bonding in, 283
Decay time, 106
Deerite, Mössbauer spectra of, 188
Defect centre, 120
Delocalized models in quantum mineralogy, 3
Delocalized valence electrons in oxides and sulphides, 257
Derivative structures, 255
Diamond, luminescence centres in, 121
Diaspore (α-AlO(OH)) X-ray emission spectra of, 48
Difference (or deformation) density *see* Electron density distributions
Diffuse reflectance, 73, 257
  spectra of haematite, 83
Digenite (Cu_9S_5) structure type, 253
Diopside, ligand field parameters for Mn^{2+} in, 117
  luminescence spectra of Mn^{2+} in, 115, 130
Discrete variation method, 3
  applied to SiO_4^{4-}, 15
Disulphide minerals
  chemical bonding in, 260
  relationships to diarsenides and sulpharsenides, 267
  spectra and properties of, 260
  structures, 252
Disiloxane molecule, application to describe Si-O bond in euclase, 21
  calculations on, 11
Distance-least-squares (DLS) approach, 13
Dolomite, luminescence spectra of, 132

Electron delocalization in minerals, application of Mössbauer spectroscopy, 184
Electron density distributions, 2, 4
  cluster calculations of valence, 19

Electron microprobe for study of
cathodoluminescence, 122
Electron-phonon coupling, 107
Electron spin resonance (ESR), 209
of calcite, 219
electron transfer reactions, 221
g-value, 211
of kaolinite, 219
linewidths, 218
of micas, 221
paramagnetic probes, 222
of powder samples, 217
of smectites, 221
spectra, 217
of transition metal ions, 223
of vermiculites, 221
Electron transfer reactions, 221
Electron traps, 106
Electronic spectra of minerals, 63
measurement of, 72
Enargite (Cu_3AsS_4) structure, 253
Enstatite, ligand field parameters for
Mn^{2+} in, 117
luminescence spectra of, 115, 128
Eskolaite (α-Cr_2O_3)
chemical bonding in, 283
optical absorption spectra of, 85
properties, 283, 286
structure, 283
thermochemical properties, 286
Euclase ($AlBeSiO_4(OH)$) difference
density maps through Be-O-H
linkage and Be-O_2-Si linkage, 20
ESR, *see* Electron spin resonance
EXAFS, *see* Extended X-ray absorption
fine structure
Exchange energy, 4
Excitons, 110
Extended X-ray absorption fine
structure, 300
Extinction coefficient, 68

F-centre, 109
Fe^{2+} ESR spectra of, 225
Fe_2O_3
conversion electron Mössbauer
spectra of, 309
on Martian surface, 92
see also Haematite
Fe_3O_4, conversion electron Mössbauer
spectra of, 309
γ-Fe_2SiO_4 crystal field spectra, 74
in the Earth's mantle, 97
$FeAs_6^{3-}$ cluster, molecular orbital
diagram for, 25
$FeAs_6^{4-}$ cluster, molecular orbital
diagram for, 25
Feldspars
Fe^{3+} centres in plagioclase, 125
ligand field parameters for Mn^{2+} in
anorthite, 117
luminescence of, 112, 113, 115, 121,
130, 136
X-ray emission spectra of, 54
X-ray photoelectron spectra of, 305
$FeNi_2S_4$, *see* Violarite
FeO_4^{5-} cluster, molecular orbitals and
application to spinel oxides, 286
FeO_4^{6-} cluster, molecular orbitals and
application to spinel oxides, 286
FeO_6^{9-} cluster, molecular orbitals and
application to haematite, 285
molecular orbitals and application to
spinel oxides, 286
FeO_6^{10-} cluster, molecular orbitals
and application to ilmenite, 285
molecular orbitals and application to
spinel oxides, 286
Fermi level, 260
Feroxyhyte (δ-FeOOH) on Mars, 92
Ferrihydrite, on Mars, 92
Ferroselite ($FeSe_2$) structure, 252
$FeCr_2S_4$, *see* Daubréelite
FeS
high-pressure behaviour, 275
magnetic and electrical properties,
275
MO/band structure models, 275, 276
molecular orbital energy levels from
SCF-X_α calculations, 275, 277
phase transformations in, 273, 275
structure, 253, 255, 273
X-ray absorption spectra, 275
X-ray emission spectra, 275
FeS_4^{5-} cluster, molecular orbital
energy level diagram, 279, 282
FeS_6^{10-} cluster, molecular orbital
energy levels, 275, 277, 282
$FeSb_2$ structure, 252
$FeTiO_3$, *see* Ilmenite
Fluorescence, 104, 107
Fluorides, calculations on M-F
distances, 10

Fluorite, luminescence of, 111, 120
Forsterite, ligand field parameters for Mn^{2+} in, 117
luminescence spectra of Mn^{2+} in, 115, 126
Frank-Condon principle, 108
Fresnel equation, 257
Full lattice calculations, 5

g-value, 211
Galena (PbS)
band structure calculations on, 271
electrical properties, 271
MO/band model for, 272
optical properties, 273
SCF-X_α calculations on, 271
structure, 252
X-ray photoelectron spectrum, 271
Garnet
in the Earth's mantle, 96
luminescence spectra of Cr^{3+} in grossularite, 118
optical absorption spectra in knorringite, 86
optical absorption spectra of Cr^{3+} in uvarovite, 85
optical absorption spectra of Fe^{3+} Andradite, 69
optical absorption spectra of pyrope-almandine series, 80
yttrium aluminium garnet (YAG), 119
Geikielite ($MgTiO_3$) structure, 254
Gibbsite ($Al(OH)_3$) X-ray emission spectra of, 48
Gillespite optical absorption spectra, 81
Glassy materials, Mössbauer spectra of, 175
Glaucophane, intervalence electronic transitions in, 71
Mössbauer spectra of, 188
Goethite (α-FeO . OH) reflectance spectrum of, 93
structure, 254
Gold, X-ray photoelectron spectra of, 301
Gold group, 252
Greenockite (CdS) structure, 253
Greigite (Fe_3S_4) chemical bonding in, 282
structure, 253
Grossularite *see* Garnet
Grunerite, Mössbauer spectra of, 178
Gudmundite (FeSbS) structure, 252

Haematite (α-Fe_2O_3)
chemical bonding in, 283
diffuse reflectance spectra of, 83, 93
properties, 82, 283, 285
structure, 254, 283
thermochemical properties, 286
see also Fe_2O_3
Hall voltage, 260
Hamiltonian operator, 4
Hartree-Fock (and Hartree-Fock-Roothaan) method, 2, 4
applied to alkali halides, 13
applied to carbonates, 8
Hauerite, MnS_2 diffuse reflectance spectrum, 265
Hausmannite (Mn_3O_4) structure, 254
Hawleyite (CdS) structure, 253
Howeite Mössbauer spectra of, 188
Huang-Rhys factor, 107, 120
Hund's rule, 113
Hydrates, calculations on M-O distances for second-row elements, 9
Hydroxides, calculations on M-O distances, 10
Hydroxyanions
calculations on difference density, 20
calculations on M-O distances for first-row elements, 9
calculations on M-O distances for second row-elements, 9
Hyperfine interactions in Mössbauer spectroscopy, 143
in nuclear magnetic resonance, 215

Idaite (Cu_3FeS_4) structure, 253
Ilmenite ($FeTiO_3$)
chemical bonding in, 283
properties, 283, 285
structure, 254, 283
thermochemical properties, 286
Ilvaite, Mössbauer spectra of, 188
Independent electron approximation, 3
Interlamellar water, investigation by NMR, 236
Intervalence (metal-metal) transitions, 70, 86
Ion scattering spectroscopy (ISS), 300
Ionic models, 255

Iron, X-ray photoelectron spectra of, 302
ISS, *see* Ion scattering spectroscopy

Jahn-Teller effect, 68

K (potassium), electronic structure at high pressure, 6
K_2CO_3 X-ray spectra of, 7
Kaolinite, ESR spectra of, 219, 226
X-ray emission spectra of, 42
Khor Temiki meteorite, luminescence spectra of orthoenstatite from, 109
Kinetics of solid state mineral reactions, investigations by Mössbauer spectroscopy, 181
Kramers-Kronig calculation, 258
Kubelka-Munk function, 258

Laplace rule, 105
Lepidocrocite (γ-FeO . OH) structure, 254
LiCl, calculated phase transition, 5
Ligand field parameter, 113
Ligand field theory, 256
Linnaeite (Co_3S_4), chemical bonding in, 281
structure, 253
Loellingite ($FeAs_2$)
application of ligand field theory to, 268
application of MO theory to, 269
relationship to other disulphides, 268
structure type, 252
localized models in quantum mechanics, 3
valence electrons in oxides and sulphides, 257
Luminescence
centres in minerals, 113
experimental techniques for study of, 121
process of, 104
Luminescope, 122
Lunar rocks and soil,
conversion electron Mössbauer spectra of, 310
luminescence of plagioclase in, 125, 131
Mössbauer spectra of, 173
optical absorption spectra of pyroxenes in, 80
remote sensed reflectance spectra of, 91
X-ray photoelectron spectra of, 306

Mackinawite $(Fe,Co,Ni,Cu)_{1+x}S)$, structure, 253
Maghaemite (γ-Fe_2O_3) on Mars, 92
structure, 254
Magic angle spinning (MAS), 209, 233
Magnesiowustite, electronic absorption spectra of, 67, 72
in the Earth's mantle, 95
Magnesite, luminescence spectra of, 132
Magnetic splitting in Mössbauer spectroscopy, 150
Magnetic susceptibility, 260
Magnetite (Fe_3O_4)
chemical bonding in, 286
on Mars, 92
photoelectron (X-ray and UV) spectra, 288
properties, 286, 288
structure, 254, 286
Mn^{2+}, ESR spectra of, 225
Manganese nodules, X-ray photoelectron spectra of, 303
Manganite (MnO . OH) structure, 254
Mantle of the Earth
electrical conduction in, 97
form of MgO in, 5
phase equilibria from electronic spectra, 94
radiative heat transport in, 95
Marcasite (FeS_2)
mechanism of formation, 270
relationship to other disulphides, 268
structure type, 252
Mars, iron oxide mineralogy of planetary surface, 92
remote sensed reflectance spectra of, 93
Melonite ($NiTe_{2-x}$) structure, 253
Metal-metal interactions in disulphide minerals, 268
in FeS, 275
Metals, 251
Meteoritic materials, Mössbauer spectra of, 175
$(Mg,Fe)_2SiO_4$, *see* Olivine
MgO
full lattice calculations on, 5

luminescence from F-centres in, 120
see also periclase
$MgSiO_3$ *see* Enstatite, Orthoenstatite
Mineral dating, application of Mössbauer spectroscopy, 183
Micas
ESR spectra of, 221
Mössbauer spectra of, 176
NMR spectra of, 236
Microcline, X-ray emission spectra of, 42
Mixed valence minerals, Mössbauer spectra of, 184
MnS_2, *see* hauerite
Modified electron gas (MEG) model, 5–6
applied to $M(OH)_n$ cluster, 11
Molecular orbital (MO) theory, 1, 31
applications in mineralogy, 1
Molybdenite (MoS_2) structure type, 253
Mössbauer spectroscopy, 141
analytical applications, 168
apparatus, 162
cationic ordering, 179
clays, 182
dating, 183
disulphides, diarsenides, sulpharsenides, 269
electron delocalization, 184
FeS at high pressure, 269
glassy materials, 175
kinetics of solid state reactions, 181
lunar rocks, 173
meteorites, 175
micas, 176
mining, 183
mixed valence minerals, 184
nodules, 177
opaque minerals, 257
pottery, 182
sediments, 177
see also Conversion electron Mössbauer spectroscopy
Muffin-tin procedure, 4
Mulliken overlap populations, 2
Mullite, X-ray emission spectra of, 2
Multiple scattering (MS)X_α methods, 2
applied to carbonates, 8

NaCl, calculated phase transition, 5
luminescence from F-centres in, 120
Neptunite, Mössbauer spectra of, 171
Niccolite (NiAs) structure type, 253, 273
NiO, studied by SCF cluster calculations, 13
NiS, 273
NiS_2 *see* Vaesite
NMR *see* Nuclear magnetic resonance
Nuclear magnetic resonance (NMR), 209, 327
applied to interlamellar water, 236
applied to micas, 236
high resolution studies, 237
of solids, 231
second moment analysis, 237

Olivine, 5
comparison between spectral data and experiment, 15
electronic spectra (optical spectra) of, 75
in the Earth's mantle, 94
luminescence spectra of, 126
transition to spinel structure, 94
X-ray photoelectron spectra of, 49
see also Forsterite
Omphacite, electronic spectra (optical spectra) of, 88
Mössbauer spectra of, 172
Optical absorption spectra, *see* Electronic spectra
Ore minerals, 251
Orgel energy level diagram, 114
Orthopyroxenes, Mössbauer spectra of, 180
Orthoenstatite, luminescence spectra of, 109, 111, 114, 116, 129
Oxides, calculations on M-O distances for transition metal, 10
X-ray photoelectron spectra of, 305

Paramagnetic probes, 222
Pauli exclusion principle, 3
PbO, X-ray photoelectron spectra of, 305
Pentlandite $(Ni,Fe)_9S_8)$ structure type, 253
Periclase (MgO)
electronic spectra of Fe^{2+} in, 66, 74
in the Earth's mantle, 95
X-ray photoelectron spectra of, 44
Perovskite in the Earth's mantle, 95

Perturbation theory, 4
application to sulphide minerals, 267
Petrological analysis, use of Mössbauer spectroscopy, 171
Phonons, 104, 108, 109
Phosphorescence, 104, 107
Photoelectron spectroscopy, 32, 294, 301
apparatus, 37
Photoluminescence, 104
Platinum metal group, 252
PO_4^{3-} anion, properties of, 16
X-ray emission spectra, 15
Polydymite (Ni_3S_4), chemical bonding in, 281
structure, 253
Pottery, Mössbauer spectra, 182
Pseudopotential approaches, 27
Pyrite (FeS_2)
absorption edge, 264
MO energy level diagram from SCF-Xα calculations, 263
properties of, 260
optical constants of, 261, 265
qualitative molecular orbital/band models, 261
S-S interactions and structure of, 267
stability in relation to marcasite, 270
structure type, 252, 260
X-ray emission spectra of, 55, 261, 263
X-ray photoelectron spectra of, 55, 261, 263
Pyrolusite (MnO_2) structure, 254
Pyroxenes
electronic spectra of Allende meteorite pyroxene, 84, 87
electronic spectra of Angra dos Reis meteorite pyroxene, 87
electronic spectra (optical spectra) of, 77, 86
in the Earth's mantle, 96
luminescence spectra of, 128
Mössbauer spectra of, 181
remote sensed spectra from lunar surface, 91
Pyrrhotite ($Fe_{1-x}S$) structure types, 253, 255, 273
see also FeS

Quadrupole splitting in Mössbauer spectroscopy, 145
Quartz (SiO_2) luminescence of, 112, 117, 120, 121, 133, 137
see also SiO_2
Quantum mechanics, applications in mineralogy, 1

Racah parameters, 65, 114, 115
Radii, atomic and ionic compared to calculated, 19
Rare-earth elements as luminescence centres in minerals, 119
RbCl, calculated phase transition, 5
Reflectance *see* Diffuse reflectance, Specular reflectance
Refractive index 96, 257
Resistivity, 259
versus temperature for sulphide minerals, 259
Rhodonite, luminescence of, 130
Riebeckite, intervalence electronic transitions in, 71
Mössbauer spectra of, 188
Rocksalt, *see* NaCl
Ruby, electronic spectra (optical spectra) of, 84
Rutile, *see* TiO_2

S_2^{2-} anion, molecular orbital energy level diagram for, 269
Safflorite ($CoAs_2$) bonding in, 270
structure, 252
Sapphire, optical absorption spectra and origin of colour, 89
see also corundum
Scattering coefficient, 258
Scheelite, luminescence in, 119
Schorlomite, Mössbauer spectra of, 190
Schrödinger equation, 4
Sediments, Mössbauer spectra of, 177
Secondary ion mass spectrometry (SIMS), 300
Seebeck voltage, 260
Siderite, conversion electron Mössbauer spectra of, 311
$(SiH_3)O$, *see* Disiloxane
Silica, *see* SiO_2
Silicates
break up of structures with electron count, 23
cluster calculations on, 11

ESR spectra of, 226
NMR spectra of, 238
X-ray photoelectron spectra of, 305
X-ray spectra of, 302, 305
SiO_2
Auger spectra of, 307
distance least squares (DLS) calculations on, 13
full lattice calculations on, 6
NMR spectra of, 239
X-ray spectra of, 53
see also Disiloxane, Quartz
SiO_4^{4-} cluster, molecular orbital diagram, 14
see also SiO_2, silicates
$Si(OH)_4$, 14
see also SiO_4^{4-}, SiO_2
Skutterudite group, qualitative MO models for, 24
Smectites, ESR spectra of, 221
Specular reflectance, 257
of haematite, ilmenite and eskolaite, 285
of PbS, PbSe, PbTe, 273
of spinel oxides, 288
Sphene glasses, X-ray photoelectron spectra of, 303
Spinel group
chemical bonding in, 286
crystal field spectra of Fe^{2+} in, 74
properties, 286, 288
structure, 254, 286
transition of olivine to spinel structure, 94
transition of spinel to periclase and perovskite, 95
X-ray spectra of, 45
Sphalerite (β-ZnS) structure, 253
see also ZnS
Spin-orbit coupling, 106
Stannite (Cu_2FeSnS_4) structure, 253
Stefan's constant, 96
Stibnite (Sb_2S_3) structure type, 253
Stishovite, 5
difference density maps, 21
Stokes shift, 107
Strontianite, luminescence in, 120
Sulphide minerals, X-ray photoelectron spectra of, 301
Sulphosalt mineral group, 254
Sulphur-sulphur interactions, 267
Surfaces of minerals, 293
Talnakhite ($Cu_9Fe_8S_{16}$) structure, 253, 255
Thermoelectric power, 260
Thermoluminescence 104, 107
Thiospinel group
electronic structure models, 281
interpretation of mineral properties, 283
structure type, 253
Thorianite (ThO_2) structure, 254
TiO_2
calculated phase transition, 6
full lattice calculations on, 6
rutile structure group, 254
X-ray photoelectron spectra of, 306
TiO_6^{8-} cluster, molecular orbitals and application to ilmenite, 285
molecular orbitals and application to spinel oxides, 286
Transition moment in luminescence process, 105
Tripuhyite, Mössbauer spectra of, 171
Troilite (FeS)
chemical bonding in, 275
magnetic and electrical properties, 275
phase transformations in, 273
structure, 253, 255, 273
see also FeS
Tungstenite (WS_2) structure, 253

Ultraviolet photelectron spectroscopy (UPS), 37, 294
Ulvöspinel ($TiFe_2O_4$)
chemical bonding in, 286
properties, 286, 288
structure, 254, 286
UPS, *see* ultraviolet photoelectron spectroscopy
Uraninite (UO_2) structure, 254

Vaesite (NiS_2)
diffuse reflectance spectrum, 265
molecular orbital energy level diagram, 266
optical constants, 265
properties, 264, 267
structure, 252
Valence bond theory, 231, 256
Vanadyl ion, ESR spectra of, 226
Vermiculites, ESR spectra of, 221
Vesuvianite, Mössbauer spectra of, 188

Violarite ($FeNi_2S_4$) chemical bonding in, 281
structure, 253
Vivianite
conversion electron Mössbauer spectra of, 311
Mössbauer spectra of, 181, 188
optical absorption spectra of, 71, 86

Wave function, 3
Wolframite group, 254
Wollastonite, luminescence spectra of, 130
Wurtzite (α-ZnS) structure, 253
see also ZnS

X_α method, *see* multiple scattering X_α methods
XPS, *see* X-ray photoelectron spectroscopy
X-ray absorption fine structure (XAFS), applications to solution and melt species, 27
X-ray emission spectroscopy, 39
applications in mineral chemistry, 42
of SiO_2 and comparison with calculation, 14
X-ray photoelectron spectroscopy, 37, 294
application in mineral chemistry, 42
calcite, 302
hydrous manganese nodules, 303
lunar rocks, 306
mineral processing, 305
silicates, 302, 305
SiO_2 and comparison with calculations, 14
sphene glass, 303
sulphide minerals, 301
surfaces of minerals, 301
X-ray spectroscopy, 31, 294
boehmite, 48
brucite, 46
diaspore, 48
feldspar, 54
gibbsite, 48
minerals containing aluminium, 42
olivine, 49
periclase, 44
pyrite, 55
silica, 53
spinel, 45

Zeolites
ESR spectra of, 226
NMR spectra of, 240
X-ray photoelectron spectra of, 305
Zero field splitting, 214
ZnS
applications of qualitative MO theory to, 23
X_α calculations on, 17
X-ray emission spectra of, 16
X-ray photoelectron spectra of, 16
ZnS_2, properties, chemical bonding and MO energy level diagram, 266
ZnS_4^{6-} cluster, *see* ZnS